Pest Risk Modelling and Mapping for Invasive Alien Species

CABI INVASIVES SERIES

Invasive species are plants, animals or microorganisms not native to an ecosystem, whose introduction has threatened biodiversity, food security, health or economic development. Many ecosystems are affected by invasive species and they pose one of the biggest threats to biodiversity worldwide. Globalization through increased trade, transport, travel and tourism will inevitably increase the intentional or accidental introduction of organisms to new environments, and it is widely predicted that climate change will further increase the threat posed by invasive species. To help control and mitigate the effects of invasive species, scientists need access to information that not only provides an overview of and background to the field, but also keeps them up to date with the latest research findings.

This series addresses all topics relating to invasive species, including biosecurity surveillance, mapping and modelling, economics of invasive species and species interactions in plant invasions. Aimed at researchers, upper-level students and policy makers, titles in the series provide international coverage of topics related to invasive species, including both a synthesis of facts and discussions of future research perspectives and possible solutions.

Titles Available

1. *Invasive Alien Plants: An Ecological Appraisal for the Indian Subcontinent*
   Edited by J.R. Bhatt, J.S. Singh, R.S. Tripathi, S.P. Singh and R.K. Kohli
2. *Invasive Plant Ecology and Management: Linking Processes to Practice*
   Edited by T.A. Monaco and R.L. Sheley
3. *Potential Invasive Pests of Agricultural Crops*
   Edited by J.E. Peña
4. *Invasive Species and Global Climate Change*
   Edited by L.H. Ziska and J.S. Dukes
5. *Bioenergy and Biological Invasions: Ecological, Agronomic and Policy Perspectives on Minimizing Risk*
   Edited by L.D. Quinn, D.P. Matlaga and J.N. Barney
6. *Biosecurity Surveillance: Quantitative Approaches*
   Edited by F. Jarrad, S.L. Choy and K. Mengersen
7. *Pest Risk Modelling and Mapping for Invasive Alien Species*
   Edited by Robert C. Venette

Pest Risk Modelling and Mapping for Invasive Alien Species

Edited by

Robert C. Venette

USDA Forest Service, USA

www.cabi.org

CABI is a trading name of CAB International

CABI
Nosworthy Way
Wallingford
Oxfordshire OX10 8DE
UK

CABI
38 Chauncy Street
Suite 1002
Boston, MA 02111
USA

Tel: +44 (0)1491 832111
Fax: +44 (0)1491 833508
E-mail: cabi@cabi.org
Website: www.cabi.org

Tel: +1 800 552 3083 (toll free)
E-mail: cabi-nao@cabi.org

© CAB International and USDA 2015. All rights reserved. No part of this publication may be reproduced in any form or by any means, electronically, mechanically, by photocopying, recording or otherwise, without the prior permission of the copyright owners.

A catalogue record for this book is available from the British Library, London, UK.

Library of Congress Cataloging-in-Publication Data

Pest risk modelling and mapping for invasive alien species / edited by Robert C. Venette, USDA Forest Service, USA.
   pages cm. -- (CABI invasives series)
 Includes bibliographical references and index.
 ISBN 978-1-78064-394-6 (alk. paper)
 1. Nonindigenous pests--Geographical distribution--Mathematical models. 2. Ecological risk assessment--Mathematical models. I. Venette, Robert C.

 SB990.P55 2015
 578.6'2--dc23

2014035632

ISBN-13: 978 1 78064 394 6

Commissioning editor: David Hemming
Assistant editor: Alexandra Lainsbury
Production editor: Claire Sissen

Typeset by Columns Design XML Ltd, Reading, UK
Printed and bound by CPI Group (UK) Ltd, Croydon, CR0 4YY

Contents

Preface *Robert C. Venette*		x
1	**The Challenge of Modelling and Mapping the Future Distribution and Impact of Invasive Alien Species** *Robert C. Venette*	1
2	**Mapping Endangered Areas for Pest Risk Analysis** *Richard Baker, Dominic Eyre, Sarah Brunel, Maxime Dupin, Philippe Reynaud and Vojtěch Jarošík*	18
3	**Following the Transportation Trail to Anticipate Human-mediated Invasions in Terrestrial Ecosystems** *Manuel Colunga-Garcia and Robert A. Haack*	35
4	**Simulation Modelling of Long-distance Windborne Dispersal for Invasion Ecology** *Hazel R. Parry, Debbie Eagles and Darren J. Kriticos*	49
5	**Using the MAXENT Program for Species Distribution Modelling to Assess Invasion Risk** *Catherine S. Jarnevich and Nicholas Young*	65
6	**The NCSU/APHIS Plant Pest Forecasting System (NAPPFAST)** *Roger D. Magarey, Daniel M. Borchert, Glenn A. Fowler and Steven C. Hong*	82
7	**Detecting and Interpreting Patterns within Regional Pest Species Assemblages using Self-organizing Maps and Other Clustering Methods** *Susan Worner, Rene Eschen, Marc Kenis, Dean Paini, Kari Saikkonen, Karl Suiter, Sunil Singh, Irene Vänninen and Mike Watts*	97
8	**Modelling the Spread of Invasive Species to Support Pest Risk Assessment: Principles and Application of a Suite of Generic Models** *Christelle Robinet, Hella Kehlenbeck and Wopke van der Werf*	115

9	**Estimating Spread Rates of Non-native Species: The Gypsy Moth as a Case Study** *Patrick C. Tobin, Andrew M. Liebhold, E. Anderson Roberts and Laura M. Blackburn*	131
10	**Predicting the Economic Impacts of Invasive Species: The Eradication of the Giant Sensitive Plant from Western Australia** *David C. Cook, Andy Sheppard, Shuang Liu and W. Mark Lonsdale*	145
11	**Spatial Modelling Approaches for Understanding and Predicting the Impacts of Invasive Alien Species on Native Species and Ecosystems** *Craig R. Allen, Daniel R. Uden, Alan R. Johnson and David G. Angeler*	162
12	**Process-based Pest Risk Mapping using Bayesian Networks and GIS** *Rieks D. van Klinken, Justine V. Murray and Carl Smith*	171
13	**Identifying and Assessing Critical Uncertainty Thresholds in a Forest Pest Risk Model** *Frank H. Koch and Denys Yemshanov*	189
14	**Making Invasion Models Useful for Decision Makers: Incorporating Uncertainty, Knowledge Gaps and Decision-making Preferences** *Denys Yemshanov, Frank H. Koch and Mark Ducey*	206
15	**Assessing the Quality of Pest Risk Models** *Steven J. Venette*	223
Index		**235**

 This book is enhanced with supplementary resources.

To access the material please visit:

www.cabi.org/openresources/43946/

Contributors

Allen, Craig R., US Geological Survey – Nebraska Cooperative Fish and Wildlife Research Unit, University of Nebraska–Lincoln, Lincoln, NE 68583-0984, USA. E-mail: allencr@unl.edu

Angeler, David G., Department of Aquatic Sciences and Assessment, Swedish University of Agricultural Sciences, Uppsala, Sweden. E-mail: david.angeler@slu.se

Baker, Richard, Department for Environment, Food and Rural Affairs, Sand Hutton, York YO41 1LZ, UK. E-mail: Richard.Baker@defra.gsi.gov.uk

Blackburn, Laura M., USDA Forest Service, Northern Research Station, Morgantown, WV 26505, USA. E-mail: laurablackburn@fs.fed.us

Borchert, Daniel M., Plant Epidemiology and Risk Analysis Laboratory, Center for Plant Health Science and Technology, Plant Protection and Quarantine, Animal Plant Health Inspection Service, USDA, Raleigh, NC 27606, USA. E-mail: Daniel.M.Borchert@aphis.usda.gov

Brunel, Sarah, European and Mediterranean Plant Protection Organization, 21 Boulevard Richard Lenoir, 75011 Paris, France. E-mail: brunel@eppo.int

Colunga-Garcia, Manuel, Center for Global Change and Earth Observations, Michigan State University, 1405 S. Harrison Road, East Lansing, MI 48823, USA. E-mail: colunga@msu.edu

Cook, David C., Department of Agriculture and Food Western Australia, PO Box 1231, Bunbury, WA 6231, Australia. E-mail: david.cook@agric.wa.gov.au

Ducey, Mark, Department of Natural Resources and the Environment, University of New Hampshire, 114 James Hall, Durham, NH 03824, USA. E-mail: mjducey@cisunix.unh.edu

Dupin, Maxime, Previously at: INRA, UR633 Zoologie Forestière, F-45075 Orléans, France. E-mail: maxime.dupin@gmail.com

Eagles, Debbie, CSIRO Australian Animal Health Laboratory, Private Bag 24, Geelong, VIC 3220, Australia. E-mail: debbie.eagles@csiro.au

Eschen, Rene, CABI, Rue des Grillons 1, CH-2800 Delémont, Switzerland. E-mail: r.eschen@cabi.org

Eyre, Dominic, Department for Environment, Food and Rural Affairs, Sand Hutton, York YO41 1LZ, UK. E-mail: Dominic.Eyre@defra.gsi.gov.uk

Fowler, Glenn A., Plant Epidemiology and Risk Analysis Laboratory, Center for Plant Health Science and Technology, Plant Protection and Quarantine, Animal Plant Health Inspection Service, USDA, Raleigh, NC 27606, USA. E-mail: Glenn.Fowler@aphis.usda.gov

Haack, Robert A., USDA Forest Service, Northern Research Station, 3101 Technology Blvd, Suite F, Lansing, MI 48910, USA. E-mail: rhaack@fs.fed.us

Hong, Steven C., Center for Integrated Pest Management, North Carolina State University, Raleigh, NC 27606, USA. E-mail: Steve.Hong@aphis.usda.gov

Jarnevich, Catherine S., US Geological Survey, Fort Collins Science Center, 2150 Centre Avenue, Building C, Fort Collins, CO 80526, USA. E-mail: jarnevichc@usgs.gov

Jarošík, Vojtěch, Lately at: Institute of Botany, Academy of Sciences of the Czech Republic, Zámek 1, Průhonice, CZ 25243, Czech Republic.

Johnson, Alan R., School of Agricultural, Forest, and Environmental Sciences, Clemson University, Clemson, SC 29634-0310, USA. E-mail: alanj@clemson.edu

Kehlenbeck, Hella, Julius Kühn-Institute, Institute for Strategies and Technology Assessment, Kleinmachnow, Germany. E-mail: hella.kehlenbeck@jki.bund.de

Kenis, Marc, CABI, Rue des Grillons 1, CH-2800 Delémont, Switzerland. E-mail: m.kenis@cabi.org

Koch, Frank H., USDA Forest Service, Southern Research Station, Eastern Forest Environmental Threat Assessment Center, 3041 Cornwallis Road, Research Triangle Park, NC 27709, USA. E-mail: fhkoch@fs.fed.us

Kriticos, Darren J., CSIRO Biosecurity Flagship, GPO Box 1700, Canberra, ACT 2601, Australia. E-mail: Darren.Kriticos@csiro.au

Liebhold, Andrew M., USDA Forest Service, Northern Research Station, Morgantown, WV 26505, USA. E-mail: aliebhold@gmail.com

Liu, Shuang, CSIRO Land and Water Flagship, GPO Box 1700, Canberra, ACT 2601, Australia & Plant Biosecurity Cooperative Research Centre, Bruce, ACT 2617, Australia. E-mail: shuang.liu@csiro.au

Lonsdale, W. Mark, CSIRO Biosecurity Flagship, GPO Box 1700, Canberra, ACT 2601, Australia & Plant Biosecurity Cooperative Research Centre, Bruce, ACT 2617, Australia. E-mail: Mark.Lonsdale@csiro.au

Magarey, Roger D., Center for Integrated Pest Management, North Carolina State University, Raleigh, NC 27606, USA. E-mail: rdmagare@ncsu.edu

Murray, Justine V., CSIRO Biosecurity Flagship, PO Box 2583, Brisbane, QLD 4001, Australia. E-mail: Justine.Murray@csiro.au

Paini, Dean, CSIRO Biosecurity Flagship, GPO Box 1700, Canberra, ACT 2601, Australia & Plant Biosecurity Cooperative Research Centre, Bruce, ACT 2617, Australia. E-mail: Dean.Paini@csiro.au

Parry, Hazel R., CSIRO Biosecurity Flagship, GPO Box 2583, EcoScience Precinct, Brisbane, QLD 4001, Australia. E-mail: hazel.parry@csiro.au

Reynaud, Philippe, Anses – Plant Health Laboratory, CBGP, Campus International de Baillarguet CS 30016, 34988 Montferrier-sur-Lez Cedex, France. E-mail: Philippe.Reynaud@anses.fr

Roberts, E. Anderson, Department of Entomology, Virginia Polytechnic Institute and State University, Blacksburg, VA 24061, USA. E-mail: roberts@vt.edu

Robinet, Christelle, INRA, UR633 Zoologie Forestière, F-45075 Orléans, France. E-mail: christelle.robinet@orleans.inra.fr

Saikkonen, Kari, MTT Agrifood Research, FI-31600 Jokioinen, Finland. E-mail: kari.saikkonen@mtt.fi

Sheppard, Andy, CSIRO Biosecurity Flagship, GPO Box 1700, Canberra, ACT 2601, Australia & Plant Biosecurity Cooperative Research Centre, Bruce, ACT 2617, Australia. E-mail: Andy.Sheppard@csiro.au

Singh, Sunil, CSIRO Biosecurity Flagship, GPO Box 1700, Canberra, ACT 2601, Australia & Plant Biosecurity Cooperative Research Centre, Bruce, ACT 2617, Australia. E-mail: sunil.singh@csiro.au

Smith, Carl, School of Agriculture and Food Sciences, The University of Queensland, St Lucia, QLD 4072, Australia. E-mail: c.smith2@uq.edu.au

Suiter, Karl, NSF Centre for Integrated Pest Management, North Carolina State University, Raleigh, NC 27606, USA. E-mail: karl_suiter@ncsu.edu

Tobin, Patrick C., School of Environmental and Forest Sciences, University of Washington, Seattle, WA 98195, USA. E-mail: pctobin@uw.edu

Uden, Daniel R., Nebraska Cooperative Fish and Wildlife Research Unit, School of Natural Resources, University of Nebraska–Lincoln, Lincoln, NE 68503-0984, USA. E-mail: danielruden87@gmail.com

van der Werf, Wopke, Centre for Crop Systems Analysis, Wageningen University, PO Box 430, 6700 AK Wageningen, The Netherlands. E-mail: Wopke.vanderWerf@wur.nl

van Klinken, Rieks D., CSIRO Biosecurity Flagship, PO Box 2583, Brisbane, QLD 4001, Australia. E-mail: Rieks.VanKlinken@csiro.au

Vänninen, Irene, MTT Agrifood Research, FI-31600 Jokioinen, Finland. E-mail: Irene.Vanninen@mtt.fi

Venette, Robert C., USDA Forest Service, Northern Research Station, 1561 Lindig Street, St. Paul, MN 55108, USA. E-mail: rvenette@fs.fed.us

Venette, Steven J., Department of Communication Studies, The University of Southern Mississippi, Hattiesburg, MS 39406, USA. E-mail: Steven.Venette@usm.edu

Watts, Mike, Information Technology Programme, AIS St Helens, PO Box 2995, Auckland 1140, New Zealand. E-mail: michaelw@ais.ac.nz

Worner, Susan, Bio-Protection Research Centre, PO Box 84, Lincoln University, Lincoln 7647, New Zealand. E-mail: Susan.Worner@lincoln.ac.nz

Yemshanov, Denys, Natural Resources Canada, Canadian Forest Service, Great Lakes Forestry Centre, 1219 Queen Street East, Sault Ste. Marie, ON P6A 2E5, Canada. E-mail: Denys.Yemshanov@NRCan-RNCan.gc.ca

Young, Nicholas, Natural Resource Ecology Laboratory, Colorado State University, Fort Collins, CO 80526, USA. E-mail: nicholas.young@colostate.edu

Preface

The Purpose of this Book

Pest risk maps are frequently used to inform strategic and tactical decisions for the management of invasive alien species. Invasive alien species can threaten human health, drive species extinctions, lower agricultural production, impede economic growth or impair ecosystem structure and function.

The International Pest Risk Mapping Workgroup acknowledges that advanced training and a 'tool kit' of software packages are needed to produce pest risk maps that are fully fit for purpose. This book is an initial attempt to address those needs. Invited chapters emphasize specific steps and data requirements to guide users through the development of pest risk models and maps, or components thereof. Each chapter describes assumptions behind each model, briefly addresses pertinent theory and illustrates key concepts through worked examples.

The methods discussed in this book are descriptive not prescriptive. That is to say, these are examples of methods that have been used in previous assessments and have met standards for publication in peer-reviewed journals. Readers should not assume that these particular methods must be used to construct pest risk maps or that simply following the methods in this book will satisfy all scientific or regulatory concerns about invasive alien species. The Sanitary and Phytosanitary Agreement reached by the World Trade Organization acknowledges the sovereign right of any nation to protect its agricultural and natural resources from invasive alien species through biosecurity measures. Those measures are justifiable insofar as they are consistent with International Standards for Phytosanitary Measures. Pest risk models and maps to support local, regional or national decision making about the management of invasive alien species are not subject to these standards until they affect international trade. However, pest risk models and maps for domestic decision making are subject to their own standards set by the decision-making body/bodies.

My hope is that by revealing greater detail about some of the methods that are currently being used to construct pest risk maps, more scientists and modellers will be able to confidently produce pest risk models and maps and will find inspiration to improve pest risk mapping techniques. The methods in the book are not exhaustive, and more approaches than those described herein are available for pest risk modelling and mapping.

The Intended Audience

The text will be instructive to current and future land managers, regulatory officials, conservation biologists, landscape ecologists, applied economists, invasion biologists and anyone else who is interested in the development and application of pest risk models and maps for invasive alien species. The text would also be appropriate in courses for advanced undergraduates or introductory graduate students.

Scope

Most of the methods that are described herein have been used to support the development of pest risk maps or components thereof. Some of these methods have been addressed in the scientific literature or in government documents, but frequently in ways that would be difficult for someone else to reproduce the results. This text and the accompanying online support materials are intended to provide much greater transparency with respect to data sources, software packages and specific methods to produce pest risk maps. As such, the text is not intended to describe all of the methods that have been, or could be, used for pest risk mapping nor is it meant to fully describe the theory behind the methods.

The text emphasizes the application of models to terrestrial ecosystems, in particular to invasive alien species that might affect plants. Many of these approaches could be adapted to invasive alien species that affect livestock, wildlife or aquatic ecosystems.

Prerequisites

Pest risk maps typically require the integration and application of a number of skills. Despite the emphasis on maps, this book does not provide an extensive discussion of geographic information systems (GIS) or cartographic methods. Additionally, the text frequently assumes that the user has some familiarity with statistics, modelling and at least one of the core biological disciplines that support pest risk analysis (e.g. entomology, plant pathology or weed science).

The Text

The text is organized around the phases that are common to all biological invasions: arrival, establishment, spread and impact. Chapters 1 and 2 offer introductory comments about general goals and challenges of producing pest risk models and maps for invasive alien species. Chapters 3 and 4 address the arrival of new invasive alien species, either through international trade or aerial dispersal. Chapters 5, 6 and 7 describe techniques to determine where invasive alien species might establish. Chapters 8 and 9 convey methods to characterize or measure the spread of invasive alien species. Chapters 10, 11 and 12 explain specific methods to assess the potential economic or environmental impacts of biological invasions and the benefits of management. Chapters 13 and 14 describe methods to measure a common form of uncertainty in pest risk models (i.e. parametric uncertainty) and how to formally incorporate uncertainty measures into products for decision makers. The text concludes with a general discussion of measures of model validity and reliability.

Several chapters include additional information in the online supplement to this book. The supplement can be found at www.cabi.org/openresources/43946.

Acknowledgements

This text is the outcome of discussions over several years within the International Pest Risk Mapping Workgroup (www.pestrisk.org). Special thanks are extended to the nearly 30 anonymous reviewers who evaluated these chapters for content and clarity. Several graduate students at the University of Minnesota, many in the graduate minor on risk analysis for invasive species and genotypes, commented on the utility of these chapters as educational tools, and their feedback is especially appreciated. I also express my deepest personal thanks to James Venette, Patricia Venette and Cathy Miller for their thorough assessments of each chapter during the final phases of the editorial process. I thank David Hemming and Alexandra Lainsbury at CABI for their assistance with the production of this book and its online supplement.

Robert C. Venette

1 The Challenge of Modelling and Mapping the Future Distribution and Impact of Invasive Alien Species

Robert C. Venette*

USDA Forest Service, Northern Research Station, St. Paul, Minnesota, USA

Abstract

Invasions from alien species can jeopardize the economic, environmental or social benefits derived from biological systems. Biosecurity measures seek to protect those systems from accidental or intentional introductions of species that might become injurious. Pest risk maps convey how the probability of invasion by an alien species or the potential consequences of that invasion vary spatially. These maps inform strategic and tactical decisions for invasive species management. Pest risk modellers must contend with the challenges of developing models that forecast the course or consequence of invasions and are more meaningful than could be obtained by chance, of demonstrating the validity of those models and of portraying results on maps in ways that will be useful for decision makers. Frequently, these forecasts depend on extrapolations from limited information to project how a species might be affected, for example, by changes in commerce, exposure to novel environments or associations with new dispersal vectors, or how these species might affect resident species or ecological processes. Consequently, pest risk maps often focus on one phase of the invasion process: arrival, establishment, spread or impact. Risk assessors use different analytical tools and information sources to address each phase. To be certain that pest risk models and maps are fully fit for purpose, models and maps must be critically evaluated at each stage of the development process. Invariably, errors will be revealed. The International Pest Risk Mapping Workgroup has offered a number of suggestions to improve the development of pest risk models and maps. In addition, short-term improvements are likely to be achieved through critical, objective assessments of model performance and greater transparency about model development.

Introduction to Pest Risk Maps

Abraham Maslow (1943) proposed a hierarchy of human needs to explain preconditions for certain human behaviours. Higher-level needs (e.g. self-actualization from which reason, creativity and morality emerge) cannot be met until more fundamental needs are satisfied. Modern concepts of biosecurity intertwine the most basal need for food and water to support life with the next most basic need to have safety and security. Indeed, biosecurity describes the measures taken 'to manage risks of infectious disease, quarantined pests, invasive alien species, living modified

* E-mail: rvenette@fs.fed.us

© USDA 2015. *Pest Risk Modelling and Mapping for Invasive Alien Species* ed. R.C. Venette) organisms, and biological weapons. ... [M]any of these problems are a subset of the issue of invasive alien species' (Meyerson and Reaser, 2002). Ultimately, biosecurity is intended to protect human health, the environment or the economy from such biological threats. Pest risk maps for invasive alien species are pivotal tools for biosecurity.

Alien species (also known as exotic, non-native, non-indigenous or introduced species) are those species that have been accidentally or intentionally introduced to one or more areas outside their native geographic range but are by no means extraterrestrial. Often, these species are pathogens, plants or animals with a history of being problematic elsewhere in the world. Three dimensions (i.e. space, time and impact) affect whether a species is considered an invasive alien. In the USA, Executive Order 13112 defines an invasive alien species as 'with respect to a particular ecosystem, any species, including its seeds, eggs, spores, or other biological material capable of propagating that species, that is not native to that ecosystem' and 'whose introduction does or is likely to cause economic harm or harm to human health' (The White House, 1999). International Standards for Phytosanitary Measures similarly recognize a quarantine pest as 'any species, strain or biotype of plant, animal, or pathogenic agent injurious to plants or plant products' that is 'of potential economic [or environmental] importance to the area endangered thereby and not yet present there, or present but not widely distributed and being officially controlled' (FAO, 2012). Both definitions implicitly acknowledge that being alien is insufficient evidence by itself to consider a species an invasive pest.

Pest risk for invasive alien species refers to both: (i) the probability that a species will arrive, establish and spread (i.e. successfully invade) within an area; and (ii) the magnitude of harm should the invasion be successful (Orr et al., 1993; Ebbels, 2003). Economic harms result from lowered yields, reduced marketability, lost trade opportunities or increased management costs. Environmental harms include altered ecosystem functioning (e.g. fire regimes or nutrient cycling) or reductions in the abundance or diversity of resident taxa. Environmental harms can also occur if management activities affect non-target species (e.g. through drift of pesticides or predation by non-specific biological control agents). Environmental harms are considered particularly severe if threatened or endangered species might be affected. Social harms can occur if an invasive species or management activities interfere with benefits people draw from an ecosystem. However, it is not reasonable to assume an invasive alien species will be present in all places at all times; thus, risks posed by invasive alien species have spatial and temporal contexts.

Pest risk maps convey how risks from invasive alien species vary spatially within an area of concern and reflect underlying models of the factors that govern the course of invasion and the effects of invasive alien species on the structure or function of ecosystems (Venette et al., 2010). Some of these models are derived from heuristic descriptions of conditions necessary for an alien species to complete each phase of an invasion or have an impact. Other empirically based, statistical models infer quantitative relationships between a response variable (e.g. the probability of pest arrival) and a number of covariates (i.e. independent or predictor variables). Conceptual mathematical models follow a logical formalism to deduce relationships among variables (e.g. factors that affect species' spread rates). Many models for pest risk maps are based on more general ecological theories about factors that affect species' distributions, rates and patterns of spread, or the outcome of species' interactions. Although the goal of a pest risk map is to characterize how the probability and consequences of invasion by an alien species vary within an area of concern, in practice, pest risk maps frequently address just one or a few components of pest risk. For example, a map could focus on the suitability of the climate for pest establishment within an area of concern, with the rationale that a species which fails to find a suitable climate cannot establish, spread or have a lasting impact.

A well-crafted risk map serves a number of purposes. From a pragmatic perspective, risk maps can be powerful tools to help managers (e.g. foresters, farmers, pest-survey coordinators and some policy makers) select appropriate strategies and tactics with which to mitigate species' risks. Such risk mitigation (i.e. biosecurity) strategies can be classified broadly as prevention, eradication, suppression and restoration, which correspond generally with the arrival, establishment, spread and impact of invasive alien species (Venette and Koch, 2009). Pest risk maps also inspire critical thought about: (i) the adequacy of current theory, models and data to characterize risks from biological invasions; and (ii) alternative explanations for the course of an invasion or the impacts that have been realized, as was suggested by Koch (2011) for maps of human disease. This chapter describes long-standing goals for pest risk maps and introduces the general process by which pest risk maps are created. General models that have been applied to address different stages of the invasion process are briefly discussed. Types of errors associated with many pest risk models are presented and discussed with respect to measures of model performance. The chapter concludes with a series of recommendations that would help to improve the future development of pest risk models and maps.

An Historical Example of a Pest Risk Map

No consensus exists about when the first modern pest risk map was created. Figure 1.1 is likely among the first maps that begin to address contemporary concepts of pest risk, although the map was not created with the formal definition of pest risk in mind. In the 1870s, the San Jose scale, currently *Quadraspidiotus perniciosus*, was detected in North America for the first time in California's San Jose Valley (Howard and Marlatt, 1896). The pernicious insect, now recognized as native to parts of Asia, feeds on several deciduous fruit trees, such as peaches, plums, apples and pears, and is easily moved on nursery stock. Few details have been published about the creation of this map, but two pragmatic questions seem to have motivated its production: where was the insect likely to spread within the USA and what major fruit production regions might be affected?

This map (Fig. 1.1), published in 1896, shows the distribution of San Jose scale up to that time relative to 'life zones' in the conterminous USA. The life zones had been proposed by C.H. Meriam to distinguish areas that were especially suitable, or unsuitable, for many plants and animals. Meriam's map identifies five major zones in North America: boreal, transition, upper austral, lower austral and tropical. Known occurrences of San Jose scale seemed to occur 'within or near the so called austral life zones' (Howard and Marlatt, 1896, p. 33). The supposition at the time was that San Jose scale should be able to continue to spread within these regions wherever suitable hosts occurred. The map was developed before modern quarantine regulations were in place, so it was intended to reassure fruit producers in New England and portions of Pennsylvania, New York, Michigan and Wisconsin that the insect would 'not establish itself to any serious extent' (Howard and Marlatt, 1896, p. 35). At the time the map was published, Howard and Marlatt (1896) cautioned about the uncertainty in this forecast by acknowledging that 'its possibility is suggested by what we know up to the present time. Against its probability may be urged the fact that, in general, scale insects ... are seldom restricted by geographical limitations which hold with other insects' (Howard and Marlatt, 1896, p. 35). The precautionary note has proven justified. San Jose scale is now established in all conterminous states except Wyoming, North Dakota, South Dakota and Maine (CABI, 1986), but is often kept under control by a suite of natural enemies (Flanders, 1960).

Fig. 1.1. An early 'pest risk' map – the historical distribution of San Jose scale (black dots) in the conterminous USA relative to C.H. Meriam's life zones. (Reproduced from Howard and Marlatt, 1896.)

The General Challenge for Pest Risk Assessors

Pest risk modellers and mappers face a three-part challenge. The first part of the challenge is to develop a model that gives a more meaningful forecast of the course and consequence of a biological invasion than would be obtained by random chance or from obvious, intuitive models (e.g. a plant pest will occur and cause damage anywhere its host plants occur). Models are often needed for species that have not yet arrived in an area of concern. In these cases, pest risk modellers cannot develop or test models with empirical observations on the distribution, biology or behaviour of a species within the area of concern. They must extrapolate from what is known about a species in its native or adventive (i.e. areas where it has invaded) range, from studies in biosecure laboratories or from inferences drawn from taxonomically related species. Further, most pest modellers must rely on a simplifying assumption that individuals in an invading population are equivalent to individuals from the native range and will be equivalent to future generations (i.e. no significant genotypic or phenotypic changes have occurred or will occur). If an invasive alien species has arrived within an area of concern, initial observations of distribution, dynamics or impact may test the robustness of current knowledge or provide the foundation for a new model.

The second part of the challenge is to demonstrate the validity of the pest risk map and the underlying model. All models (physical, conceptual, statistical or mathematical) are an abstraction of reality. They are never intended to incorporate all of reality. Rather, models are intended to capture enough reality to be useful. What constitutes enough or useful is often a matter of debate. Venette (Chapter 15 in this volume) provides a typology of validity, drawn from the social sciences, to apply to pest risk maps. One might reasonably consider the map to be a hypothesis, so the validity of the model would be demonstrated though empirical testing (i.e. comparisons of model outputs with observations that are independent of the model). Such evaluation would also formally confirm that the first challenge was met. However, in many cases, relevant empirical observations are likely to be rare or non-existent, at least in the short term. So, pest risk modellers must use other lines of reasoning to argue for model validity. In some cases, arguments for or against a model reduce to so-called first principles with respect to content or construct validity. First principles are axiomatic statements about forces that drive biological invasions, affect population dynamics or affect invasion outcomes. Given the imperfect state of knowledge about biological invasions, debates based on first-principle arguments are seldom resolved, except in the most extreme cases.

Once a model is created and its validity established, the third part of the challenge is to portray results in a way that will be useful for decision making. The risk mapper must consider the required geographic extent (e.g. continent, country, region or parcel) and resolution of the map. Resolution (i.e. grain) refers to the size of grid cells (i.e. pixels) that comprise the map; smaller grid cells provide higher resolution. High-resolution maps can be visually appealing but come with additional uncertainty as fine-scale information is often interpolated from distantly neighbouring observations. Because all maps have some degree of distortion, a consequence of plotting the Earth's curved surface in two dimensions, thought should be given to the appropriate map projection that accurately represents area or distances (DeMers, 1997). Model results should be reported with sufficient precision to support decision making, but should not be so precise as to visually overwhelm the end user (Smans and Estève, 1996). Consider a model output, such as the Ecoclimatic Index from CLIMEX (Sutherst and Maywald, 1985; Sutherst et al., 2007), with values from 0 to 100. At the extreme, each model output could be associated with a unique colour and that colour scheme applied to the map, but subtle variations among 101 colours may be difficult to distinguish. Typically, all possible model results are divided into classes. For example, Vera et al. (2002) interpret an Ecoclimatic Index of 0 as a climate that is unsuitable for pest establishment; values of 1–10 are marginal; values of 11–25 are suitable; and values >25 are very suitable. A unique colour, stippling or shading is assigned to each class and that classification scheme is applied to each grid cell. For many pest risk maps, red designates the highest-risk areas (i.e. red zones or hot spots).

The Production of Pest Risk Maps

Despite calls for more pluralistic approaches to ecological risk analysis (NRC, 1996), the creation of pest risk maps regularly follows a technocratic approach with distinct roles for assessors, managers and stakeholders. The assessor (i.e. pest risk modeller/mapper or analyst) typically has advanced academic training in entomology, plant pathology, weed science or one of the other core biological disciplines that provide foundational knowledge about taxa that might become invasive alien species (Worner et al., 2014). Assessors may also have experience with computer science or geographic information systems or be asked to collaborate with individuals who do. Formal training in pest risk assessment is rare but is slowly increasing (Worner et al., 2014). Individual assessors or assessment teams have the technical knowledge to produce pest risk models and maps or to evaluate such products from others.

Risk managers are those decision makers, end users or land managers who use pest risk models to mitigate the likelihood or impacts of pest invasion. Risk managers are often senior personnel within governmental agencies. In the ideal case, distinctions between risk assessors and managers are maintained to prevent undue outside pressures from influencing the pest risk model or map. Likewise, risk managers use pest risk models and maps as components of a much broader decision-making framework.

Stakeholders are those individuals with an interest in the outcome of the risk mitigation decision and are most likely to be affected by that decision. Stakeholders are often given the opportunity to comment on interim or proposed/final pest risk models, maps or mitigation decisions. Stakeholders also have the opportunity to sue in court a government agency if a pest risk model, map or mitigation decision can be shown to cause demonstrable harm. Alternative approaches to ecological risk analysis seek more inclusive roles for stakeholders throughout the process, especially during problem formulation and risk mitigation.

An event, such as a request to import commodities or an incursion by an invasive alien species (as illustrated in the historical example presented above), typically triggers the development of a pest risk map, but the work of the risk assessor begins with problem formulation (Fig. 1.2). In this phase, risk managers and assessors articulate the purpose of the map, identify practical limitations (e.g. budgets and deadlines) and discuss consequences of particular errors. The purpose of the map may dictate whether it is especially important to analyse the arrival, establishment or spread phase of an invasion or to attempt a more integrative analysis across phases. Errors of commission, when some sites are classified as having higher risk than they do in reality, might be acceptable in certain contexts, for example, when risk managers want to know the maximum possible geographic extent of risk. Errors of omission, when some sites are classified as having less risk than they do in reality, might be acceptable to severely resource-constrained decision makers, for example, who are able only to expend resources where needs are greatest. As part of problem formulation, risk assessors evaluate the extent and quality of information about the invasive alien species and the endangered area and identify additional questions for research.

Problem formulation is the most important phase in the production process but is often most neglected because decision makers are not always able to articulate fully how they intend to use a risk map or what would constitute an acceptable end product. The challenge for scientists who are responsible for the production of pest risk maps is to 'balance rigor and timeliness in their work to obtain an acceptable degree of accuracy' in their map for decision makers; for decision makers, 'the challenge is to describe clearly what information is needed to support time-critical decision making' (Venette et al., 2013, p. 1). The urgency for a pest risk map can become especially high when an invasive alien species has been detected within an area of concern and decision makers contemplate needs for quarantine, eradication or containment and consider the potential consequences if no, or ineffective, action is taken.

Once the problem has been fully described, the pest-risk-mapping process moves to the analytical phase (Fig. 1.2). This phase begins with the selection of a model or suite of models appropriate to the task. Numerous software packages, described in the next section, exist to support model development. These packages typically rely on information about an invasive alien species and spatially explicit covariates. This information is more accessible now than it has ever been. For example, several online databases provide current and historical species' distributions, climatological records, elevation data, land-use classifications and population censuses. Nevertheless, not all desired data may be available, so a number of models may be considered but ultimately rejected if the requisite data to forecast the outcome of interest cannot be obtained.

Next, the assessor calibrates the model(s) by estimating key parameters to account for unique aspects of the invasive alien species under consideration or the qualities of the endangered area that has been or might be invaded. In some situations, the assessor calibrates the model by fitting it to a training data set. In the verification step, the assessor checks for coding errors and confirms that the model is giving outputs that are consistent with the data that were used to develop the model. A

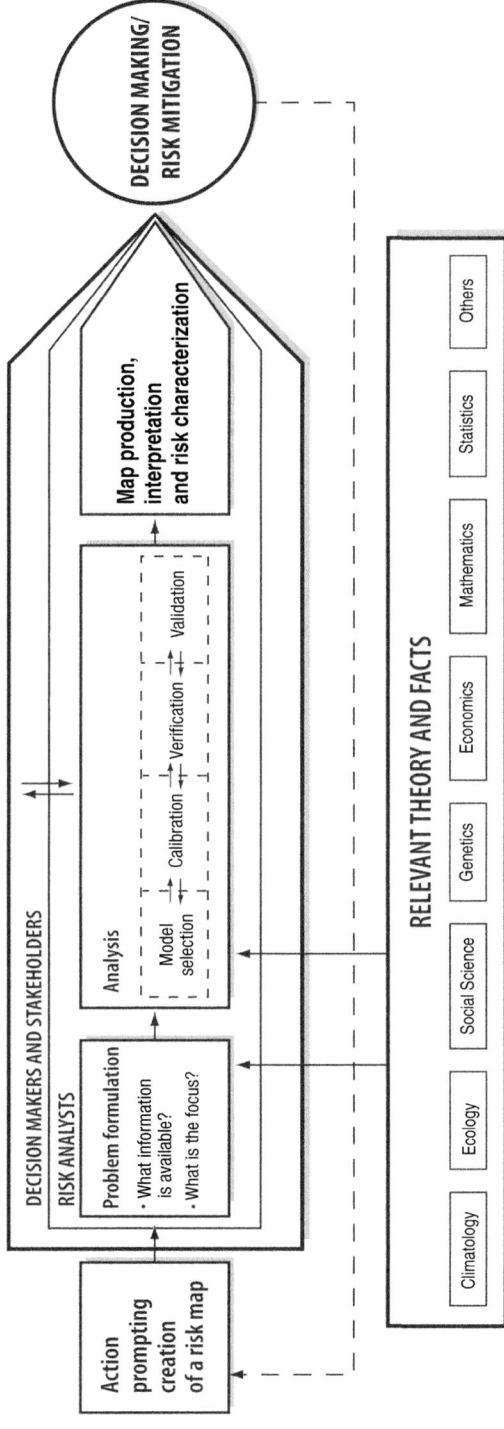

Fig. 1.2. Diagram of the events that lead to the development of a pest risk map. Pest risk analysts communicate with decision makers and stakeholders throughout the process. As the process concludes, models and maps are given to decision makers to select risk mitigation options.

validation step follows. Here, model outputs are compared with a completely independent set of data, sometimes called a test data set. Validation exercises are meant to gauge the reliability of the model. The analysis phase culminates in the production of a pest risk map. Advanced geographic information systems now exist to allow users to create visually compelling map products. These products are used by decision makers to select strategies and tactics to mitigate the risk posed by the invasive alien species. The mitigation decision itself may prompt the development of a new risk map.

An aspirational component to include with the map is a representation of the uncertainty associated with the forecast for each grid cell. Methods to characterize uncertainty in support of decision making have been proposed (e.g. Koch and Yemshanov, Chapter 13 and Yemshanov et al., Chapter 14 in this volume) but are not yet routine. Uncertainty stems from imprecise uses of language (i.e. linguistic uncertainty), a lack of knowledge (i.e. epistemic uncertainty) and inherent variation in a system (i.e. aleatory uncertainty; Regan et al., 2002). Further study of a system can reduce epistemic uncertainty, but can only serve to characterize aleatory uncertainty. Uncertainty assessments for pest risk maps presently address aleatory uncertainty.

Overview of Models to Create Pest Risk Maps

In many respects, pest risk maps take spatially implicit concepts about the course of biological invasions or impacts from invasive alien species and make them spatially explicit. For example, Orr et al. (1993) proposed one of the first qualitative risk assessment models for invasive alien species. The model had two major components, the likelihood of pest invasion and the potential consequences of pest invasion. Each of these components had three to four sub-components. Although the model only required answers of high, medium or low for each of these sub-components, the model asked a number of intrinsically spatial questions about an alien species or a commodity with which that species might be associated. Where is the species likely to arrive within the area of concern? Where might the species move? Where could the species encounter suitable climate and host plants? Pest risk mappers use a variety of models to provide spatially specific answers to these questions.

Arrival

Arrival (i.e. entry) describes the likelihood that a species could be brought into an area of concern and the conditions under which arrival would occur. Models for arrival are generally intended to answer one or more of the following questions: where is a species most likely to arrive; by what means is it likely to arrive; and in what numbers and condition (i.e. live, moribund or dead) will it arrive? Answers to these questions are typically used to justify biosecurity measures, such as inspecting cargo, screening luggage or prohibiting imports, to prevent the arrival of highly threatening species into an area of concern. Answers may also suggest the futility of such efforts for species that are likely to be brought into an area of concern by natural processes (e.g. with wind or water). When justified, biosecurity strategies that prevent the arrival of a highly threatening species are generally considered the most effective and least costly relative to other biosecurity measures.

A number of models are used to address the arrival of invasive alien species. In general, these models are used to identify particular alien species that should be of concern or to analyse pathways by which those species might arrive within the area of concern. For example, in trait-based screening assessments, a number of characteristics that improve the likelihood that a species will successfully invade a site and cause harm are identified by expert opinion or through statistical analyses of previous invasions. For example, Pheloung et al. (1999) provide an extensive list of characteristics related to the biogeography and ecology of invasive plants (e.g. evidence that the species has naturalized beyond its native range or produces buoyant propagules). Risk assessors evaluate whether a plant species of concern has these traits and use the results to generate a weed risk assessment score. This index correlates well with experts' perceptions about the degree of risk posed by particular plant species. Similarly, Kolar and Lodge (2002) analyse previous fish invasions to identify characteristics of fish (e.g. relative growth rate, diet breadth, temperature tolerances and previous invasion history) that are associated with (un-) successful invasions and estimate the probability that new fish species will establish, spread or cause harm in the Great Lakes. Such trait-based models typically are not used to generate pest risk maps but frequently rely on risk maps, especially spatial assessments of climate suitability, in the course of the assessment. Screening assessments can also be used to select species for more in-depth analysis.

Pathway models typically focus on the means by which a species might arrive in the area of concern or on the suite or abundance of potentially pestiferous species that might be associated with a particular commodity or conveyance. Pathway models often decompose the process by which pests/commodities are moved from a country of origin to the area of concern into discrete steps. The probability of successful completion of each step is estimated. For example, Colunga-Garcia and Haack (Chapter 3 in this volume) analyse import trade statistics to determine where bark- and wood-boring insects associated with solid wood packing might first arrive in the USA and be subsequently moved within the country.

In other pathway models, each step in the pathway is characterized by a probability distribution (i.e. a probability-density function). In a process generally known as Monte Carlo analysis, values are selected repeatedly at random in proportion to their likely occurrence as defined by each probability distribution. The values are used to calculate a new probability distribution for the number of individuals that are likely to complete all of the steps and arrive in the area of concern. For example, Gould et al. (2013) studied the potential for *Copitarsia corruda* to arrive in the USA with imports of Peruvian asparagus. In the course of their studies, they developed a model using statistical distributions to describe: (i) the volume of asparagus imported at different times of the year; (ii) potential rates of infestation, i.e. eggs per spear; (iii) proportionate survival of eggs during transport; (iv) the likelihood of disposing of asparagus at importer warehouses, wholesale distributors or retail outlets; and (v) the potential for individual insects to develop into adults (i.e. moths) by feeding on discarded asparagus. Monte Carlo analysis was used to estimate the likelihood that at least one mating pair of *C. corruda* on Peruvian asparagus would arrive in the USA and escape into the wild (B. Caton, North Carolina, 2014, personal communication). The analysis suggested that the likelihood would be low because once produce moved beyond importation facilities, an insufficient volume of asparagus was present at any point in time for a mating pair to develop.

Although many invasive species have been transported into areas of concern directly through human activities (e.g. international trade), some species, particularly pathogens and small insects, may be transported via wind. Models such as HYSPLIT and PMTRAJ have been used to describe where low-level jet streams might carry species of concern and in what numbers (Parry et al., Chapter 4 in the current volume). These models may also be used to describe the passive spread of a species within the area of concern after the species has established.

Establishment

Establishment occurs when an invasive alien species sustains a population through time by local reproduction. Models for establishment are generally used to determine where a species is most likely to persist through time if it were to arrive, where a species might ultimately spread if given enough time and where a species might eventually cause economic or multiple harms (Baker *et al.*, Chapter 2 in this volume). For establishment to occur, an invasive alien species must encounter suitable climate, food and a mate and avoid local antagonists (e.g. predators, pathogens and competitors). Maps of potential establishment are useful, for example, to determine the need to eradicate an invading population or to keep the invasive alien species from spreading by enacting quarantines.

Several spatially explicit models to characterize where a pest might establish are known synonymously as bioclimatic envelopes, habitat models, species distribution models or ecological niche models. Frequently these models focus on an analysis of climate because many invasive alien species are ectothermic (i.e. poikilothermic) and temperature directly affects developmental, reproductive and survival rates. Different schemes have been proposed to classify these models. Inductive modelling approaches relate information about a species' geographic distribution, either presence-and-absence or presence-only data, to any number of environmental covariates to infer statistically what factors might explain where a species occurs. Inductive models do not depend on knowledge of mechanisms by which environmental covariates might affect distribution. In contrast, deductive modelling approaches specify factors a priori that should shape species' range limits and abundances and rely on results from appropriately designed studies to evaluate, for example, how changes in temperature will affect population growth rate. Information about the climate at a site (i.e. grid cell) and the relationship between the response and the covariates is used collectively to forecast if a species might persist in the area represented by the grid cell if it were to arrive.

Software to create models for establishment varies considerably. Some software is devoted to a single modelling approach and provides a clear structure for the analysis. Venette *et al.* (2010) identify 13 software packages that have been used to evaluate environmental suitability for establishment. Two popular inductive models, GARP (genetic algorithm for rule-set prediction) and MAXENT (maximum entropy; e.g. Jarnevich amd Young, Chapter 5 in this volume), ask the user to provide latitudes and longitudes for known occurrences of a species and to select ecologically relevant, geo-referenced covariates, commonly climatological data.

Structured deductive models (e.g. NAPPFAST; Magarey *et al.*, Chapter 6 in this volume) ask the user to provide estimates of key parameters (e.g. upper and lower temperature thresholds for development) from published literature or from appropriately designed experiments. Another software tool, CLIMEX (Sutherst and Maywald, 1985; Sutherst *et al.*, 2007), may be used inductively, deductively or through a combined approach. The software assumes that broken stick models will describe a species' response to temperature or moisture gradients. The challenge is to estimate key parameters for these models. Estimates may be derived from field or laboratory studies (i.e. deductively) or by iteratively altering parameters in the model until a qualitatively satisfactory fit between the model outputs and the known distribution is reached (i.e. inductively).

Other statistical software packages (e.g. R or SAS) provide powerful analytical tools to assess the relationship between species occurrences and covariates but offer little structure to guide the analyses. Venette *et al.* (2010) identify six classes of statistical models that have been used for this purpose.

Another type of inductive model, a self-organizing map (Worner *et al.*, Chapter 7 in this volume), uses information about communities of pest organisms in an area of concern and their similarity to communities around the world to determine which species (and from where) are most likely to establish within the area of concern.

A significant assumption behind many inductive species distribution models is that the species is in equilibrium with its environment and geographic range boundaries are stable. Such a condition is more likely to be true for a species in its native range than in its adventive range.

Inductive models developed from occurrence data in the native range may underestimate the potential distribution of an invasive alien species in the adventive range because the models typically do not explicitly account for the effects of natural enemies. Natural enemies are more likely to constrain a species in its native range than in the adventive range. Freedom from natural enemies is a commonly cited reason for the success of many biological invasions. The constrained distribution from natural enemies in the native range might be misattributed to the effects of climate. Conversely, inductive models developed from a species' adventive range may overestimate the distribution in a species' native range.

Spread

Spread describes the means by which a species redistributes itself in an area of concern after it has established. Spread models are often used to answer general questions such as: where is the species likely to move through time and when is it likely to get there? Spread is either active (i.e. by flying, walking or swimming) or passive (e.g. wind-, animal- or water-dispersed). Passive, anthropogenic spread occurs when humans intentionally or accidentally move a species to new areas. If spread of an invasive alien species has been extensive before detection occurs or if a species is so highly dispersive that quarantines are unlikely to be effective (e.g. certain wind-borne pathogens), biosecurity measures begin to focus on managing and mitigating damage from the species.

A number of quantitative models have been developed to measure and forecast spread by invasive alien species. This extensive body of literature will not be reviewed here; see Shigesada and Kawasaki (1997) and Hastings et al. (2005) for excellent reviews. Many spread models have not been organized into software packages as has been done with species distribution models (but see Robinet et al., Chapter 8 in this volume); most spread models are derived mathematically and key parameters are estimated statistically from empirical observations (e.g. Tobin et al., Chapter 9 in this volume). Geographic information systems can be useful to estimate distances moved over periods of time, the results of which are further analysed in statistical software, and to project the location of the expanding invasion front over time.

In general, many spread models require information about the probability that an individual will move, or be moved, a particular distance (i.e. the dispersal kernel) and quantitative information about life history parameters, particularly population growth rates. Such information is frequently unavailable for species that have only recently been detected in areas outside their native range. Rare long-distance dispersal events have a significant impact on patterns and rates of spread but are difficult to forecast. Gravity models and individual-based models have been used to describe where invasive alien species might be moved, for example, based on an understanding of flows of people and goods through transportation corridors (Prasad et al., 2010; Crespo-Perez et al., 2011; Koch et al., 2011).

An alternative approach, applicable to invasive alien species that have already arrived within the area of concern and started to spread but do not seem to have reached the limits of their distribution, relies on statistical analyses of realized spread. The rationale is that future spread is likely to be similar to previous patterns of spread. Generalized linear models (GLMs) such as logistic regression have been used for this purpose. Care must be taken to account for spatial and temporal autocorrelations in these data or risk misestimating the statistical significance and explanatory power of the resulting model.

Many spread models assume that the propensity of a species to disperse is spatially independent. However, tests of this assumption have found that the likelihood that an individual will move depends on the environment into which it is moving. This phenomenon poses a significant complication for pest risk modellers because models of spread with parameters estimated from data from one location may have little relevance to spread behaviour in another location (Hastings et al., 2005). For pest risk modellers, extrapolations based on experiences with an invasive alien species in another location or at another time seem inescapable, so results must be interpreted with great caution.

Integrative models of invasion or impact

Integrative models attempt to synthesize results from different data sources or individual models into a more complete characterization of pest risk. For each grid cell, models estimate the probability of an invasive alien species arriving, either directly as a beachhead population or indirectly from other infested sites within the area of concern, and establishing and the associated magnitude of impact. Complete, spatially explicit characterizations of risk are exceptionally difficult to prepare and, as a result, are rare (but see Murray and Brennan, 1998).

Rule-based models (Meentemeyer et al., 2004; Seybold and Downing, 2009), often implemented in geographic information systems, have been used to describe spatial differences in relative degrees of risk. The models are forms of multi-criteria decision models. Rules, often expressed as 'if ..., then ...' statements, are typically provided by experts to reflect their knowledge and opinions about factors that might contribute to invasion risk or impact. The 'if' describes a condition that would increase or reduce risk and the 'then' typically results in the assignment of a risk score. Often, one gridded data layer relevant to each rule is queried and scores assigned to each grid cell on the map. Resultant scores from each rule for the same grid are summed and the final summation used as an indicator of the degree of risk in that cell. A simple summation indicates that each rule contributes equally to the final risk score. In more complex situations, a weighting factor can be applied to reflect the relative importance of each rule. Weights can be assigned directly by experts or can be elicited through structured questions. For example, the analytical hierarchy process, a form of multi-criteria decision model, provides an elicitation approach to generate these weights.

The weighted sums are quantitative, but ordinal, data and reflect the correct rank order of cells in a data set. The (weighted) sums do not necessarily capture the correct ratio between observations. So, for example, if sites 1, 2 and 3 get sums of 4, 16 and 32, respectively, site 3 has greater risk than sites 1 and 2, but the risk at site 3 is not eight times the risk at site 1 or twice the risk at site 2. Some authors prefer to refer to such scores as semi-quantitative assessments of pest risk.

Periodically, rule-based models will incorporate the results from an arrival, establishment or spread model. A rule is formed for the model output and applied just as it would have had the data been based on empirical observations. However, because the models generate an estimate of the actual value, the model outputs have some degree of error. This error propagates through the model, but only rarely is such error propagation formally measured.

Individual-based models have also been constructed to generate a more comprehensive representation of pest risk. For example, Koch and Smith (2008) describe the potential spread of the ambrosia beetle *Xyleborus glabratus* in the south-eastern USA after accounting for the geographic distribution of hosts, the density of hosts, the degree of climate similarity to regions in Asia where the beetle is native, the expected spread rate and the effect of host density on spread. This well-integrated model does not attempt to forecast the potential impact of the beetle and its fungal symbiont *Raffaelea lauricola* on potentially affected host plants.

Impacts from invasive alien species remain exceptionally difficult to forecast quantitatively (Venette et al., 2010). In general, impacts depend on the response of resident species or genotypes to the invasive alien, and those responses are likely to vary depending on the densities the invasive alien pest can achieve. For example, the detection of a single individual may trigger international quarantines on potentially affected hosts and cause severe economic impacts. Partial budgeting, which can be performed in any spreadsheet, is a relatively simple technique to estimate the net effect on farm incomes from value of lost yield (e.g. volume of timber or tonnes of grain) and increased control costs, less any additional revenue or cost savings (Soliman et al., 2010). Such impact assessments work well at the micro scale, but do not capture potential macroeconomic impacts, such as shifts in market prices or consumer demand. However, it is not yet clear whether these effects can be mapped.

Model Performance

Risk modellers can rigorously evaluate the performance of a model by comparing outputs with independent observations from the field. In the calibration phase of model development (see 'The Production of Pest Risk Maps' above), such comparisons are useful to determine if model parameter estimates are appropriate and during the validation phase, to evaluate the quality of a model's forecasts. A 2 × 2 decision matrix (i.e. confusion matrix; Table 1.1) is one approach to compare model outputs and field observations when the model gives a binary result (e.g. species presence or absence). Cell A includes the true negatives, cases when a model suggests that an event will not happen and it does not. Cell B describes instances where the model suggests an event will occur, but in actuality it does not. These circumstances are known as type I errors, false positives or commission errors. Cell C describes instances when the model indicates no event will occur but it does. These are type II errors, false negatives or omission errors. Cell D includes the true positives, cases where the model suggests the event will occur and it does. Overall model accuracy is the proportion of all cases in which the model correctly forecast the outcome: $(A + D)/(A + B + C + D)$. Specificity is the proportion of observed non-events that were correctly indicated by the model: $A/(A + B)$. Sensitivity is the proportion of events that were correctly identified by the model: $D/(C + D)$.

Sensitivity can receive considerable attention in the development of pest risk models and maps for two primary reasons. First, greater confidence is placed in the observance of events than non-events (i.e. not seeing anything). Detecting and recognizing an event (e.g. a species is present) is empirically testable. For example, specimens can be examined for the accuracy of identification and new specimens can be collected to confirm initial reports. For non-events, the adage 'you can never prove a negative' is appropriate. To conclude that no event of interest has occurred and will not ever occur at a location implies a great deal of knowledge about current and future events. For newly arrived alien invasive species, in particular, it may be more accurate to conclude that a species has not arrived at a site yet than to conclude that it has not arrived. Second, decision makers frequently consider false positives to be more acceptable than false negatives. The reason for this bias is not completely clear, but studies of risk perceptions in humans consistently indicate that the fear of loss is greater than the fear of a missed gain.

If events are trusted over non-events and false positives are preferable to false negatives in pest risk analysis, pest risk

Table 1.1. Confusion matrix for the comparison of binary model outputs with independent field results.

	Field results	
Model output	No event	Event
No event	A. True negatives	C. False negatives
Event	B. False positives	D. True positives

models and maps may overestimate the potential distribution and impact from invasive alien species. Models that forecast a more widespread occurrence of an alien invasive species will be judged superior to models that forecast a more limited distribution. In fact, the simplest model, 'the invasive alien species will be everywhere', will have perfect sensitivity now and in the future. Overestimation increases the likelihood that an area will be covered by biosecurity measures. If resources for biosecurity are allowed to vary 'as needed', overestimation may lead to unnecessary biosecurity measures and a suboptimal allocation of resources to other valued goods and services. Alternatively, if resources are fixed, overestimation may lead to resources being spread too thin.

In cases where model specificity is a concern, the potential for overfitting a model becomes an issue (Peterson et al., 2011). Overfitting describes the case where an excessive number of covariates, more than can be justified statistically, are included in a model. Overfit models will often suggest that an invasive alien species has specific environmental requirements and will not be able to survive in areas other than those it currently occupies. Overfitting is not easily recognized a priori but is revealed during model validation. Models found to have low error during the verification stage but high error when tested on an independent data set are likely to be overfit (Peterson et al., 2011). Such models lack robustness and are likely to underestimate the area potentially affected by an invasive alien species over time.

Many risk models describe the potential for future events as a probability, not as an absolute yes or no. In these cases, the analyst must select a threshold to describe when an event is probable (e.g. when the probability is >50%) or when it is not. At very low threshold values, the model will have perfect sensitivity but no specificity. Conversely, at high threshold values, the model will have perfect specificity but no sensitivity. A receiver-operating characteristic (ROC) curve describes the trade-off between model sensitivity (i.e. the true positive rate) and the false positive rate (i.e. 100 − specificity). If a model has no discriminatory power, the ROC curve will fall along the diagonal and the area under the ROC curve (AUC) will be 0.5. As the discriminatory power of the model improves, the AUC will approach 1. AUC is not appropriate to evaluate models of potential distribution (Jimenez-Valverde, 2012). Alternatives to AUC have been proposed when only presence data are available (Phillips and Elith, 2010; Li and Guo, 2013).

Conclusions

The International Pest Risk Mapping Workgroup (IPRMW) offered pragmatic recommendations to address pressing issues for the production of pest risk maps (Venette et al., 2010). The IPRMW now includes nearly 90 scientists, modellers and decision makers from around the world who specialize in aspects of pest risk analysis. Most members have formal or informal affiliations with decision-making bodies that regulate the movement of plants, plant products or pests that might affect plants. Some recommendations were intended to support ongoing pest risk analysis efforts. Species' distributions and environmental covariates are still frequently lacking, so a call was made to expand the availability and accessibility of primary data sources. In addition, greater communication between pest risk analysts and decision makers is needed to clarify interpretation and uses of risk maps. Some recommendations called for changes of practice within the pest risk modelling and mapping community. For example, calls were made for pest risk modellers to more fully document model development and validation, improve representations of uncertainty, increase international collaborations and work towards pest risk maps that include impacts. Some recommendations were intended to support new individuals coming into the discipline. General needs were recognized for a best-practice guide, a modelling tool kit and advanced training in risk modelling practice. Lastly, a call was made to incorporate global climate change and studies of human behaviour into pest risk models.

Pest risk analysts must also contend with some pressing issues in basic science. Foremost, many risk assessment models do not account for evolution in the invading population or recipient communities. In fact, the assumption is that phenotypic traits remain constant. This simplifying assumption constrains most risk models to relatively short time horizons (e.g. <30 years). Critical questions pertain to the heritability and conservation of the fundamental niche, selection for resilience to environmental stresses and the evolution of increased competitive ability (Felker-Quinn *et al.*, 2003; Wiens *et al.*, 2010; Morey *et al.*, 2013). Pest risk analysis has been criticized for failing to account for post-invasion evolution by alien species (Whitney and Gabler, 2008).

Pest risk modelling and mapping occurs at the interface of science and policy. Some nations may perceive governmental actions to protect human, animal and plant health from risks posed by invasive alien species as non-tariff barriers to free trade. The World Trade Organization's Agreement on the Application of Sanitary and Phytosanitary Measures (also known as the SPS Agreement) defines principles for the appropriate implementation of such protective measures (WTO, 1994). Devorshak (2012) provides a thorough discussion of the history of the SPS Agreement and its current and future ramifications for plant health protection. A chief principle of the SPS Agreement is that nations have the sovereign right to implement biosecurity policies and practices as they deem necessary insofar as those measures are supported by scientific evidence, are consistent with international standards and are not a disguised barrier to trade. The sovereignty principle acknowledges that no nation or body can dictate the biosecurity measures taken by another nation. Similarly, although the methods described in this text have been scrutinized for scientific credibility during the peer review process, no nation or organization is obligated to use these approaches.

Common challenges for pest risk modellers and mappers are outlined in this chapter and some solutions are offered throughout the remaining text of this book. These solutions (i.e. modelling approaches and software tools) were developed out of necessity and have proven to be scientifically credible. However, these tools could be refined or new tools developed to address long-standing concerns over risks posed by invasive alien species. By making the conceptual and logistical challenges that underlie pest risk mapping more transparent, my hope is that others may see new opportunities for scientific and technical advancements.

References

CABI (1986) *Quadraspidiotus perniciosus. Distribution Maps of Plant Pests, Map 7.* CAB International, Wallingford, UK.

Crespo-Perez, V., Rebaudo, F., Silvain, J.F. and Dangles, O. (2011) Modeling invasive species spread in complex landscapes: the case of potato moth in Ecuador. *Landscape Ecology* 26, 1447–1461.

Demers, M.N. (1997) *Fundamentals of Geographic Information Systems*. John Wiley and Sons, New York.

Devorshak, C. (2012) *Plant Pest Risk Analysis, Concepts and Application*. CAB International, Wallingford, UK.

Ebbels, D.L. (2003) *Principles of Plant Health and Quarantine*. CAB International, Wallingford, UK.

FAO (2012) *Glossary of Phytosanitary Terms. International Standards for Phytosanitary Measures Publication No. 5*. Food and Agriculture Organization of the United Nations, Rome. Available at: https://www.ippc.int/publications/glossary-phytosanitary-terms (accessed 29 December 2013).

Felker-Quinn, E., Schweitzer, J.A. and Bailey, J.K. (2003) Meta-analysis reveals evolution in invasive plant species but little support for Evolution of Increased Competitive Ability (EICA). *Ecology and Evolution* 3, 739–751.

Flanders, S.E. (1960) The status of San Jose scale parasitization (including biological notes). *Journal of Economic Entomology* 53, 757–759.

Gould, J., Simmons, R. and Venette, R. (2013) *Copitarsia* spp.: biology and risk posed by potentially invasive Lepidoptera from Central and South America. In: Pena, J.E. (ed.) *Potential Invasive Pests of Agricultural Crops*. CAB International, Wallingford, UK, pp. 160–182.

Hastings, A., Cuddington, K., Davies, K.F., Dugaw, C.J., Elmendorf, S., Freestone, A., Harrison, S., Holland, M., Lambrinos, J., Malvadkar, U., Melbourne, B.A., Moore, K., Taylor, C. and Thomson, D. (2005) The spatial spread of invasions: new developments in theory and evidence. *Ecology Letters* 8, 91–101.

Howard, L.O. and Marlatt, C.L. (1896) *The San Jose Scale: Its Occurrence in the United States with a Full Account of its Life History and the Remedies to be Used Against It. Bulletin No. 3*. US Department of Agriculture, Division of Entomology, Government Printing Office, Washington, DC.

Jimenez-Valverde, A. (2012) Insights into the area under the receiver operating characteristic curve (AUC) as a discrimination measure in species distribution modelling. *Global Ecology and Biogeography* 21, 498–507.

Koch, F.H. and Smith, W.D. (2008) Spatio-temporal analysis of *Xyleborus glabratus* (Coleoptera: Circulionidae (*sic*): Scolytinae) invasion in eastern US forests. *Environmental Entomology* 37, 442–452.

Koch, F.H., Yemshanov, D., Colunga-Garcia, M., Magarey, R.D. and Smith, W.D. (2011) Potential establishment of alien-invasive forest insect species in the United States: where and how many? *Biological Invasions* 13, 969–985.

Koch, T. (2011) *Disease Maps*. The University of Chicago Press, Chicago, Illinois.

Kolar, C.S. and Lodge, D.M. (2002) Ecological predictions and risk assessment for alien fishes in North America. *Science* 298, 1233–1236.

Li, W.K. and Guo, Q.H. (2013) How to assess the prediction accuracy of species presence–absence models without absence data? *Ecography* 36, 788–799.

Maslow, A.H. (1943) A theory of human motivation. *Psychological Review* 50, 370–396.

Meentemeyer, R., Rizzo, D., Mark, W. and Lotz, E. (2004) Mapping the risk of establishment and spread of sudden oak death in California. *Forest Ecology and Management* 200, 195–214.

Meyerson, L.A. and Reaser, J.K. (2002) A unified definition of biosecurity. *Science* 295, 44.

Morey, A.C., Venette, R.C. and Hutchison, W.D. (2013) Could natural selection change the geographic range limits of light brown apple moth (Lepidoptera, Torticidae) in North America? *NeoBiota* 18, 151–156.

Murray, G.M. and Brennan, J.P. (1998) The risk to Australia from *Tilletia indica*, the cause of Karnal bunt of wheat. *Australasian Plant Pathology* 27, 212–225.

NRC (1996) *Understanding Risk: Informing Decisions in a Democratic Society*. National Academy Press, Washington, DC.

Orr, R.L., Cohen, S.D. and Griffin, R.L. (1993) *Generic Nonindigeous Pest Risk Assessment Process (for Estimating Pest Risk Associated with the Introduction of Nonindigenous Organisms)*. US Department of Agriculture, Animal and Plant Health Inspection Service, Riverdale, Maryland.

Peterson, A.T., Soberón, J., Pearson, R.G., Anderson, R.P., Martínez-Meyer, E., Nakamura, M. and Araújo, M.B. (2011) *Ecological Niches and Geographic Distributions*. Princeton University Press, Princeton, New Jersey.

Pheloung, P.C., Williams, P.A. and Halloy, S.R. (1999) A weed risk assessment model for use as a biosecurity tool evaluating plant introductions. *Journal of Environmental Management* 57, 239–251.

Phillips, S.J. and Elith, J. (2010) POC plots: calibrating species distribution models with presence-only data. *Ecology* 91, 2476–2484.

Prasad, A.M., Iverson, L.R., Peters, M.P., Bossenbroek, J.M., Matthews, S.N., Sydnor, T.D. and Schwartz, M.W. (2010) Modeling the invasive emerald ash borer risk of spread using a spatially explicit cellular model. *Landscape Ecology* 25, 353–369.

Regan, H.M., Colyvan, M. and Burgman, M.A. (2002) A taxonomy and treatment of uncertainty for ecology and conservation biology. *Ecological Applications* 12, 618–628.

Seybold, S.J. and Downing, M. (2009) What risk do invasive bark beetles and woodborers pose to forests of the western US?: a case study of the Mediterranean pine engraver, *Orthotomicus erosus*. In: Hayes, J.L. and Lundquist, J.E. (eds) *The Western Bark Beetle Research Group: A Unique Collaboration with Forest Health Protection. General Technical Report PNW-GTR-874*. US Department of Agriculture, Forest Service, Pacific Northwest Research Station, Portland, Oregon, pp. 111–134.

Shigesada, N. and Kawasaki, K. (1997) *Biological Invasions: Theory and Practice*. Oxford University Press, Oxford.

Smans, M. and Estève, J. (1996) Practical approaches to disease mapping. In: Elliott, P., Cuzick, J., English, D. and Stern, R. (eds) *Geographical and Environmental Epidemiology: Methods for Small Areas*. Oxford University Press, Oxford, pp. 141–150.

Soliman, T., Mourits, M.C.M., Oude Lansink, A.G.J.M. and van der Werf, W. (2010) Economic impact assessment in pest risk analysis. *Crop Protection* 29, 517–524.

Sutherst, R.W. and Maywald, G.F. (1985) A computerized system for matching climates in ecology. *Agriculture, Ecosystems & Environment* 13, 281–299.

Sutherst, R.W., Maywald, G.F. and Kriticos, D. (2007) *CLIMEX Version 3. Users Guide*. Hearne Scientific Software Pty Ltd, Melbourne, Australia.

The White House (1999) Executive Order 13112 of February 3, 1999: Invasive species. *Federal Register* 64, 6183–6186.

Venette, R.C. and Koch, R.L. (2009) IPM for invasive species. In: Radcliffe, E.B., Hutchison, W.D. and Cancelado, R.E. (eds) *Integrated Pest Management*. Cambridge University Press, New York, pp. 424–436.

Venette, R.C., Kriticos, D.J., Magarey, R.D., Koch, F.H., Baker, R.H.A., Worner, S.P., Raboteaux, N.N.G., Mckenney, D.W., Dobesberger, E.J., Yemshanov, D., De Barro, P.J., Hutchison, W.D., Fowler, G., Kalaris, T.M. and Pedlar, J. (2010) Pest risk maps for invasive alien species: a roadmap for improvement. *Bioscience* 60, 349–362.

Venette, R.C., Kriticos, D.J., Koch, F.H., Rafoss, T., van der Werf, W. and Baker, R. (2013) *Summary of the International Pest Risk Mapping Workgroup Meeting Sponsored by the Cooperative Research Program on Biological Resource Management for Sustainable Agricultural Systems*. Organisation for Economic Cooperation and Development, Paris. Available at: http://www.oecd.org/tad/crp/CRP%20Conference%20Summary-Advancing%20risk%20assessment%20models%20for%20invasive%20alien%20species.pdf (accessed 29 December 2013).

Vera, M.T., Rodriguez, R., Segura, D.F., Cladera, J.L. and Sutherst, R.W. (2002) Potential geographical distribution of the Mediterranean fruit fly, *Ceratitis capitata* (Diptera: Tephritidae), with emphasis on Argentina and Australia. *Environmental Entomology* 31, 1009–1022.

Whitney, K.D. and Gabler, C.A. (2008) Rapid evolution in introduced species, 'invasive traits' and recipient communities: challenges for predicting invasive potential. *Diversity and Distributions* 14, 569–580.

Wiens, J.J., Ackerly, D.D., Allen, A.P., Anacker, B.L., Buckley, L.B., Cornell, H.V., Damschen, E.I., Davies, T.J., Grytnes, J.A., Harrison, S.P., Hawkins, B.A., Holt, R.D., Mccain, C.M. and Stephens, P.R. (2010) Niche conservatism as an emerging principle in ecology and conservation biology. *Ecology Letters* 13, 1310–1324.

Worner, S.P., Venette, R.C., Braithwaite, M. and Dobesberger, E. (2014) The importance of core biological disciplines in plant biosecurity. In: Gordh, G. and McKirdy, S. (eds) *The Handbook of Plant Biosecurity*. Springer, New York, pp. 73–117.

WTO (1994) *The WTO Agreement on the Application of Sanitary and Phytosanitary Measures*. World Trade Organization, Geneva. Available at: http://www.wto.org/english/tratop_e/sps_e/spsagr_e.htm (accessed 17 December 2013).

2 Mapping Endangered Areas for Pest Risk Analysis

Richard Baker,[1]* Dominic Eyre,[1] Sarah Brunel,[2] Maxime Dupin,[3] Philippe Reynaud[4] and Vojtěch Jarošík[5]

[1]*Department for Environment, Food and Rural Affairs, York, UK;* [2]*European and Mediterranean Plant Protection Organization, Paris, France;* [3]*Previously at: INRA, UR633 Zoologie Forestière, Orléans, France;* [4]*Anses – Plant Health Laboratory, CBGP, Montferrier-sur-Lez, France;* [5]*Lately at: Institute of Botany, Academy of Sciences of the Czech Republic, Průhonice, Czech Republic*

Abstract

This chapter provides guidance on mapping risks posed by invasive alien species to support pest risk analysis (PRA), the process required to justify phytosanitary measures. Because pest risk mapping can be challenging and resource-intensive, the situations in which risk maps are particularly useful are highlighted. The procedures described focus on mapping areas where the pest can establish and potentially cause the greatest harm. In the first stage of risk mapping, the factors that might influence the potential distribution and impacts of an invasive alien species are identified and the data are assembled and mapped. In the second stage, the maps of each factor are combined using matrix rules to generate areas of potential establishment and highest risk. These general procedures are illustrated with two examples. Risk maps for the western corn rootworm, a maize pest that has invaded Europe, are based on the combination of maps of climatic suitability, the presence of sandy soils, the distribution of grain and forage maize and the value of these commodities in Europe. Uncertainty is estimated by varying the classification of climatic suitability to obtain the worst, best and most likely scenarios. Risk maps for the common water hyacinth, an invasive plant on the Iberian Peninsula, are based on maps of climatic suitability, the distribution of suitable wetland habitats and areas of conservation importance. The chapter concludes by summarizing some of the major challenges that remain to enhance the production of risk maps for PRA and their interpretation by risk managers.

Background: The Context for Pest Risk Mapping

Pest risk analysis (PRA) is fundamental to plant biosecurity because it requires a review of the biology and ecology of an invasive alien invertebrate, pathogen or plant to evaluate its potential to enter, establish, spread and cause harm to plants in an area where it is, or might become, established. If the risk is unacceptable, appropriate phytosanitary measures to prevent entry and establishment are identified.

PRAs that might affect trade between two or more countries should follow International Standards for Phytosanitary

---

* Corresponding author. E-mail: Richard.Baker@defra.gsi.gov.uk

Measures (ISPMs), principally ISPM 11 (FAO, 2013), because ISPMs have been formulated by the International Plant Protection Convention and are recognized by the World Trade Organization (WTO, 1994; Devorshak, 2012). Although international standards clearly describe the elements that need to be assessed to evaluate the likelihood of entry and establishment together with the magnitude of spread and impacts, they do not provide guidance on the methods to be used. A number of schemes (i.e. methodologies) have been created to assist pest risk analysts with the production of PRAs based on expert judgement supported by documented evidence. EPPO (2011) provides one of the best-known examples. Analysts need additional guidance on when and how to produce maps to support PRA.

This chapter is principally based on work undertaken by PRATIQUE, a project funded by the European Union (EU) 7th Framework Programme in 2008–2011 to enhance PRA techniques (Baker et al., 2009; Baker, 2012) by developing decision support schemes (or systems; hereafter DSSs) for mapping endangered areas. DSSs provide organization, structure and assistance in evaluating alternative courses of action.

Definitions of critical terms: the area of potential establishment, the endangered area and the area at highest risk

Within the geographical area of interest to the pest risk analyst, three risk areas can be distinguished: (i) the area of potential establishment; (ii) the area at highest risk; and (iii) the endangered area. The area of potential establishment is where it is likely that there is 'perpetuation, for the foreseeable future, of a pest within an area after entry' (FAO, 2012). This area encompasses the geographical range in which climate is suitable, hosts (or suitable soils for weeds) are available and no other biotic or abiotic constraints prevent establishment. The area at highest risk does not have an official definition but is where, independent of economic loss, impacts are likely to be greatest (e.g. because particularly valuable or vulnerable hosts occur where abiotic and biotic factors are most suitable for an invasive alien species). The endangered area is where 'ecological factors favour the establishment of a pest whose presence in the area will result in economically important loss' (FAO, 2012). The endangered area is therefore within and smaller than or equal to the area at highest risk. Likewise, the area at highest risk is always within and almost always smaller than the area of potential establishment. Although these three areas can be described by listing geographical regions or sub-national boundaries (e.g. states, provinces or counties) that might be affected, maps generally convey a clearer message.

Mapping areas of potential establishment and highest risk for pests of plants is difficult, but mapping endangered areas creates additional challenges because endangered area maps need to show where plants have the greatest economic, environmental and social value and where the pest population density is likely to cause an economically important loss. The threshold for such a loss has been named the economic injury level (EIL) and defined by Stern et al. (1959) as 'the lowest population density of a pest that will cause economic damage; or the amount of pest injury which will justify the cost of control'. The concept has been developed further by Pedigo et al. (1986) and Pedigo and Higley (1992). Although the EIL concept has been used in PRA (Lammers and MacLeod, 2007), the authors of this chapter are not aware of PRAs that include maps showing where the EIL would be expected to be exceeded. The EIL is very difficult to determine because, although yield loss data are available for some crops (e.g. Oerke et al., 1994; CABI, 2013), these losses need to be extrapolated to the current crop, pest management and market conditions in the PRA area. The EIL concept can be extended beyond yield to include marketability. For example, high quality standards demanded for fresh produce in Europe can mean that a relatively low level of pest damage leads to a large drop in value of fresh fruit and vegetables.

Even in the rare cases where an appropriate EIL is available, it is difficult to make accurate estimates of population abundance because population dynamics models are complex to construct and difficult to parameterize reliably even for well-known species. Although EILs can be estimated for particular crops in particular locations in particular years, they will rarely be sufficient to account for spatial or temporal variations in agronomic practices or climate within the large areas that often need to be assessed in PRAs.

Scope: aspects of pest risk mapping in this chapter

Maps of the area at highest risk are often sufficient to support written descriptions of the endangered area in a PRA. The type of map utilized, whether the area at highest risk or the endangered area, should be explicitly stated and justified. This chapter addresses three key questions: (i) when is it appropriate to map areas at highest risk or endangered areas; (ii) what data and information are required for mapping areas at highest risk; and (iii) how can maps that represent key factors be integrated to identify endangered areas or areas at highest risk? Examples demonstrate how these questions can be answered for invasive alien pests of plants.

Guidance and Needs for Mapping Highest-Risk Areas

Guidance is needed to help inexperienced pest risk analysts, students and even more advanced practitioners both to correctly apply modelling and mapping techniques and to accurately interpret risk and uncertainty. This guidance includes assistance in deciding whether it is necessary or even appropriate to construct a map given the information available and selecting from the large number of modelling methods that can be used to create the maps. The lack of such important guidance was a key incentive for the establishment of the International Pest Risk Mapping Workgroup in 2007 which identified the principal challenges in pest risk mapping and set them out in a roadmap (Venette *et al.*, 2010).

The simplest method to describe the highest-risk area or endangered area, as recommended by the European and Mediterranean Plant Protection Organization (EPPO) DSS for PRA, is to refer to existing ecoclimatic zones, geographic areas, crop distributions, production systems or ecosystems within a geographic area of concern. While such descriptions can be effective, maps generally provide a clearer method for visualizing, summarizing and communicating risk in PRA. However, maps need to be constructed, interpreted and communicated with caution as they may convey a false sense of certainty. Risk maps have been deployed in a number of detailed PRAs. In some cases, these maps have been limited to assessments of climatic suitability with or without host or habitat distributions. Such maps may provide an indication of the area of potential establishment, but because they omit critical economic, agronomic or environmental factors, they are not sufficient to identify the endangered area or the area at highest risk.

DSSs are now available to guide the construction of pest risk maps. For example, NAPPFAST (Magarey *et al.*, 2007, 2011, Chapter 6 in this volume) provides a suite of models that allows risk assessors to model pest phenology and pathogen infection, apply a technique for climatic matching, generate pathway risk maps and utilize a Pareto dominance method (Yemshanov *et al.*, 2010) to combine climate suitability analyses with host distribution records. Although the methods available in NAPPFAST are widely applicable, NAPPFAST is particularly designed to support the requirements of the US Department of Agriculture, uses customized climatic data sets and is thus difficult for other countries to apply. For the EU, the PRATIQUE project has developed DSSs for mapping climatic suitability (Eyre *et al.*, 2012), the area of potential establishment and the area at highest risk (Baker et al., 2012). When linked to models of spread (Kehlenbeck et al., 2012; Robinet et al., 2012, Chapter 8 in this volume) and economic impact (Soliman et al., 2012), the PRATIQUE DSSs can help to map invasion dynamics and illustrate endangered areas. The DSSs are independent of models used and the area of concern. Examples of the application of the DSSs provided in this chapter are for the EU.

The Structure of the PRATQUE DSSs to Map Areas at Highest Risk and Endangered Areas

A flow diagram (Fig. 2.1) shows how the different stages of the DSSs are linked. The Introduction identifies situations when pest risk mapping is most appropriate, the resources required and prior steps that need to be taken. Four stages follow the Introduction. In Stage 1, key factors that might affect the course or impact of invasion are identified, data are assembled and maps are generated. In Stage 2, mapped layers of spatial data are combined to determine the areas at highest risk. In Stage 3, areas are identified where the EIL might be exceeded by the pest and economically important losses could occur (Soliman et al., 2012). During Stage 3, potential relationships between modelled climatic suitability and measures of impact are explored. This stage is difficult and requires a detailed modelling approach that will rarely be possible within the PRA process. In Stage 4, generic spread models are applied to portray invasion dynamics (Kehlenbeck et al., 2012; Robinet et al., Chapter 8 in this volume).

The PRATIQUE DSSs and this chapter emphasize Stages 1 and 2 to identify the area of potential establishment and the area at highest risk. Maps of the area at highest risk often suffice to describe endangered areas. In practice, maps of the areas at highest risk provide evidence to support the PRA process, surveillance programmes, contingency plans and remedial actions.

Application of the PRATIQUE DSS

Before commencing the process of mapping areas of highest risk, it is important to think through options and implications. Detailed mapping may require considerable resources including time and expertise. For example, models and maps of climatic suitability from computer programs such as CLIMEX (Sutherst et al., 2007) and MAXENT (Phillips et al., 2006; Phillips and Dudík, 2008) require extensive inputs. Guidance on options and implications is provided in the introduction to the PRATIQUE DSS.

The DSSs help users determine if risk maps are needed and feasible. Mapping may be only necessary once a qualitative PRA has been completed, at least in draft, and may be only practical after careful consideration of the information available and resources required. When risk maps are desirable but insufficient resources are available, published maps, simpler models or individual maps of key risk components may still be useful in communicating risk (Baker et al., 2013).

Maps may be particularly important when there is great uncertainty concerning the likelihood of establishment; but, if establishment occurred, the magnitude of impact would be expected to be high. Maps may also be critically important when the magnitude of impacts is likely to vary spatially, significant areas of crop production or special areas of conservation are particularly vulnerable or targeted surveillance and protection measures are required. If PRAs include phytosanitary measures likely to have a significant impact on trading partners, maps can provide valuable justification for the PRA conclusions.

Low priorities for mapping

Pests that clearly pose little risk to an area are generally considered low priorities for mapping. Maps are also a low priority for pests already considered likely to become

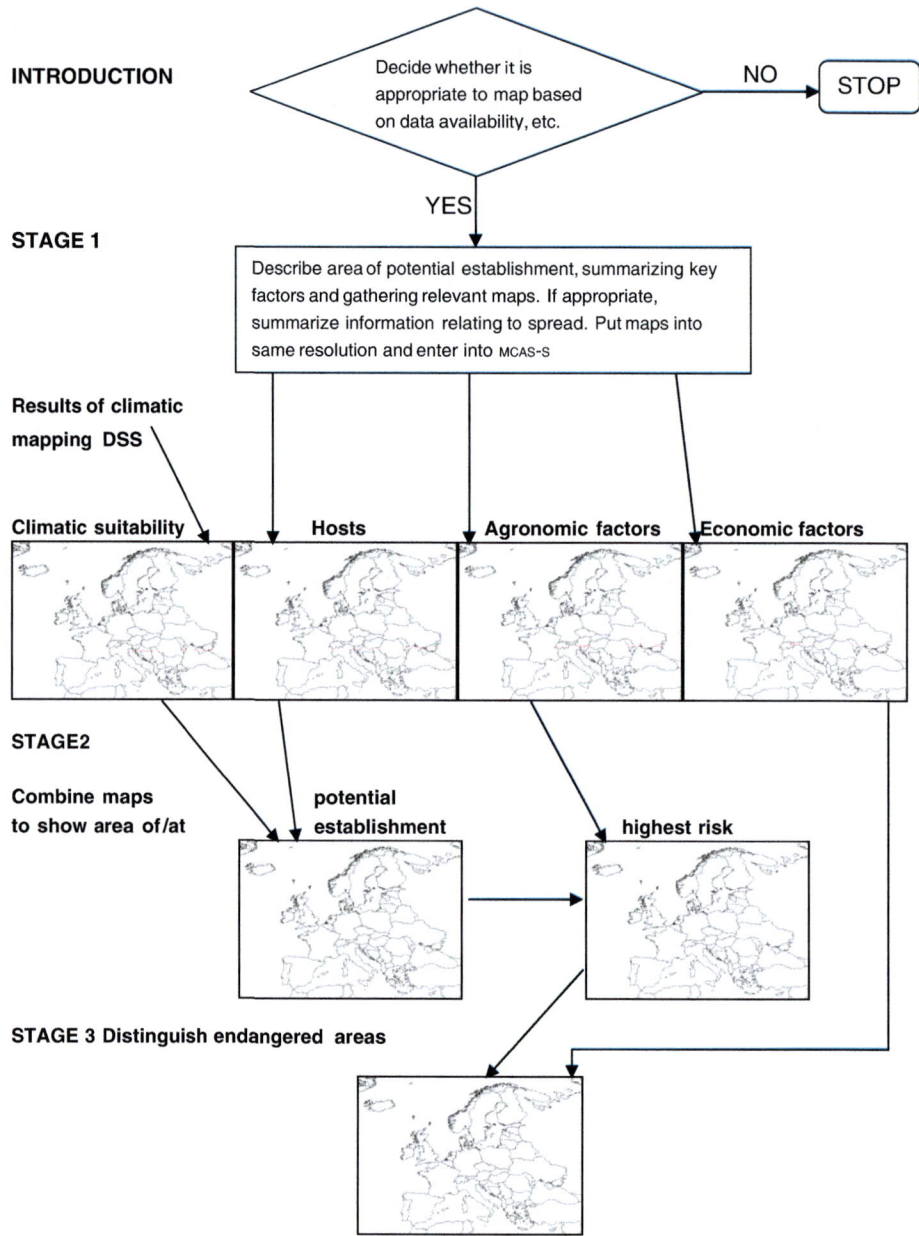

Fig. 2.1. Mapping endangered areas decision support scheme (DSS) – an example of its use.

widely established (e.g. because the pest is common in neighbouring countries with similar climates and host distributions). For example, the pine wood nematode (*Bursaphalencus xylophilus*) clearly posed a risk to Spain once outbreaks were discovered in Portugal (Vicente *et al.*, 2012). Sometimes, an area of potential establishment can be identified without risk mapping, especially for pests with obvious risks (e.g. primary tree killers), if analysts can refer to existing national boundaries, climatic zones, eco-climatic zones, host distributions, habitats or geographic coordinates. Lastly, maps may be considered low priority when extant data are too unreliable.

Information and resources for the PRATIQUE DSSs

Before starting the DSSs, analysts need to assemble three critical pieces of information: (i) a list of key factors that constrain establishment; (ii) a description of the area of potential establishment and suitability for population growth therein; and (iii) an assessment of potential economic, environmental and social impacts. Information about the species' biology, distribution and climatic responses can also be useful. Analysts can obtain this information from a draft or final detailed PRA.

Locating relevant data sets can be time-consuming for the PRA in general and for mapping pest risk. Useful pest distribution, host distribution and economic data sets have been collated and rated as part of the PRATIQUE project, and these are now listed online (EPPO, 2012a). PRATIQUE also provides assistance through: (i) links to models for assessing spread (Kehlenbeck et al., 2012; Robinet et al., Chapter 8 in this volume) and impacts (Soliman et al., 2012); (ii) a database of online data sources for PRA (EPPO, 2012a) such as crop maps (based on Monfreda et al., 2008); (iii) automation of several common data manipulation procedures by using the R software language (R Core Team, 2012); and (iv) a user-friendly method for combining maps using open source software, the Multi-Criteria Analysis Shell for Spatial Decision Support (MCAS-S; ABARES, 2012), as an alternative to expensive and complex geographic information systems (GISs). MCAS-S requires all data sets to be at the same resolution and adjustments may require some GIS skills, but it does allow different spatial data sets to be displayed and combined to represent risk using a variety of arithmetical, logical and matrix methods that can be set by the user. Guidance, data and tools are provided by the DSS to ensure that all the key data sets are available at 10 km × 10 km resolution for the EU by appropriate re-projection, up-scaling and down-scaling.

Linked to the DSS for mapping highest-risk area is a separate climatic DSS that assists with selection of an appropriate climatic risk mapping model. Selection is based on: (i) the pest's biology, its known climatic responses and the extent of knowledge of its distribution at the centre and the periphery of its geographical range; and (ii) relevant attributes of each model, such as functionality, ease of use and quality assurance (Eyre et al., 2012).

Stage 1: Assembling and Mapping Data for Key Factors

Stage 1 starts by assembling data for the key factors that determine the suitability of an area for establishment. These factors are described in detail in the EPPO PRA scheme (EPPO, 2011) and relate to the suitability of both the biotic environment, e.g. host availability, and the abiotic environment, e.g. climatic suitability. The next task is to describe what data are available for each of the relevant factors that could potentially be put into a map. An example is provided in Table 2.1.

Data and any existing maps then need to be assembled for the factors that determine where hosts or habitats are at highest risk from impacts (excluding the factors that enhance establishment that have already been selected). Determining the factors that lead to highest risk requires substantial knowledge of the host, its cultivation or habitat, and its ecological or commercial value. Key factors may be related to host genotype or crop husbandry. Some cultivated plants are inherently more susceptible to damage by particular pests than other plants. This difference in susceptibility can be for a variety of reasons including: the crop cultivar; cropping patterns, especially grower reliance on one cultivar; growing conditions that favour pest impacts; high quality standards in crop

Table 2.1. Summary of data and maps for key factors that determine the area of potential establishment for *Diabrotica virgifera virgifera* in the EU.

Factor	Assessment and uncertainty score	Maps for the PRA area
Host plants and suitable habitats	Abundant Low uncertainty	EU grain and forage maize maps
Climatic suitability	Largely similar to native range Low uncertainty	CLIMEX maps (Kriticos *et al.*, 2012)
Other abiotic factors	Sandy soil has a negative effect Low uncertainty	Soil maps

EU, European Union; PRA, pest risk analysis.

products; and, if the host is killed, long replacement times. For example, development of the chestnut leaf miner (*Cameraria ohridella*) varies considerably between different chestnut species and hybrids. *Aesculus* × *hemiacantha* is the most susceptible, but no damage is observed on *A.* × *carnea* (Rämert *et al.*, 2011). Debilitating damage to plants has a greater potential to cause economic damage to crops with long replacement times, such as forest and orchard trees.

Growing conditions and production methods can make some crops more vulnerable to damage; for example, maize grown in sandy soil is less likely to be damaged by western corn rootworm (*Diabrotica virgifera virgifera*) than when the crop is grown in other soil types (Turpin and Peters, 1971). Western corn rootworm is also much less likely to be a significant pest in first year maize fields than in fields which have been in continuous maize production for several years (Dillen *et al.*, 2010; Krügener *et al.*, 2011).

Certain conditions increase the likelihood that an invasive alien species will have major environmental impacts. The most severe impacts are anticipated when the pest damages populations of keystone, rare or endemic species or those that provide important ecosystem services. For example, the invasive harlequin ladybird, *Harmonia axyridis*, is thought to have led to a decline of a natural insect predator in northern Europe, the native two-spotted ladybird, *Adalia bipunctata* (Roy *et al.*, 2012). Environmental impacts are also likely to be high in special areas for conservation such as national nature reserves. Islands are known to be particularly vulnerable to the impacts of invasive species, e.g. invasive species have had a major impact on the people and economy of Hawaii (Jang, 2007).

Techniques for classifying variables representing different risk factors

Once relevant data sets have been converted to the same resolution, uploaded to MCAS-S and mapped, thresholds for data classification are then selected. Each factor has a range of values and these values are divided into a number of classes, with the number and definition of each class specified by experts on a case-by-case basis (Stephens *et al.*, 2007). Six classes (e.g. absent, very low, low, medium, high and very high) or two classes (e.g. absent and present) are common. Classification is undertaken before factors are combined because assessors generally find it easier to attempt a judgement about the relationship between a single factor (e.g. climate suitability or host abundance) and the topic of interest (e.g. the likelihood of establishment) than to consider combinations of variables simultaneously. It can also be worthwhile to explore the effect of different classifications on the outcome once factors are combined because combining mapped variables that have been pre-classified will lead to a loss of some information (Dupin *et al.*, 2011). An alternative is to combine variables arithmetically and then classify the results of this combination, thus eliminating one source of error.

Representing uncertainty

Since the difficulties of representing risk and uncertainty are magnified when mapping combinations of risk factors, analysts must ensure that maps displaying the different risk factors are examined before combination. Where possible, maps showing the most likely outcome and the extremes should also be provided to risk managers. One way of doing this is to use lower, most likely and upper estimates of factors that contribute to a calculation of the endangered area. To calculate the best case, most likely case and worst case extent of the endangered area, the lower, most likely and upper estimates of different variables (e.g. crop values) can be combined. For example, a best case scenario could be calculated by using the lower estimates of pest damage crop value. If there is any evidence to support a probability distribution of the factors contributing to the endangered area, it is then possible to create maps representing the 'best case' (e.g. 10% percentile outcome), median scenario (50%) and 'worst case' scenario (90%). In some cases, it will be appropriate to investigate different scenarios, especially when considerable changes in risk are expected over time. Further discussion about the representation of uncertainty is provided by Koch and Yemshanov (Chapter 13) and Yemshanov et al. (Chapter 14) in this volume.

Stage 2: Combining Maps

In Stage 2 the maps of the key risk factors prepared in Stage 1 are combined to: (i) illustrate the area of potential establishment; (ii) map the area of potential impacts; and (iii) define the area(s) at highest risk. The assumptions and combination rules used by the assessor should be documented. Seven matrix rules (Fig. 2.2; see colour plate section) have been implemented using MCAS-S to cover the map combinations required:

1. The *minimum* rule matrix is used when both factors, e.g. climatic suitability and host presence, are required so that the factor with the most severe constraint (lowest classification) dominates and the other factor is ignored. Equivalent scores for each factor impose equivalent levels of constraint.
2. The *maximum* rule matrix is the inverse of the minimum rule matrix. The factor with the highest classification is what matters and the classification of the other factor is ignored. For example, if protected environments are present, the climate, as measured at weather stations, may not be important.
3. The *addition* rule matrix sums both factors. This rule can be applied, for example, when combining production maps for two host crops, when production is expressed on ordinal scales from none to high. This rule was used in a PRA for western corn rootworm to combine the lack of crop rotation (which may increase population densities) with the amount of crop production in the area of potential establishment.
4. The *limiting factor* matrix can be used when the absence of one factor prevents pest establishment but, if present, the value of the factor does not have any influence on the potential for pest establishment. For example, a pathogen may only become established in soils below pH 7; however, there may be no difference in the likelihood of establishment between soils of pH 6 or 5.
5. The *modified average* matrix increases the classification by one class if favourable or decreases the classification by one if unfavourable. For example, an early harvest date may prevent some pests from completing their life cycle whereas a late harvest date may allow time for all individuals to develop.
6. The *high risk* matrix can be used to identify locations where there is a coincidence of high classes, e.g. when combining the suitability of the area of potential establishment with factors influencing impact, to produce a five-level risk classification, in addition to a zero-risk level. This differs from the maximum rule matrix in that both factors have to be at maximum to give the highest risk output, whereas for the maximum rule matrix only one factor has to be at maximum. The high risk matrix is recommended, for example, to produce a map of the areas at highest risk from the

combination of a map of potential establishment with a map of economic, environmental or social impacts.

7. *Cox's risk* matrix is a 5 × 5 classification scheme that is useful to characterize risk as the product of probability and consequence, when these two factors are measured qualitatively. The results are categorized into three classes (in addition to zero).

To produce a classified map of potential establishment, the analyst begins with maps of host distribution, other biotic constraints (e.g. presence of vectors), climatic suitability and other abiotic constraints (e.g. soil suitability, irrigation usage or protection from weather). All calculations occur within each grid cell, so all cells must have a common resolution (e.g. 10 km × 10 km). If an invasive alien species has more than one host, host densities in each cell are added together and that sum is given an ordinal rating (e.g. 0–5). The ordinal ranking of total host density can be combined with maps of the presence/absence of other biotic factors required by the invasive alien species by using the limiting factor matrix. The combined map shows areas where biotic factors are suitable for establishment. This map can be integrated with a map of climatic suitability, also described with ordinal rankings, by using the minimum rule matrix, to generate a map of the area where climatic factors and hosts are suitable for establishment. This map can be joined to maps of the presence/absence of additional abiotic requirements for establishment (e.g. soil types) by using the limiting factor matrix to produce a map of the area of potential establishment without additional management factors. The resulting map can be combined with maps of management activities that would favour the establishment of the pest (e.g. irrigation or protection of sites from weather) by using the maximum rule matrix to create a map of the area of potential establishment with additional management.

To produce a classified map of the areas at highest risk, the map of potential establishment is combined with a map of economic impacts. This part of the process assumes crop values are known. For each affected crop, the distribution (i.e. crop hectares per 10 km × 10 km grid), yield (e.g. tonnes per crop hectare) and value (e.g. Euros per tonne) are multiplied together, and the resulting crop production value (Euros per 10 km × 10 km grid) is given an ordinal rating. (We recommend six classes divided by equal area.) When two or more host crops are affected, productivity value maps are combined by using the addition matrix to generate a map of total crop production value. If an additional factor places the host crops at higher risk (e.g. a lack of crop rotation increases the potential damage caused by western corn rootworm), this factor can be combined with the total crop production value by using the modified average matrix. If crop production values are not known, the areas at highest risk still can be identified. The map of potential establishment is combined with a map of environmental, social or other economic impact maps. We recommend using the high risk matrix for this integration.

Stage 3: Mapping Endangered Areas

Before attempting to map an endangered area, it is important to have a clear understanding of economically important loss in terms of the PRA. For producers, as explained above in the section 'Definitions of critical terms', economically important loss occurs when the EIL is exceeded. EILs can be expressed in several ways such as the number of pests recorded over a unit of time on traps or the proportion of plants infected with a disease. When the EIL is exceeded for a crop pest, an intervention such as treating the crop with a plant protection product gives greater financial benefits than the cost of the treatment. Defining an EIL for a pest that does not attack a commercially important crop, but rather attacks a resource viewed as a common good, ecosystem service or plant of conservation concern, may require other strategies.

To forecast the area where economically important losses are likely to occur, it is important to identify the areas at highest risk (see 'Stage 1: Assembling and Mapping Data for Key Factors' and 'Stage 2: Combining Maps' above) and to know the extent to which the conditions necessary for pest populations to exceed the economic injury level are present in the PRA area. For example, MacLeod et al. (2005) created a model to determine the potential area in which the western corn rootworm might cause economic damage in England and Wales. In the absence of models predicting pest population densities and the extent to which they are likely to exceed the EIL, yield and quality loss scenarios can still be explored. For example, the analyst could create a worst case scenario, where it is assumed that the pest has reached its maximum geographical extent and the maximum pest density.

With sufficient data it would not only be possible to map the area where economic loss may occur but also the estimated losses per unit area, e.g. in a 10 km × 10 km grid cell. For example, if $100,000 worth of a susceptible crop is grown in a particular grid cell and losses due to a pest are likely to be about 10%, then losses could be mapped in units of $10,000. However, it is unlikely that such estimates can be made with confidence because changes in pest populations, relationships between population densities and yield or quality loss, the climate, crop variety, agronomic practices, other crops grown in the area, surrounding non-crop habitat and market values can influence the magnitude of economic loss.

In cases where yield and quality loss are caused by polyphagous species, decision rules for combining maps of economic impact for different hosts are provided.

Worked Examples

To illustrate how the DSS functions, the western corn rootworm is used as the principal example. Methods for mapping the risks posed by the water hyacinth (*Eichhornia crassipes*), a highly invasive freshwater plant, are also described to illustrate the procedures for an organism that primarily causes environmental impacts.

Example 1: the western corn rootworm

The western corn rootworm is a highly destructive pest of maize in the USA. Beetles and lodged maize plants were first detected in Serbia (Yugoslavia) in 1992 near the Belgrade airport. The western corn rootworm is univoltine, overwintering as an egg in the soil. Larvae injure maize plants by feeding on root tissue. Adults feed on all above-ground parts of maize plants, especially maize pollen and silk.

For Stage 1, information from PRAs for France (EPPO, 1997), The Netherlands (Lammers, 2006) and the UK (CSL, 2007) and other sources obtained by the PRATIQUE project were used to complete Table 2.1.

In Stage 2, maps were produced for each key factor. A map of climatic suitability for the western corn rootworm was prepared by using climatology from 1961–1990 gridded to a resolution of 0.5° latitude × 0.5° longitude in the software CLIMEX to generate an Ecoclimatic Index (EI) (Kriticos et al., 2012), a measure of climate suitability that ranges from 0 (unsuitable) to 100 (ideal suitability) for a species. The CLIMEX model for western corn rootworm assumed, among other parameters, that 666 degree-days (base 9°C) would be required annually to complete a generation (modified from Baufeld et al., 1996). This map was rescaled to a 10 km × 10 km grid and divided into six classes for entry into MCAS-S (Fig. 2.3a; see colour plate section).

A map of sandy soil, defined as clay <18% and sand >65%, from the European Soil Database version 2 (JRC, 2010) was rescaled to a 10 km × 10 km grid and divided into presence or absence classes (Fig. 2.3b; see colour plate section). The importance of sandy soil is based on observations from Hungary that the damage caused by larvae of the western corn rootworm on sandy soil is insignificant.

Maize was considered to be the only host and the western corn rootworm was assumed only to be able to establish where grain or forage maize is present. Maps of grain maize (Fig. 2.3c) and forage maize data (Fig. 2.3d; see colour plate section) at

5 arcminutes latitude × 5 arcminutes longitude resolution from Monfreda et al. (2008) were rescaled to provide the percentage of hectares in a 10 km × 10 km grid cell with maize and divided into five classes based on equal area, with a sixth class representing the absence of maize. Maize production in tonnes per hectare at 5 arcminutes latitude × 5 arcminutes longitude resolution from Monfreda et al. (2008) was rescaled to a 10 km × 10 km grid cell and divided into six classes (with one absent class) based on equal area. Maize production maps (in tonnes per 10 km × 10 km grid cell) were generated for grain maize (Fig. 2.3e) and forage maize (Fig. 2.3f; see colour plate section) by multiplying the harvested area (expressed as the percentage of the grid square area) by 100 (to convert to hectares) and then by the grain maize yield (in tonnes per hectare). (Note: As each 10 km × 10 km grid cell contains 10,000 ha, the calculation of harvested hectares is equivalent to multiplying the harvested proportion of the area of the grid cell by 10,000.) To obtain the production value per crop per grid cell, the grain maize and forage maize production were multiplied by the estimated price per tonne for each crop (€50 per tonne for forage maize and €250 per tonne for grain maize). To map the area at highest risk of impacts based on economic data, the grain maize and the forage maize production value maps were then combined to generate the total maize production value map. Maps showing the proportion of maize that is not rotated and present a much higher risk are available only for a few countries, e.g. France, and so cannot be mapped at a European scale. The environmental and social impacts of western corn rootworm were considered to be negligible.

To take soil type into account, the map of total maize production was combined with the map of soil textures. Where the soil texture is defined as sandy, damage is expected to be low, so the modified average matrix rule was applied and the impact classification was reduced by one. Where the soil texture is not sandy, the impact level was unchanged.

The key factors that were considered to influence the area at highest risk of damage by the western corn rootworm were: (i) crop value; (ii) whether maize is grown for forage or grain production; (iii) yield; (iv) sandy soils; and (v) the extent to which maize is rotated with other crops.

An area-of-potential-establishment map for the western corn rootworm was produced by using the minimum matrix rule to combine the climatic suitability map based on the CLIMEX EI with the host distribution map based on the areas harvested for grain and forage maize (in hectares per 10 km × 10 km grid cell). The maps of host production (i.e. value per grid cell) and the area where the soils are not sandy were combined by using a modified average matrix and then combined with the area of potential establishment to map the area at highest risk. The area in each risk category can also be summarized in a table using standard GIS procedures. The MCAS-S software allows the user to easily export data and statistics from the combined maps. Table 2.2 shows the number of 10 km × 10 km grid cells in each of the six levels of risk for each EU member state.

Uncertainties arise when classifying data and model outputs before or after combining data sets. For example, in classifying the CLIMEX EI for the climatic suitability of western corn rootworm, different interpretations of the EI might reasonably be applied. For example, 'very highly suitable' cells might have an EI $\geq$ 12 (i.e. the worst case), an EI $\geq$ 20 (i.e. the most likely case) or an EI $\geq$ 28 (i.e. the best case). As the threshold for 'very high' suitability increases, the total area classified as very highly suitable will decline. This reduction in turn reduces the area of potential establishment and the area at highest risk. Figure S2.1 in the online supplement to Chapter 2 shows where the areas of climatic suitability, potential establishment and highest risk differ according to the three scenarios.

The choice of scenario can have important ramifications. For example, if the numbers of grid cells in the high and very high risk ratings are tallied for each EU country under the best case scenario, the

Table 2.2. Number of 10 km × 10 km grid cells in each EU country classified by risk level for *Diabrotica virgifera virgifera*. Countries are ranked based on the sum of high and very high cells.

		Number (proportion) of cells at each risk level					
Rank	Region[a]	Absent	Very low	Low	Moderate	High	Very high
1	France	700 (0.13)	1356 (0.25)	897 (0.17)	864 (0.16)	1099 (0.20)	486 (0.09)
2	Italy	896 (0.31)	656 (0.22)	569 (0.20)	274 (0.09)	310 (0.11)	212 (0.07)
3	Hungary	5 (0.01)	92 (0.10)	126 (0.14)	288 (0.32)	213 (0.23)	190 (0.21)
4	Romania	327 (0.14)	775 (0.33)	391 (0.16)	642 (0.27)	228 (0.10)	8 (<0.01)
5	Spain	1141 (0.23)	3276 (0.67)	317 (0.06)	130 (0.03)	55 (0.01)	0 (0.00)
6	Bulgaria	92 (0.08)	728 (0.66)	177 (0.16)	64 (0.06)	46 (0.04)	0 (0.00)
7	Portugal	336 (0.40)	235 (0.28)	148 (0.18)	71 (0.09)	38 (0.05)	3 (<0.01)
8	Greece	255 (0.22)	687 (0.59)	149 (0.13)	49 (0.04)	25 (0.02)	7 (0.01)
9	Austria	540 (0.64)	126 (0.15)	73 (0.09)	80 (0.10)	23 (0.03)	0 (0.00)
10	Slovenia	74 (0.39)	71 (0.38)	17 (0.09)	12 (0.06)	15 (0.08)	0 (0.00)
11	Germany	188 (0.05)	2695 (0.75)	575 (0.16)	119 (0.03)	11 (<0.01)	0 (0.00)
12	Netherlands	0 (0.00)	190 (0.63)	61 (0.20)	43 (0.14)	8 (0.03)	0 (0.00)
13	Slovakia	104 (0.22)	248 (0.53)	64 (0.14)	47 (0.10)	5 (0.01)	0 (0.00)
14	Czech Republic	37 (0.05)	668 (0.86)	65 (0.08)	7 (0.01)	1 (<0.01)	0 (0.00)
15	Belgium	6 (0.02)	203 (0.68)	67 (0.22)	24 (0.08)	0 (0.00)	0 (0.00)
16	Denmark	5 (0.01)	373 (0.99)	0 (0.00)	0 (0.00)	0 (0.00)	0 (0.00)
17	Estonia	407 (1.00)	0 (0.00)	0 (0.00)	0 (0.00)	0 (0.00)	0 (0.00)
18	Finland	3278 (1.00)	0 (0.00)	0 (0.00)	0 (0.00)	0 (0.00)	0 (0.00)
19	Ireland	636 (1.00)	0 (0.00)	0 (0.00)	0 (0.00)	0 (0.00)	0 (0.00)
20	Latvia	617 (0.99)	8 (0.01)	0 (0.00)	0 (0.00)	0 (0.00)	0 (0.00)
21	Lithuania	110 (0.16)	557 (0.84)	0 (0.00)	0 (0.00)	0 (0.00)	0 (0.00)
22	Luxembourg	0 (0.00)	25 (1.00)	0 (0.00)	0 (0.00)	0 (0.00)	0 (0.00)
23	Poland	182 (0.06)	2929 (0.94)	8 (<0.01)	0 (0.00)	0 (0.00)	0 (0.00)
24	Sweden	4379 (1.00)	0 (0.00)	0 (0.00)	0 (0.00)	0 (0.00)	0 (0.00)
25	UK	1246 (0.54)	1047	0 (0.00)	0 (0.00)	0 (0.00)	0 (0.00)

[a]Cyprus and Malta are missing.

five countries at greatest risk are France (342 high cells and 345 very high cells), Italy (300 and 25), Hungary (31 and 0), Portugal (27 and 3) and Spain (22 and 0). Similar tallies under the most likely scenario change five countries at greatest risk to France (1099 and 486), Italy (310 and 212), Hungary (213 and 190), Romania (228 and 8) and Spain (55 and 0). Under the worst case scenario, the five countries at greatest risk change yet again to France (1442 and 571), Romania (979 and 66), Hungary (431 and 224), Italy (305 and 227) and Germany (315 and 43). Although the largest area under threat is in France under each scenario, the worst case scenario showed Romania and Hungary are more threatened than Italy.

Example 2: the common water hyacinth

Risk mapping for water hyacinth provides an example of identifying areas at high risk from environmental impacts. A PRA for the common water hyacinth, a highly invasive floating freshwater weed native to South America, was produced by EPPO (2008) and included a CLIMEX model to identify areas in Europe that are climatically suitable or unsuitable for invasion. Water hyacinth is sensitive to frost, which kills the leaves and upper petioles that normally protect rhizomes. At prolonged cold temperatures (<5°C), rhizomes may be killed (Owens and Madsen, 1995). Water hyacinth's low tolerance to frost limits distribution of this species to southern Europe. The common water hyacinth colonizes still or slow-moving water in estuarine habitats, lakes, urban areas, watercourses and wetlands. Although the species can tolerate extremes of water-level fluctuation and seasonal variations in flow velocity, nutrient availability, pH, temperature and toxic substances (Gopal, 1987), seawater concentrations >20–25% are lethal (Muramoto et al., 1991).

The area of potential establishment was identified by mapping habitats suitable for establishment based on the EU Corine Land Cover Map (i.e. inland marshy areas, watercourses and water bodies; EEA, 2012) and climatic suitability as characterized by CLIMEX and combining these maps by using a limiting factor matrix. The process is illustrated in Fig. S2.2 in the online supplement to Chapter 2. The areas at highest risk were considered to be the habitats of highest conservation importance. These were represented by the Natura 2000 sites, special areas of conservation established under the 1993 EU habitats directive (EEA, 2011), that could be invaded by the common water hyacinth.

Based on an expert interpretation of the evidence, five Natura 2000 habitat types were selected as being particularly suitable for water hyacinth invasion: (i) natural eutrophic lakes with *Magnopotamion* or *Hydrocharition*; (ii) constantly flowing Mediterranean rivers with *Glaucium flavum*; (iii) watercourses of plain to montane levels with *Ranunculion fluitantis* and *Callitricho–Batrachion* vegetation; (iv) rivers with muddy banks with *Chenopodion rubri* p.p. and *Bidention* p.p. vegetation; and (v) constantly flowing Mediterranean rivers with *Paspalo–Agrostidion* species and hanging curtains of *Salix* and *Populus alba*. The area of the Iberian Peninsula at highest risk from the common water hyacinth was identified by overlaying the raster map of the area of potential establishment with a vector map of Natura 2000 sites that are particularly suitable for colonization by common water hyacinth. Figure S2.3 in the online supplement to Chapter 2 provides details about where the highest-risk locations occur.

Further Work

DSSs provide pest risk analysts with a structured process to identify key factors that affect the probability and consequence of invasion by alien species and to combine maps of those factors to identify endangered areas. PRATIQUE DSSs focus on mapping areas of highest risk due to the difficulties in identifying areas where economic loss will occur. However, considerable testing and development are required to identify best practices and further enhance guidance. Advancements are needed to: (i) identify the most appropriate methods for calculating and representing uncertainty in risk maps; (ii) accurately reflect current climate and taking climate change into account; (iii) map organisms with complex life cycles; (iv) map areas where economic loss will occur; and (v) provide guidance on the choice of models in particular situations. Additional challenges include: (i) the lack of consistently accurate and up-to-date data for all the key parameters, particularly if the PRA area includes several countries; (ii) the difficulty of combining data sets at different resolutions and formats without clear combination rules, (iii) the complexity of many GISs; and (iv) the absence of clear techniques or conventions for representing pest risk uncertainty in maps.

Priorities for enhanced pest risk mapping DSSs are already articulated in the roadmap provided by Venette et al. (2010). There is a continuing need to help risk mappers provide clear, accurate and readily interpretable maps to risk managers and policy makers. Policy makers in Europe increasingly require the use of shorter PRA schemes that can be completed quickly, e.g. the Quick Scan PRA scheme of The Netherlands (Netherlands Plant Protection Service, 2012), the Rapid PRA scheme of the UK (Fera, 2013) and the EPPO Express PRA scheme (EPPO, 2012b). This demand for rapid PRAs creates additional challenges for pest risk mapping that are partly addressed by the guidance given here and by Baker et al. (2013), who give examples of how shortcuts to the detailed DSSs can be made when information, time or resources are limited.

Acknowledgements

The EU PRATIQUE project was funded by the EU 7th Framework Programme (grant number 212459). The DSS for mapping the areas at highest risk was created by the authors of this paper in collaboration with Jan Benninga, Johan Bremmer and Tarek Soliman (LEI, The Netherlands), Zhenya Ilieva (PPI, Bulgaria), Hella Kehlenbeck (JKI, Germany), Darren Kriticos (CSIRO, Australia), David Makowski (INRA, France), Jan Pergl (IBOT, Czech Republic), Christelle Robinet (INRA, France), Wopke van der Werf (Wageningen University, The Netherlands) and Susan Worner (Bio-Protection, New Zealand). We gratefully acknowledge their contributions and the suggestions for improvement from the editor and two anonymous reviewers. Sadly, Vojtěch Jarošík died while this chapter was being written and we acknowledge the important contributions he made to this work.

This chapter is enhanced with supplementary resources.

To access the material please visit:

www.cabi.org/openresources/43946/

References

ABARES (2012) *Multi-Criteria Analysis Shell for Spatial Decision Support (MCAS-S)*. Australian Bureau of Agricultural and Resource Economics and Sciences, Canberra. Available at: http://www.daff.gov.au/abares/data/mcass (accessed 17 December 2013).

Baker, R.H.A. (2012) An introduction to the PRATIQUE Research Project. *EPPO Bulletin* 42, 1–2.

Baker, R.H.A., Battisti, A., Bremmer, J., Kenis, M., Mumford, J., Petter, F., Schrader, G., Bacher, S., De Barro, P., Hulme, P.E., Karadjova, O., Lansink, A.O., Pruvost, O., Pysek, P., Roques, A., Baranchikov, Y. and Sun, J.H. (2009) PRATIQUE: a research project to enhance pest risk analysis techniques in the European Union. *EPPO Bulletin* 39, 87–93.

Baker, R.H.A., Benninga, J., Bremmer, J., Brunel, S., Dupin, M., Eyre, D., Ilieva, Z., Jarosik, V., Kehlenbeck, H., Kriticos, D.J., Makowski, D., Pergl, J., Reynaud, P., Robinet, C., Soliman, T., van der Werf, W. and Worner, S. (2012) A decision-support scheme for mapping endangered areas in pest risk analysis. *EPPO Bulletin* 42, 65–73.

Baker, R.H.A., Eyre, D. and Brunel, S. (2013) Matching methods to produce maps for pest risk analysis to resources. *NeoBiota* 18, 25–40. Available at: http://www.pensoft.net/journals/neobiota/article/4056/ (accessed 17 December 2013).

Baufeld, P., Enzian, S. and Motte, G. (1996) Establishment potential of *Diabrotica virgifera* in Germany. *EPPO Bulletin* 26, 511–518.

CABI (2013) *Crop Protection Compendium*. CAB International, Wallingford, UK. Available at: http://www.cabi.org/cpc/ (accessed 17 December 2013).

CSL (2007) *CSL Pest Risk Analysis for Western Corn Rootworm (Diabrotica virgifera virgifera)*. Central Science Laboratory, York, UK. Available at: http://fera.co.uk/plantClinic/documents/factsheets/csldiab.pdf (accessed 27 February 2015).

Devorshak, C. (2012) *Plant Pest Risk Analysis, Concepts and Application*. CAB International, Wallingford, UK.

Dillen, K., Mitchell, P.D. and Tollens, E. (2010) On the competitiveness of *Diabrotica virgifera virgifera* damage abatement strategies in Hungary: a bio-economic approach. *Journal of Applied Entomology* 134, 395–408.

Dupin, M., Brunel, S., Baker, R., Eyre, D. and Makowski, D. (2011) A comparison of methods for combining maps in pest risk assessment: application to *Diabrotica virgifera virgifera*. *EPPO Bulletin* 41, 217–225.

EEA (2011) *Natura 2000 Data – the European Network of Protected Sites*. European Environment Agency, Copenhagen. Available at: http://www.eea.europa.eu/data-and-maps/data/natura-2000 (accessed 17 December 2013).

EEA (2012) *Corine Land Cover 2006 Raster Data*. European Environment Agency, Copenhagen. Available at: http://www.eea.europa.eu/data-and-maps/data/corine-land-cover-2006-raster-2 (accessed 16 December 2013).

EPPO (1997) *Pest Risk Assessment for* Diabrotica virgifera *prepared by the French Plant Protection Service (translation). EPPO Document 97/6445*. European and Mediterranean Plant Protection Organization, Paris.

EPPO (2008) *Pest Risk Analysis for* Eichhornia crassipes. European and Mediterranean Plant Protection Organization, Paris. Available at: http://www.eppo.int/QUARANTINE/Pest_Risk_Analysis/PRAdocs_plants/08-14407%20PRA%20record%20Eichhornia%20crassipes%20EICCR.pdf (accessed 17 December 2013).

EPPO (2011) *Decision-support Scheme for Quarantine Pests. EPPO Standard PM 5/3(5)*. European and Mediterranean Plant Protection Organization, Paris. Available at: http://archives.eppo.int/EPPOStandards/PM5_PRA/PRA_scheme_2011.doc (accessed 17 December 2013).

EPPO (2012a) *Useful Datasets for PRA*. EPPO Capra Network, Paris. Available at: http://capra.eppo.org/dataset/ (accessed 17 December 2013).

EPPO (2012b) *Decision-support Scheme for an Express Pest Risk Analysis. EPPO Standard PM 5.5(1)*. European and Mediterranean Plant Protection Organization, Paris. Available at: http://dx.doi.org/10.1111/epp.2591 (accessed 17 December 2013).

Eyre, D., Baker, R.H.A., Brunel, S., Dupin, M., Jarošík, V, Kriticos, D.J., Makowski, D., Pergl, J., Reynaud, P., Robinet, C. and Worner, S. (2012) Rating and mapping the suitability of the climate for pest risk analysis. *EPPO Bulletin* 42, 48–55.

FAO (2012) *Glossary of Phytosanitary Terms. International Standards for Phytosanitary Measures Publication No. 5*. Food and Agriculture Organization of the United Nations, Rome. Available at: https://www.ippc.int/publications/glossary-phytosanitary-terms (accessed 17 December 2013).

FAO (2013) *Pest Risk Analysis for Quarantine Pests including Analysis of Environmental Risks. International Standards for Phytosanitary Measures Publication No. 11, Rev. 1*. Food and Agriculture Organization of the United Nations, Rome. Available at: https://www.ippc.int/publications/pest-risk-analysis-quarantine-pests (accessed 17 December 2013).

Fera (2013) *PRAs for Consultation on Pest Risk Management*. Food and Environment Research Agency, York, UK. Available at: http://www.fera.defra.gov.uk/plants/plantHealth/pestsDiseases/praTableNew.cfm (accessed 16 December 2013).

Gopal, B. (1987) *Water Hyacinth*. Elsevier, Amsterdam.

Jang, E.B. (2007) Fruit flies and their impact on agriculture in Hawaii. *Proceedings of the Hawaiian Entomological Society* 39, 117–119.

JRC (2010) *European Soils Database*. Joint Research Centre, European Commission, Brussels. Available at: http://eusoils.jrc.ec.europa.eu/esdb_archive/ESDB/Index.htm (accessed 14 December 2013).

Kehlenbeck, H., Robinet, C., van der Werf, W., Kriticos, D., Reynaud, P. and Baker, R. (2012) Modelling and mapping spread in pest risk analysis: a generic approach. *EPPO Bulletin* 42, 74–80.

Kriticos, D.J., Reynaud, P., Baker, R.H.A. and Eyre, D. (2012) Estimating the global area of potential establishment for the western corn rootworm (*Diabrotica virgifera virgifera*) under rain-fed and irrigated agriculture. *EPPO Bulletin* 42, 56–64.

Krügener S., Baufeld, P. and Unger, J.G. (2011) Modellierung der populationsentwicklung des westlichen maiswurzelbohrers (*Diabrotica virgifera virgifera*) – Betrachtung verschiedener Eingrenzungsoptionen. *Journal für Kulturpflanzen* 63, 69–76.

Lammers, W. (2006) *Report of a Pest Risk Analysis for Diabrotica virgifera virgifera*. Netherlands Plant Protection Service, Utrecht, The Netherlands. Available at: http://www.nvwa.nl/txmpub/files/?p_file_id=2000886 (accessed 14 December 2013).

Lammers, W. and MacLeod, A. (2007) *Report of a Pest Risk Analysis for Helicoverpa armigera (Hübner, 1808)*. Netherlands Plant Protection Service, Utrecht, The Netherlands and Central Science Laboratory, York, UK. Available at: https://secure.fera.defra.gov.uk/phiw/riskRegister/plant-health/documents/helicoverpa.pdf (accessed 26 February 2015).

MacLeod, A., Baker, R.H.A. and Cannon, R.J.C. (2005) Costs and benefits of European Community (EC) measures against an invasive alien species – current and future impacts of *Diabrotica virgifera virgifera* in England & Wales.

In: Alford, D. and Backhaus, G. (eds) *Introduction and Spread of Invasive Species: BCPC Symposium Proceedings No. 81*. British Crop Production Council, Alton, UK, pp. 167–172.

Magarey, R.D., Fowler, G.A., Borchert, D.M., Sutton, T.B., Colunga-Garcia, M. and Simpson, J.A. (2007) NAPPFAST: an internet system for the weather-based mapping of plant pathogens. *Plant Disease* 91, 336–345.

Magarey, R.D., Borchert, D.M., Engle, J.S., Colunga-Garcia, M., Koch, F.H. and Yemshanov, D. (2011) Risk maps for targeting exotic plant pest detection programs in the United States. *EPPO Bulletin* 41, 46–56.

Monfreda, C., Ramankutty, N. and Foley, J.A. (2008) Farming the planet: 2. Geographic distribution of crop areas, yields, physiological types, and net primary production in the year 2000. *Global Biogeochemical Cycles* 22, 1–19.

Muramoto, S., Aoyama, I. and Oki, Y. (1991) Effect of salinity on the concentration of some elements in water hyacinth (*Eichhornia crassipes*) at critical levels. *Journal of Environmental Sciences and Health* 26, 205–215.

Netherlands Plant Protection Service (2012) *Evaluation of Pest Risks*. Netherlands Plant Protection Service, Wageningen, The Netherlands. Available at: http://www.vwa.nl/onderwerpen/english/dossier/pest-risk-analysis/evaluation-of-pest-risks (accessed 17 December 2013).

Oerke, E.-C., Dehne, D.-W., Schonbeck, F. and Weber, A. (1994) *Crop Production and Crop Protection: Estimated Losses in Major Food and Cash Crops*. Elsevier, Amsterdam.

Owens, C.S. and Madsen, J.D. (1995) Low temperature limits of water hyacinth. *Journal of Aquatic Plant Management* 33, 63–68.

Pedigo, L.P. and Higley, L.G. (1992) The economic injury level concept and environmental quality: a new perspective. *American Entomologist* 38, 12–21.

Pedigo, L.P., Hutchins, S.H. and Higley, L.G. (1986) Economic injury levels in theory and practice. *Annual Review of Entomology* 31, 341–368.

Phillips, S.J., Anderson, R.P. and Schapire, R.E. (2006) Maximum entropy modeling of species geographic distributions. *Ecological Modelling* 190, 231–259.

Phillips, S.J. and Dudík, M. (2008) Modeling of species distributions with Maxent: new extensions and a comprehensive evaluation. *Ecography* 31, 161–175.

R Core Team (2012) *R: A Language and Environment for Statistical Computing*. R Foundation for Statistical Computing, Vienna. Available at: http://www.r-project.org (accessed 17 December 2013).

Rämert, B., Kenis, M., Kärnestam, E., Nyström, M. and Rännbäck, L.-M. (2011) Host plant suitability, population dynamics and parasitoids of the horse chestnut leafminer *Cameraria ohridella* (Lepidoptera: Gracillariidae) in southern Sweden. *Acta Agriculturae Scandinavica, Section B – Soil & Plant Science* 61, 480–486.

Robinet, C., Kehlenbeck, H., Kriticos, D.J., Baker, R.H.A., Battisti, A., Brunel, S., Dupin, M., Eyre, D., Faccoli, M., Ilieva, Z., Kenis, M., Knight, J., Reynaud, P., Yart, A. and van der Werf, W. (2012) A suite of models to support the quantitative assessment of spread in pest risk analysis. *PLoS One* 7, e43366.

Roy, H.E., Adriaens, T., Isaac, N.J.B., Kenis, M., Onkelinx, T., Martin, G.S., Brown, P.M.J., Hautier, L., Poland, R., Roy, D.B., Comont, R., Eschen, R., Frost, R., Zindel, R., Van Vlaenderen, J., Nedvěd, O., Ravn, H.P., Grégoire, J.-C., de Biseau, J.-C. and Maes, D. (2012) Invasive alien predator causes rapid declines of native European ladybirds. *Diversity and Distributions* 18, 717–725.

Soliman, T., Mourits, M.C.M., van der Werf, W., Hengeveld, G.M., Robinet, C. and Oude Lansink, A.G.M. (2012) Framework for modelling economic impacts of invasive species, applied to pine wood nematode in Europe. *PLoS One* 7, e45505.

Stephens, A.E.A., Kriticos, D.J. and Leriche, A. (2007) The current and future potential geographical distribution of the oriental fruit fly, *Bactrocera dorsalis* (Diptera: Tephritidae). *Bulletin of Entomological Research* 97, 369–378.

Stern, V.M., Smith, R.F., Van Den Bosch, R. and Hagen, K.S. (1959) The integrated control concept. *Hilgardia* 29, 81–101.

Sutherst, R.W, Maywald, G.F. and Kriticos, D. (2007) *CLIMEX Version 3. Users Guide*. Hearne Scientific Software Pty Ltd, Melbourne, Australia.

Turpin, F.T. and Peters, D.C. (1971) Survival of southern and western corn rootworm larvae in relation to soil texture. *Journal of Economic Entomology* 64, 1448–1451.

Venette, R.C., Kriticos, D.J., Magarey, R.D., Koch, F.H., Baker, R.H.A., Worner, S.P., Gomez. R., Nadilia, N., McKenney, D.W., Dobesberger, E.J., Yemshanov, D., De Barro, P. J., Hutchison, W.D., Fowler, G., Kalaris, T.M. and Pedlar, J. (2010) Pest risk maps for invasive alien species: a roadmap for improvement. *Bioscience* 60, 349–362.

Vicente, C., Espada, M., Vieira, P. and Mota, M. (2012) Pine wilt disease: a threat to European forestry. *European Journal of Plant Pathology* 133, 89–99.

WTO (1994) *The WTO Agreement on the Application of Sanitary and Phytosanitary Measures (SPS Agreement)*. World Trade Organization, Geneva. Available at: http://www.wto.org/english/tratop_e/sps_e/spsagr_e.htm (accessed 17 December 2013).

Yemshanov, D., Koch, F.H., Ben-Haim, Y. and Smith, W.D. (2010) Detection capacity, information gaps and the design of surveillance programs for invasive forest pests. *Journal of Environmental Management* 91, 2535–2546.

3 Following the Transportation Trail to Anticipate Human-mediated Invasions in Terrestrial Ecosystems[1]

Manuel Colunga-Garcia[1]* and Robert A. Haack[2]

[1]*Center for Global Change and Earth Observations, Michigan State University, East Lansing, Michigan, USA;* [2]*USDA Forest Service, Northern Research Station, Lansing, Michigan, USA*

Abstract

Freight transportation is an important pathway for introduction and dissemination of invasive alien species. Identifying the final destination of imports is critical in determining the likelihood of introduction by invasive alien species. In this chapter, we used the Freight Analysis Framework (FAF) database to model freight transport of imports to the USA with urban areas as their final destination. We also used FAF-based projections for the year 2040 for 20 pathways consisting of four commodities from five world regions of origin. After modelling the final distribution of imports among urban areas in the USA, we characterized the distribution patterns of imports and assessed the introduction potential of invasive alien species to urban forests. Freight pattern analyses were conducted for one category of agricultural products and three categories of imports whose products or packaging materials are associated with invasive alien species: wood products, non-metallic mineral products and machinery. We found that the type of import and the world region of origin greatly influenced the final distribution of imported products. A simple pest risk assessment model for invasive alien species in urban forests was built based on the percentage of forestland and volume of imports for each urban area. We found that several urban areas are potentially highly vulnerable to introductions from invasive alien species, regardless of import pathway. We conclude that inclusion of freight movement information is critical for proper risk assessment of invasive alien species.

Introduction to Trade Analysis for Invasive Alien Species

Economic losses by invasive alien species in agricultural and forest ecosystems of the USA have been estimated at US$37.1 billion per annum (Pimentel *et al.*, 2005). International transport of goods has been widely acknowledged as one of the leading causes of introductions of invasive alien species (Meyerson and Mooney, 2007; Westphal *et al.*, 2008; Hulme, 2009; Perrings *et al.*, 2009). Manufactured and agricultural goods, including their associated wood packaging material and cargo containers, can harbour a number of invasive alien species (NACEC, 2003; Caton *et al.*, 2006; McCullough *et al.*, 2006; Liebhold *et al.*, 2012; Haack and Rabaglia, 2013). Wood-based crating, dunnage and pallets have been implicated as major pathways by which bark- and wood-infesting insects have

* Corresponding author. E-mail: colunga@msu.edu

© CAB International and USDA 2015. *Pest Risk Modelling and Mapping for Invasive Alien Species* (ed. R.C. Venette) moved with commerce among countries (IPPC, 2009). Imports such as tiles, machinery, marble, steel and ironware have been commonly associated with borer-infested wood packaging material (Haack, 2006).

Port inspections can help detect invasive alien species and reduce their frequency of entry. When high-risk pests are intercepted at ports in the USA, reports of those interceptions are stored in electronic databases maintained by the US Department of Agriculture, Animal and Plant Health Inspection Service (NRC, 2002; McCullough et al., 2006; PPQ, 2011). This information allows federal and state agencies to implement detection surveys and possible mitigation measures at ports if warranted (McCullough et al., 2006). Inspection rates of imports entering the USA are under 2% (NRC, 2002); and thus a non-zero probability exists that some infested cargo will be missed at the ports and be transported to the cargo's final destination. Therefore, it would be valuable to identify the principal final destinations of selected imports that are commonly associated with invasive alien species to aid in regional risk assessments and detection surveys.

The Freight Analysis Framework (FAF) database shows promise in helping to predict the movement of invasive alien species via imports. FAF consists of several data tables for US imports, exports and within-country flow of 43 commodity groups (Southworth et al., 2011). FAF is compiled from multiple data sources in which data gaps are filled using a combination of log-linear modelling and iterative proportional fitting (Southworth et al., 2011). There are eight world regions of origin for the imports in FAF and 63 FAF regions within the USA. The latter consist of Core Based Statistical Areas (CBSAs), which cover entire US states or portions of states.

To obtain an initial idea of the potential usefulness of the FAF data to understand freight transportation patterns in the USA, we can look at one record in the database (Table 3.1). In this record, we see that the

Table 3.1. Summary variables from Freight Analysis Framework's Origin–Destination Database. This example shows codes and values for machinery imports from Asia (southern, central or western regions) that arrived at Seattle–Tacoma–Olympia CSA[a] ports in the USA via maritime transportation and were delivered to Los Angeles–Long Beach–Riverside CSA via truck transportation in 2007, with projections for 2040.

Variable	Description	Code or value	Description of code or value
TRADE_TYPE	Trade type (import, export, domestic)	2	Import
SCTG2	Two-digit SCTG[b] commodity (43 categories)	34	Machinery
FR_ORIG	World region of origin (eight regions)	806	Southern–Central–Western Asia
FR_INMODE	Transportation mode from world region of origin to US point of entry (seven modes)	3	Water
DMS_ORIG	US point of entry (123 regions)	531	Seattle–Tacoma–Olympia CSA
ST_ORIG	US state of point of entry	53	Washington
DMS_MODE	Transportation mode from point of entry to final destination (seven modes)	1	Truck
DMS_DEST	US regional final destination (123 regions)	061	Los Angeles–Long Beach–Riverside CSA
ST_DEST	US state of regional final destination	06	California
TONS_2007	Volume of imports in 2007	0.3759	Thousand short tons[c]
VALUE_2007	Value of imports in 2007	1.565	US$ million
TONS_2040	Projected weight of imports in 2040	2.4805	Thousand short tons
VALUE_2040	Projected value of imports in 2040	10.329	US$ million

[a]CSA, Combined Statistical Area.
[b]SCTG, Standard Classification of Transported Goods.
[c]One short ton = 0.9072 tonnes (metric tons).

Combined Statistical Area (CSA) of Los Angeles–Long Beach–Riverside in the state of California imported 376 short tons (~341 tonnes), with a total value of US$2.5 million, of machinery goods that originated somewhere in the southern, central or western regions of Asia during 2007. These imports initially arrived via maritime transport at ports located in the Seattle–Tacoma–Olympia CSA in the state of Washington. After arrival, imports were then transported via truck to their final destination in the Los Angeles–Long Beach–Riverside CSA. In the same record, we can see that, by the year 2040, annual imports are projected to increase to 2480 short tons (~2250 tonnes) with a total value of US$10.3 million. The full record includes projected import figures every five years from 2015 to 2040. With more than half a million import records, the FAF database provides a comprehensive view of freight flow in the USA.

FAF data have been used to enhance pest risk assessment and survey efforts for invasive alien species (Koch et al., 2011; Magarey et al., 2011). To illustrate the usefulness of the FAF database from an invasive alien species perspective, we used projections of regional freight transport in the USA to characterize the potential for introductions into urban forests. We focused on urban areas because as hubs of economic activity (McGranahan and Satterhwaite, 2003; Rodrigue et al., 2009), urban areas are likely the final destination for most imports and thus are at greater risk of invasive alien species introductions. The initial introductions of the Asian longhorned beetle (*Anoplophora glabripennis* (Coleoptera: Cerambycidae)) in Illinois, New York, New Jersey and Toronto (Haack et al., 1997, 2010) and the emerald ash borer (*Agrilus planipennis* (Coleoptera: Buprestidae)) in Michigan (Poland and McCullough, 2006) are prime examples. As with the emerald ash borer, once invasive alien species become established in urban areas they are likely to continue spreading into rural areas.

Before continuing, we provide some concepts defined by the US Census Bureau (USCB, 2012) that are important in understanding this chapter:

1. *Combined Statistical Areas* (CSAs) consist of two or more adjacent core-based statistical areas that have substantial economic connections, as measured by the number of commuters.
2. *Core Based Statistical Areas* (CBSAs) refer collectively to metropolitan and micropolitan statistical areas and consist of the county or counties associated with at least one core population of at least 10,000, plus adjacent counties having a high degree of social and economic integration with the core, as measured through commuting ties with the counties associated with the core.
3. *Metropolitan Statistical Areas* are CBSAs associated with at least one urbanized area.
4. *Micropolitan Statistical Areas* are CBSAs associated with at least one urban cluster.
5. *Urbanized area* consists of densely developed territory that contains ≥50,000 people.
6. *Urban cluster* consists of densely developed territory that has ≥2500 people but <50,000 people.
7. *Urban area* refers collectively to urbanized areas and urban clusters.

In a nutshell, from the above definitions: (i) CSAs are clusters of CBSAs; (ii) CBSA is the generic term for metro/micropolitan areas; and (iii) metro/micropolitan areas differ from each other by whether their main populated centres involve an urbanized area or urban cluster. Of relevance from a biological invasion perspective is that the terms metropolitan and micropolitan are not equivalent to urban land use. Both metropolitan and micropolitan concepts imply the presence of urban–rural gradients in their territory (Office of Information and Regulatory Affairs, 2010). In fact, in the USA, only 11% of the combined metro/micropolitan area is classified as developed land, while forest, agriculture and shrub/scrub account collectively for 70% of the rest of the land (Colunga-Garcia et al., 2013). Moreover, the socio-economic interactions between urban and rural areas that are characteristic of metro/micropolitan areas make them highly suitable for the establishment and spread of invasive alien species.

In the remainder of this chapter, we show a methodology to use the FAF database to model freight transport of US imports with urban areas as their final destination. We use FAF projections for the year 2040 for 20 pathways consisting of combinations of four commodities and five world regions of origin. After modelling the final distribution of imports among urban areas in the USA, we: (i) characterize the distribution patterns of imports; and (ii) assess the introduction potential of invasive alien species to urban forests.

Modelling Urban Areas as Destinations of US Imports

Modelling the final distribution of US imports among urban areas was done in two phases (Fig. 3.1). In the first phase, we used the international FAF origin–destination data set to determine the regional destination (i.e. CBSAs and other territories in the contiguous USA) of commodities and the total volume imported as projected for the year 2040. In phase two, the regional imported volumes were disaggregated

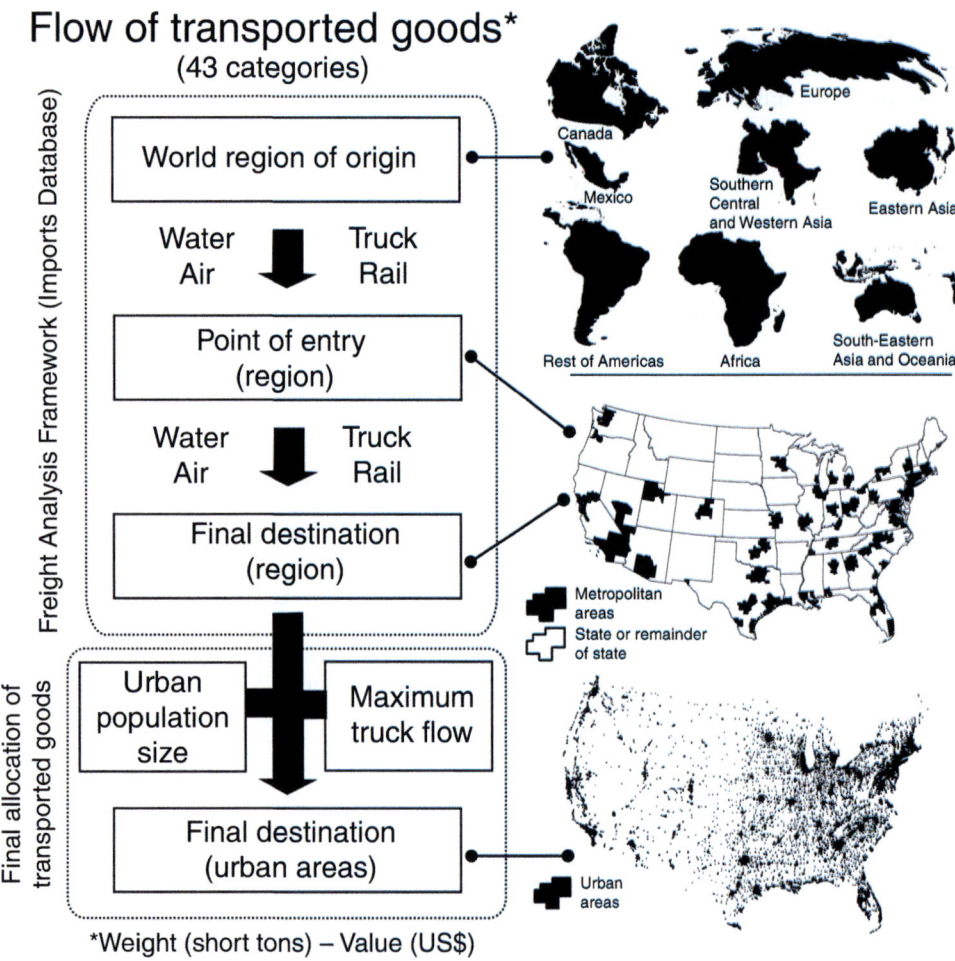

Fig. 3.1. Conceptual model for the flow of transported goods from selected world regions to US urban areas. (Modified from Colunga-Garcia et al., 2009.)

among the urban areas within a region. The combination of urban population size and long-distance truck volume was used as disaggregation criteria.

We focused our analyses on four commodity groups derived from the two-digit Standard Classification of Transported Goods (SCTG). Three of the commodity groups, including wood products (SCTG 26), non-metallic mineral products (including marble and ceramic tiles; SCTG 31) and machinery (SCTG 34), have been associated with wood-infesting insects (Haack, 2006). Our concern with the last two is the wood packaging material used to protect and handle the products. The fourth commodity was general agricultural products (SCTG 3), which does not include cereal grains, seed, animal feed, or live animals and fish. The five world regions of origin included in the analysis were Eastern Asia, Central and South America, Europe, Canada and Mexico. These world regions were selected because, according to preliminary calculations we made with the FAF database, they were projected to be major regions of origin for US imports of agricultural products (92%), wood products (97%), non-metallic mineral products (92%) and machinery (95%) in 2040.

Regional destination of imports (Phase 1)

For each selected pathway (i.e. combination of world region × commodity) we used data projections for the year 2040 from the FAF origin–destination data set (Box 3.1) to compute the total volume of imports at their projected regional destinations as follows:

$$R_i = \sum_{j=1}^{m} E_{ij} \qquad (3.1)$$

where R_i is the total imported volume in a selected pathway to reach the ith regional destination; E_{ij} is the total imported volume in a selected pathway arriving at the jth US point of entry to be transported to the ith regional destination; and m is the number of all US points of entry involved in a selected pathway for the ith regional destination.

Volume estimates were converted from FAF's original values in short tons to tonnes or metric tons (1 short ton = 0.9072 tonnes). We pooled the FAF data as to the pathway by which the imports arrived in the USA (i.e. via water, air and land). We also pooled the FAF data for the following modes of domestic transportation: truck, rail, water, air and truck–rail.

Disaggregation of regional imports to urban areas (Phase 2)

For this analysis, we used the Urban Areas Boundary Layer as delimited by the US Census Bureau (Box 3.1). Because some urban areas overlap the boundaries of more than one FAF region, we had to split those urban areas by FAF region. To do this, we intersected the Urban Areas Boundary Layer with the FAF Regions Boundary Layer (Box 3.1) in ArcGIS 10.1 to produce an FAF Region-Urban Areas Layer.

The disaggregation of regional import volumes (see R_i in Eqn 3.1) among the urban areas within a region was conducted as a function of urban population size and maximum long-distance truck volume using the following equation:

$$U_{ij} = R_i \times \frac{P_{ij} T_{ij}}{\sum_{j=1}^{n} P_{ij} T_{ij}} \qquad (3.2)$$

where U_{ij} is the total imported volume in a selected pathway to reach the jth urban area of the ith regional destination; R_i is the total imported volume in a selected pathway to reach the ith regional destination (from Eqn 3.1); P_{ij} is the size of the urban population in the jth urban area of the ith regional destination; T_{ij} is the maximum long-distance truck volume in the jth urban area of the ith regional destination; and n is the number of urban areas in the ith regional destination.

To compute T_{ij} in Eqn 3.2, we joined the FAF Network Data to the FAF Network Layer (Box 3.1) using ArcGIS 10.1. The resulting FAF Network Layer was intersected

Box 3.1. Sources of data sets used in this chapter

Source: US Department of Transportation – Federal Highway Administration
       Name: Freight Analysis Framework (FAF) Version 3.4
Link: http://www.ops.fhwa.dot.gov/freight/freight_analysis/faf/
            Data set: FAF Origin–Destination Data
            Data set: FAF Network Layer
            Data set: FAF Network Data
            Data set: FAF Regions Boundary Layer
Source: US Department of Commerce–US Census Bureau
       Name: TIGER/Line® Shapefiles and TIGER/Line® Files (web interface)
       Link: http://www.census.gov/geo/maps-data/data/tiger-line.html
            Data set: 2010 Urban Areas Boundary Layer
       Name: 2010 Census Gazetteer Files
       Link: http://www.census.gov/geo/maps-data/data/gazetteer2010.html
            Data set: 2010 Urban Areas Gazetteer Files (2010 population counts)
       Name: TIGER/Line® Shapefiles and TIGER/Line® Files (FTP site)
       Link: ftp://ftp2.census.gov/geo/tiger/TIGER2010/TABBLOCK/2010/
            Data set: 2010 Census Blocks Layers (including 2010 population counts)
Source: Multi-Resolution Land Characteristics (MRLC) consortium
       Name: National Land Cover Database 2006
       Link: http://www.mrlc.gov/nlcd06_data.php
            Data set: NLCD2006 Land Cover

with the FAF Region–Urban Areas Layer to obtain for each FAF Region–Urban Area polygon the maximum value of the 'FAF40' data field of the FAF Network Layer. This data field corresponded to the projected long-distance truck volume data for the year 2040. FAF Region–Urban Area polygons whose FAF40 values were equal to zero were discarded from the analysis. To obtain P_{ij} in Eqn 3.2, we used the 2010 Census Gazetteer File (Box 3.1). When urban areas were completely part of an FAF region, P_{ij} values were taken directly from the Census Gazetteer File. For urban areas that overlapped more than one FAF region, we had to disaggregate the urban area population values in the Census Gazetteer File among FAF Region–Urban Area polygons. To do this, we intersected the FAF Region–Urban Areas Layer described above with the Census Blocks Layers (Box 3.1). Then we summed the urban population for all the blocks that intersected the same FAF Region–Urban Area polygon. Summed values for each FAF Region–Urban Area polygon were then used as weights to disaggregate proportionally the urban area population values in the Census Gazetteer Files among the urban area–FAF region polygons.

Analysing distribution patterns of imports among urban areas

The purpose of the analysis was to determine if the 20 selected pathways produced different patterns in the distribution of imports among urban areas. Following the Pareto principle (i.e. the 80–20 rule), we selected urban areas that contributed to the upper 80% of total imported volume for each pathway. Two spatial statistics were computed using ArcGIS 10.1: (i) the weighted mean centre; and (ii) the weighted standard distance (Mitchell, 2005). The mean centre is the centre of concentration of the urban areas in a selected pathway and is computed as follows:

$$\bar{x} = \frac{\sum_i^n x_i}{n}, \quad \bar{y} = \frac{\sum_i^n y_i}{n} \quad (3.3)$$

where, for each selected pathway, $\bar{x}$ and $\bar{y}$ are, respectively, the centre mean x and y coordinates; x_i and y_i are the x and y coordinates of the ith urban area; and n is the number of urban areas. The weighted mean centre is a mean centre adjusted (or weighted) for the volume imported by each urban area for a selected pathway as follows:

$$w\bar{x} = \frac{\sum_i^n (U_i x_i)}{\sum_i^n U_i}, \quad w\bar{y} = \frac{\sum_i^n (U_i y_i)}{\sum_i^n U_i} \quad (3.4)$$

where, for each selected pathway, $w\bar{x}$ and $w\bar{y}$ are, respectively, the weighted centre mean x and y coordinates; x_i and y_i are the x and y coordinates of the ith urban area; U_i is the total imported volume in a selected pathway to reach the ith urban area (Eqn 3.2); and n is the number of urban areas.

The standard distance (SD) measures the degree to which urban areas were concentrated or dispersed around the mean centre and is computed as follows:

$$SD = \sqrt{\frac{\sum_i^n (x_i - \bar{x})^2}{n} + \frac{\sum_i^n (y_i - \bar{y})^2}{n}} \quad (3.5)$$

where x_i, $\bar{x}$, y_i, $\bar{y}$, i and n are as in Eqns 3.3 and 3.4. In the weighted standard distance we adjusted (i.e. weighted) the computations using the volume imported by each urban area for a selected pathway as follows:

$$WSD = \sqrt{\frac{\sum_i^n U_i (x_i - \bar{x})^2}{\sum_i^n U_i} + \frac{\sum_i^n U_i (y_i - \bar{y})^2}{n \sum_i^n U_i}} \quad (3.6)$$

where, for each selected pathway, WSD is the weighted standard distance and the other variables are as in Eqns 3.3 and 3.4. The greater the standard distance value, the more widely dispersed the urban areas were around the centre (Mitchell, 2005).

We also used two non-spatial indicators of concentration of import volume among urban areas. First, we measured the number of urban areas and US states that contributed to import volume. The smaller the numbers, the more concentrated the imports were among urban areas. Second, we measured the largest percentage contribution made by any individual urban area for each pathway. The smaller the number, the more evenly distributed were the imports among urban areas.

Assessing the introduction potential of invasive alien species to urban forests

The purpose of this analysis was to explore if pathways differed in their effect on the vulnerability of urban areas to potential introductions of invasive alien species. Introduction, for the purposes of this chapter, is defined as the entry of an invasive alien species resulting in its establishment (IPPC, 2012). To measure introduction potential, we built a simple model based on the volume of imports and the percentage of forested land within each urban area for each selected pathway. The model assumed that if a pathway was carrying invasive alien species, then urban areas with higher percentage of forest cover that were also the destination for high import volumes would have higher introduction potential for invasive alien species. Because the focus of the analysis for this chapter was primarily on trade patterns, we did not consider phytosanitary, environmental or biological factors other than forest cover (see Koch et al., 2011 or Magarey et al., 2011 for examples of comprehensive analyses of introduction risk involving trade patterns). The equation used to calculate the introduction potential was:

$$I_j = \frac{U_j}{U_{max}} \times \frac{F_j}{F_{max}} \quad (3.7)$$

where I_j is the introduction potential of invasive alien species in the jth urban forest; U_j is the total imported volume in a selected pathway to reach the jth urban area (i.e.

from Eqn 3.2); U_{max} is the highest U_j observed among all urban areas in a selected pathway; F_j is the percentage of forest land cover within the jth urban area; and F_{max} is the highest F_j observed among all urban areas.

To estimate the percentage of forest land cover in each urban area (F_j), we used ArcGIS 10.1 to cross-tabulate areas between the 2006 USA National Land-Cover Data Layer and the Urban Areas Boundary Layer (Box 3.1). Then we combined the areas of the deciduous, evergreen and mixed forest classes. Finally, we computed the percentage of forest land cover with respect to the total area of the urban area. After we computed I_j values for all 20 pathways, we selected a subset of urban areas for further analysis. To do this we computed I_j' as:

$$I_j' = \frac{I_j}{\sum_{j=1}^{n} I_j} \times 100 \qquad (3.8)$$

and selected the urban areas within the upper 95% of I' values for each pathway. The urban areas within each selected pathway subset were ranked for introduction potential by using a scale of 1–6, where 1 was the highest value. To implement the ranking, we first used the boxplot method (Tukey, 1977) to divide the subset of urban areas into four quartiles (i.e. Q1, Q2, Q3, Q4) based on their I_j values. We assigned the introduction potential rank values of 3, 4, 5 and 6 respectively to Q4, Q3, Q2 and Q1. Then we computed the interquartile range (IQR; i.e. Q3–Q1) to identify urban areas with outlier values (i.e. $I_j > Q3 + 1.5 \times IQR$) and extreme outlier values (i.e. $I_j > Q3 + 3 \times IQR$). Urban areas identified as outliers and extreme outliers were assigned, respectively, the introduction potential rank values of 2 and 1. We counted the number of urban areas included in each ranking category for each pathway to determine if the number of urban areas with introduction potential varied with pathway type. Then we selected the urban areas with a rank of 1 (i.e. with the highest introduction potential) and computed their mean centre (Eqn 3.3) and standard distance (Eqn 3.5) by using ArcGIS 10.1 to characterize invasion potential patterns under different pathways.

Distribution Patterns of Imports among Urban Areas

We modelled the final distribution of US imports to 3156 urban areas via 20 pathways (Fig. 3.2). Considering all 20 pathways combined, 211 urban areas (i.e. 6.7% of all urban areas in the USA) across 49 states (including Hawaii) contributed to the upper 80% of the US imported tonnage in at least one pathway. There were on average 50 ± 6 urban areas across 23 ± 2 states for the 20 pathways analysed (Table 3.2). The weighted mean centre of urban areas for each pathway showed strong influence by the world region of origin in the distribution of tonnage among urban areas (Table 3.2, Fig. 3.3a). In general, mean centres were in the north for products from Canada, south-west for products from Mexico, west for products from Eastern Asia, south-east for products from Central and South America, and east for products from Europe. Mean centres for pathways involving machinery were relatively clustered, suggesting this commodity has a strong influence on the distribution pattern. Weighted standard distances varied from 1000 km for machinery from Canada to over 1700 km for non-metallic mineral products from Eastern Asia (Table 3.2). This indicates that urban areas within each pathway are in general geographically widespread. The model results indicated that the geographic concentration of tonnage among urban areas varied depending on the pathway involved. At one end of the spectrum, agricultural products from Mexico went to 23 urban areas in seven states (i.e. 0.7% of all US urban areas) with the largest percentage contribution by an individual urban area to all US imports of 27.9% (Table 3.2). At the other end of the spectrum, wood products from Canada went to 119 urban areas in 43 states (i.e. 3.8% of all US urban areas) with the largest percentage contribution by an individual urban area to all US imports of 8.6%.

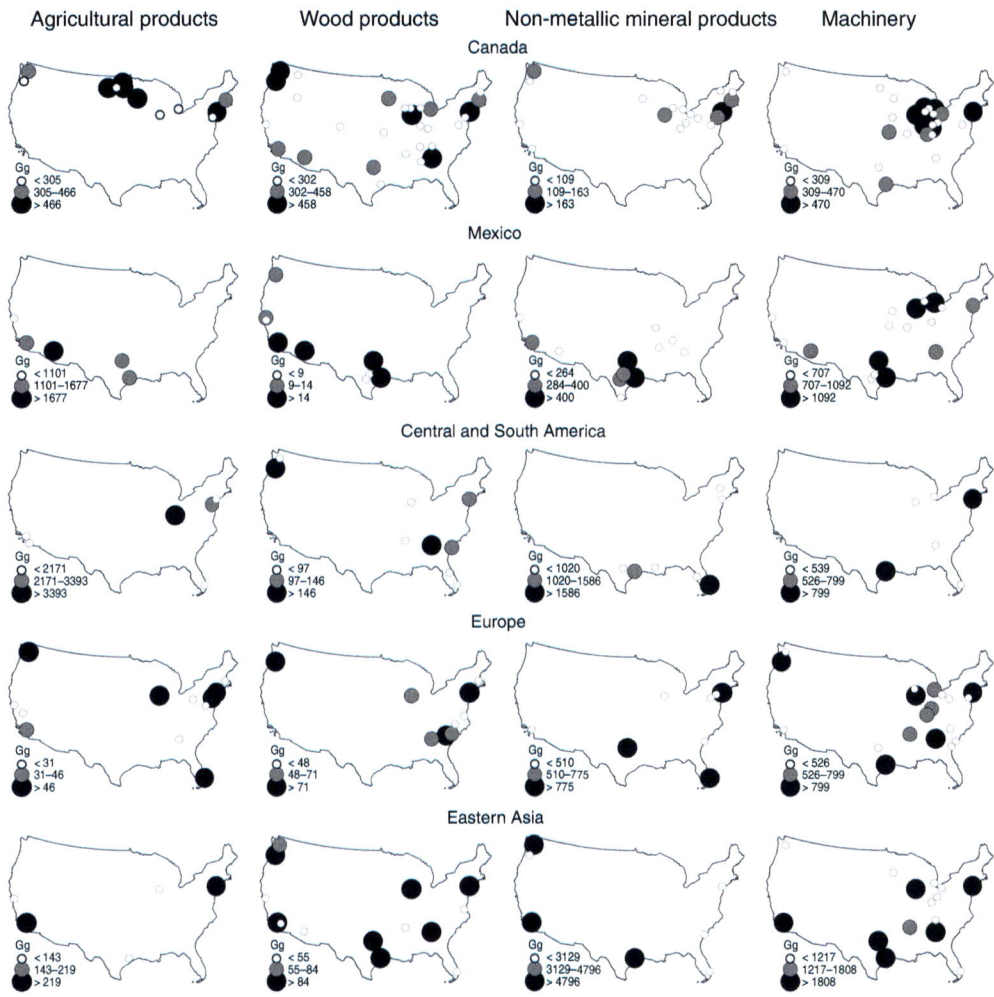

Fig. 3.2. Modelled destinations (urban areas) and projected imported volume (tonnes × 10^3, or Gg) for 2024 for 20 pathways (i.e. combinations of commodity type and world region of origin) in the contiguous USA. Only the fourth quartile of urban areas that contributed to the upper 80% of the US import volume in each pathway are shown. The two upper size classes (larger circles) of import volume in the labels for each of the 20 maps represent destination hot spots, i.e. respectively, the 'outliers' and the 'extreme outliers'.

Introduction Potential of Invasive Alien Species to Urban Forests

The geographic patterns resulting from the introduction potential model (Eqn 3.5) for each of the 20 pathways are shown in Fig. 3.3b. Overall, compared with the import 'cloud' of mean centres in Fig. 3.3a, the introduction cloud of mean centres in Fig. 3.3b was centred more in the north-eastern USA (i.e. the mean centre of the mean centres in Fig. 3.3b moved 1°N and 5.5°E in relation to Fig. 3.3a) and was slightly more clustered (i.e. the standard deviation of all mean centres in Fig. 3.3a was 734 km versus 595 km in Fig. 3.3b). This likely reflects the effect of forest vegetation abundance patterns in the contiguous USA. In general,

Table 3.2. Number of urban areas that contributed to the upper 80% of the US import volume via 20 pathways (combinations of commodity and world region of origin). Also included are number of US states involved, imported tonnage, largest percentage contribution made by an individual urban area, weighted mean centre and weighted standard distance.

SCTG[a]	Region of origin	Imported volume[b]	States	Urban areas	Largest percentage	WMC[c]	WSD[d] (km)
Agricultural products (SCTG 03)	Canada	8,881	25	48	15.2	45.1 N, 92.9 W	1180
	Mexico	14,390	7	23	27.9	34.2 N, 109.1 W	1000
	C&S America[e]	16,790	16	24	17.9	38.4 N, 89.9 W	1626
	Europe	1,104	24	49	29.3	39.4 N, 85.7 W	1558
	Eastern Asia	1,389	16	23	21.9	38.5 N 101.2 W	1742
Wood products (SCTG 26)	Canada	15,291	43	119	8.6	41.9 N, 96.7 W	1611
	Mexico	200	12	39	13.9	34.4 N, 105.6 W	1175
	C&S America	1,847	22	39	22.7	36.6 N, 90.0 W	1483
	Europe	1,208	21	39	21.3	37.8 N, 86.7 W	1425
	Eastern Asia	1,617	29	56	13	38.7 N, 99.4 W	1650
Non-metallic mineral products (SCTG 31)	Canada	5,542	35	77	15.2	42.6 N, 86.7 W	1489
	Mexico	6,521	15	48	15.2	34.1 N, 98.5 W	1099
	C&S America	10,756	14	27	28.7	31.6 N, 84.8 W	1070
	Europe	10,641	21	34	23.1	35.4 N, 83.9 W	1351
	Eastern Asia	39,272	16	27	18.9	37 N, 103.5 W	1627
Machinery (SCTG 34)	Canada	14,657	40	101	9.3	40.3 N, 88.5 W	1000
	Mexico	19,648	24	70	7.8	37.8 N, 92.4 W	1141
	C&S America	6,594	20	26	21.2	37 N, 85.6 W	1122
	Europe	18,855	31	79	9.2	38.8 N, 88.3 W	1210
	Eastern Asia	44,753	29	59	12.8	38.2 N, 92.1 W	1317

[a]SCTG, two-digit Standard Classification of Transported Goods.
[b]Imported volume in thousand tonnes.
[c]WMC, weighted mean centre; coordinates (latitude, longitude) are in the World Geodetic System 84 (WGS 84).
[d]WSD, weighted standard distance; equals one standard deviation.
[e]C&S America, Central and South America.

mean introduction centres were similar to those for imports (Fig. 3.3a). However, mean centres for introduction from Central and South America shifted primarily to the north-east, and for the wood products pathway shifted towards the mid-west. Differences among pathways were observed in the number of urban areas within each ranking level (Table 3.3). For instance, imports from Central and South America went to an average of 6 ± 0.5 urban areas where the introduction potential was 1 (i.e. highest ranking). On the other hand, imports from Mexico and Canada went to an average of 14 ± 2 and 15 ± 1 urban areas, respectively, where the introduction potential was 1. On average, 44.3 ± 1.8% of the total tonnage of US imports from 20 pathways went to urban areas where invasive alien species had the best opportunity to establish.

The potential for introduction of invasive forest species among some urban areas remained high (i.e. ranking category 1) regardless of the pathway involved (Fig. 3.3c). For instance, Atlanta, GA; New York–Newark, NY/NJ/CT; Houston, TX; Chicago, IL; Cincinnati, OH/KY/IN; Seattle, WA; and Washington DC/VA/MD were ranked in category 1 for ten or more pathways.

Discussion

The role of trade and commerce in the dispersal of plant pests is well recognized; however, research on this topic has been limited. The use of FAF data to model the final destination of imports to urban areas in the USA was useful to identify relevant patterns of potential introductions of invasive alien species. Our analysis in this

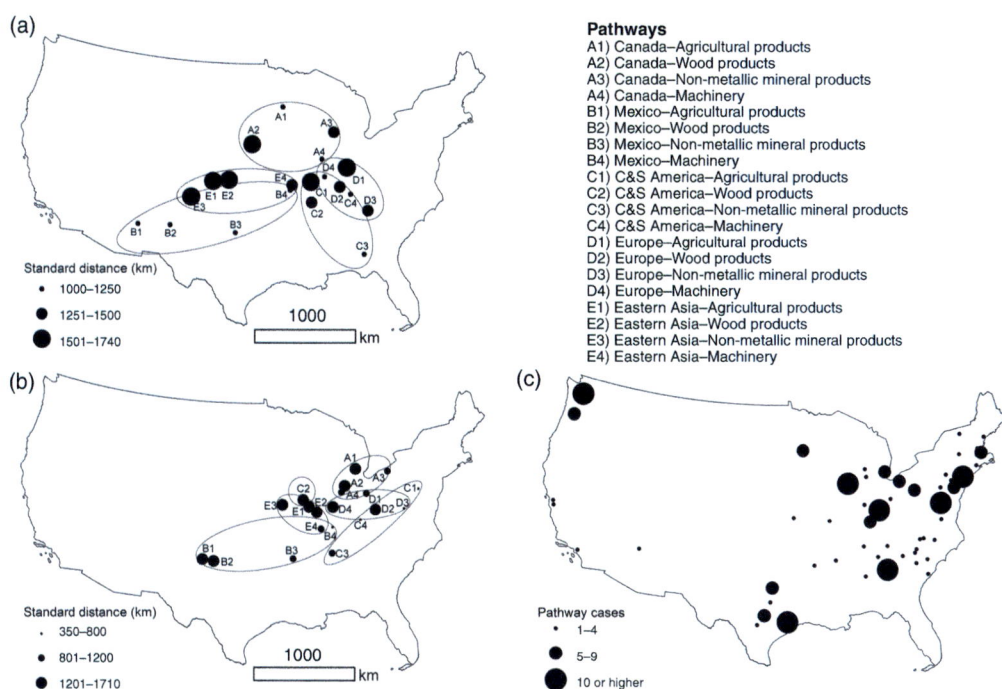

Fig. 3.3. (a) Spatial statistics for the final destination (urban areas) of US imports via 20 pathways. Only urban areas that contributed to the upper 80% of the US import volume in at least one pathway are shown. Dot locations represent the weighted mean centre and dot size is the weighted standard distance (one standard deviation). (b) Spatial statistics for urban areas in the contiguous USA with the highest introduction potential for invasive forest pest species via 20 pathways. Each dot location represents the mean centre and dot size represents the standard distance (one standard deviation). (c) Urban areas with the highest introduction potential for invasive forest pest species via 20 pathways. Dot size indicates the number of pathways where a particular urban area was ranked. Introduction potential was estimated based solely on volume of imports (i.e. rate of entry of invasive species was assumed to be proportional to the amount of tonnage) and percentage of forest cover in an urban area. Highest introduction potential means they were in the ranking category 1 on a scale of 1–6 (with 1 being the highest ranking). Pathways were combinations of commodity and world region of origin (C&S America, Central and South America) as listed in the upper right. Ellipses were drawn as a visual reference to indicate values belonging to the same world region of origin (with the exception of C2 in Fig. 3.3b for clarity).

chapter showed that the type of imports and the world region of origin affect the final distribution of the imported products to urban areas and thereby can influence the initial distribution of introduced invasive alien species. Incorporating such trade information into risk assessments should help plant health specialists prioritize areas for detection and monitoring programmes.

Some adaptations to the methodology described in this chapter have already been implemented in the analysis of invasive alien species. For instance, Koch et al. (2011) and Magarey et al. (2011) modified Eqn 3.1 to include the domestic movement of goods. Presumably, the flow of imported goods could continue domestically after delivery to the final destination specified in the bill of lading, which is assumed to be the final destination in the present chapter. Eqn 3.1 can also be modified to provide rates of flow (i.e. percentages or proportions) instead of tonnages. Such modification would enable the integration of other trade databases such as the port import database in USA Trade® Online (https://usatrade.census.gov/).

Table 3.3. Number of urban areas in the contiguous USA (and percentage contribution to US imports) in different ranking categories for the introduction potential of invasive forest species via 20 pathways (combinations of world region of origin and commodity)[a].

SCTG	World region of origin	Introduction potential ranking[c]					
		1	2	3	4	5	6
Agricultural products (SCTG 03)	Canada	13 (31)	6 (4)	26 (17)	46 (21)	45 (8)	45 (4)
	Mexico	10 (39)	9 (5)	11 (2)	31 (3)	32 (4)	30 (1)
	C&S America[d]	5 (45)	3 (7)	8 (4)	17 (10)	16 (3)	16 (7)
	Europe	11 (51)	4 (2)	15 (9)	30 (6)	29 (7)	30 (2)
	Eastern Asia	7 (38)	2 (2)	11 (33)	20 (6)	19 (1)	20 (3)
Wood products (SCTG 26)	Canada	17 (38)	11 (9)	29 (13)	59 (10)	57 (8)	58 (3)
	Mexico	13 (49)	7 (12)	17 (6)	39 (6)	37 (4)	38 (1)
	C&S America	7 (47)	5 (6)	9 (10)	22 (9)	22 (5)	21 (3)
	Europe	9 (51)	1 (2)	15 (16)	25 (11)	26 (4)	25 (4)
	Eastern Asia	17 (44)	9 (20)	22 (6)	49 (8)	49 (3)	48 (2)
Non-metallic mineral products (SCTG 31)	Canada	14 (48)	7 (6)	21 (10)	43 (11)	42 (5)	42 (3)
	Mexico	14 (53)	6 (5)	17 (8)	38 (11)	38 (5)	37 (2)
	C&S America	6 (26)	5 (5)	18 (13)	30 (7)	29 (38)	29 (1)
	Europe	10 (38)	7 (7)	18 (8)	35 (10)	35 (3)	35 (24)
	Eastern Asia	7 (43)	5 (7)	16 (21)	30 (8)	28 (2)	29 (1)
Machinery (SCTG 34)	Canada	17 (44)	9 (9)	35 (15)	61 (9)	62 (6)	61 (4)
	Mexico	20 (53)	6 (5)	22 (9)	48 (11)	47 (4)	48 (3)
	C&S America	7 (58)	7 (8)	12 (7)	27 (7)	27 (3)	26 (2)
	Europe	15 (44)	12 (14)	20 (11)	48 (10)	48 (5)	47 (4)
	Eastern Asia	14 (47)	13 (10)	21 (17)	49 (7)	49 (4)	48 (2)

[a]Introduction potential was estimated based solely on percentage of forest cover in an urban area and on volume of imports (rate of entry of invasive species was assumed to be proportional to the amount of imported tonnage).
[b]SCTG, two-digit Standard Classification of Transported Goods.
[c]A ranking of 1 indicates the greatest likelihood of introduction.
[d]C&S America, Central and South America.

This database provides data on port imports at a finer resolution in terms of the origin of imports (i.e. country) and commodity type (i.e. six-digit Harmonized Code Classification). Import volume data from USA Trade *Online* can be used in a modified Eqn 3.1 for rates to determine transportation flow of those imported goods. If knowing the actual flow of imports is more important than simply knowing the final destinations, it would be worthwhile to take an in-depth look at the FAF Network Data (Box 3.1) to model the transportation of goods (i.e. truck mode) on US roads. If knowing the point of entry is not a priority (i.e. only the origin and final destination are desired), other approaches could be implemented. For example, Colunga-Garcia et al. (2013) used the State Import Destination database of USA Trade *Online* to obtain information about the country of origin and US state of destination of imports and then used geocoded data on wholesale or retail establishments and sales within the US states to disaggregate the state import data.

In working with these data sources, it is important to keep in mind that databases such as FAF or USA Trade *Online* were designed to be used at the scale provided. Attempts to disaggregate the data at finer resolutions are as good as the assumptions behind the disaggregation approach and therefore caution should be used when deriving conclusions from their analysis. That being said, these types of analyses should be encouraged because they provide important insights into the potential dynamics between international trade, freight movement and biological invasions. Colunga-Garcia et al. (2010) modelled the potential destination of imports within urban areas based on human population numbers and commercial land use. Their goal was to explore the potential interactions between propagule pressure and forest cover and the potential establishment of a hypothetical invasive forest pest. When they delimited hot spots for its potential establishment, they found that such hot spots encompassed historical initial detections of actual invasive alien species in those urban areas.

An important finding of the analysis in this chapter is that some urban areas were highly ranked in terms of invasion potential across several pathways (Fig. 3.3b). The fact that some of those urban areas (e.g. Chicago, New York–Newark) have experienced introductions of invasive alien pests in the past is significant because our analysis was conducted with FAF data projections for the year 2040. This means that urban areas that have experienced invasions in the past are also vulnerable to future invasions. In fact, Colunga-Garcia et al. (2010) argue that hot spots for invasion due to international trade or freight transport are exposed to propagule pressure from multiple pathways. The combination of multiple pathways with high volumes of propagule pressure increases the likelihood of entry and establishment by invasive alien species where environmental conditions are favourable (Haack and Rabaglia, 2013; Brockerhoff et al., 2014). Although we may not be able to predict which species will be introduced next nor when, our analysis indicates that we can anticipate where introductions will likely occur by following the transportation trail.

Acknowledgements

This work was partially supported by the NRI–USDA/CSREES/Plant Biosecurity Program (grant number 2006-55605-16658).

Notes

[1] Material for this chapter was adapted from Colunga-Garcia, M., Haack, R.A. and Adelaja, A.O. (2009) Freight transportation and the potential for invasions of exotic insects in urban and periurban forests of the United States. *Journal of Economic Entomology* 102, 237–246 with permission by the Entomological Society of America.

References

Brockerhoff, E.G., Kimberley, M., Liebhold, A.M., Haack, R.A. and Cavey, J. (2014) Predicting how altering propagule pressure changes establishment rates of biological invaders across species pools. *Ecology* 95, 594–601.

Caton, B.P., Dobbs, T.T. and Brodel, C.F. (2006) Arrivals of hitchhiking insect pests on international cargo aircraft at Miami International Airport. *Biological Invasions* 8, 765–785.

Colunga-Garcia, M., Haack, R.A. and Adelaja, A.O. (2009) Freight transportation and the potential for invasions of exotic insects in urban and periurban forests of the United States. *Journal of Economic Entomology* 102, 237–246.

Colunga-Garcia, M., Haack, R.A., Magarey, R.A. and Margosian, M.L. (2010) Modeling spatial establishment patterns of exotic forest insects in urban areas in relation to tree cover and propagule pressure. *Journal of Economic Entomology* 103, 108–118.

Colunga-Garcia, M., Haack, R.A., Magarey, R.D. and Borchert, D.M. (2013) Understanding trade pathways to target biosecurity surveillance. *NeoBiota* 18, 103–118.

Haack, R.A. (2006). Exotic bark- and wood-boring Coleoptera in the United States: recent establishments and interceptions. *Canadian Journal of Forest Research* 36, 269–288.

Haack, R.A. and Rabaglia, R.J. (2013) Exotic bark and ambrosia beetles in the USA: potential and current invaders. In: Peña, J.E. (ed.) *Potential Invasive Pests of Agricultural Crop Species*. CAB International, Wallingford, UK, pp. 48–74.

Haack, R.A., Law, K.R., Mastro, V.C., Ossenbruggen, H.S. and Raimo, B.J. (1997) New York's battle with the Asian long-horned beetle. *The Journal of Forestry* 95, 11–15.

Haack, R.A., Hérard, F., Sun, J. and Turgeon, J.J. (2010) Managing invasive populations of Asian longhorned beetle and citrus longhorned beetle: a worldwide perspective. *Annual Review of Entomology* 55, 521–546.

Hulme, P.E. (2009) Trade, transport and trouble: managing invasive species pathways in an era of globalization. *Journal of Applied Ecology* 46, 10–18.

IPPC (International Plant Protection Convention) (2009) *International Standards for Phytosanitary Measures: ISPM No. 15. Regulation of Wood Packaging Material in International Trade*. Food and Agriculture Organization of the United Nations, Rome.

IPPC (International Plant Protection Convention) (2012) *International Standards for Phytosanitary Measures: ISPM No. 5. Glossary of*

Phytosanitary Terms. Food and Agriculture Organization of the United Nations, Rome.

Koch, F.H., Yemshanov, D., Colunga-Garcia, M., Magarey, R.D. and Smith, W.D. (2011) Potential establishment of alien-invasive forest insect species in the United States: where and how many? *Biological Invasions* 13, 969–985.

Liebhold, A.M., Brockerhoff, E.G., Garrett, L.J., Parke, J.L. and Britton, K.O. (2012) Live plant imports: the major pathway for forest insect and pathogen invasions of the United States. *Frontiers in Ecology and the Environment* 10, 135–143.

Magarey, R.D., Borchert, D.M., Engle J.S., Colunga-Garcia, M., Koch, F.H. and Yemshanov, D. (2011) Risk maps for targeting exotic plant pest detection programs in the United States. *EPPO Bulletin* 41, 46–56.

McCullough, D.G., Work, T.T., Cavey, J.F., Liebhold, A.M. and Marshall, D. (2006) Interceptions of nonindigenous plant pests at US ports of entry and border crossings over a 17-year period. *Biological Invasions* 8, 611–630.

McGranahan, G. and Satterthwaite, D. (2003) Urban centers: an assessment of sustainability. *Annual Review of Environment and Resources* 28, 243–274.

Meyerson, L.A. and Mooney, H.A. (2007) Invasive alien species in an era of globalization. *Frontiers in Ecology and the Environment* 5, 199–208.

Mitchell, A. (2005) *ESRI Guide to GIS Analysis. Volume 2: Spatial Measurements and Statistics*. ESRI Press, Redlands, California.

NACEC (North American Commission for Environmental Cooperation) (2003) Invasive species, agriculture and trade: case studies from the NAFTA context. *Second North American Symposium on Assessing the Environmental Effects of Trade*. North American Commission for Environmental Cooperation, Center for International Environmental Law, and Defenders of Wildlife, Mexico City.

NRC (National Research Council) (2002) *Predicting Invasions of Nonindigenous Plants and Plant Pests*. National Academies Press, Washington, DC.

Office of Information and Regulatory Affairs (2010) 2010 standards for delineating metropolitan and micropolitan statistical areas. US Office of Management and Budget. *Federal Register* 75, 123 (28 June 2010), 37246–37252.

Perrings C., Fenichel, E. and Kinzig, A. (2009) Globalization and invasive alien species: trade, pests, and pathogens. In: Perrings, C., Mooney, H. and Williamson, M. (eds) *Bioinvasions and Globalization: Ecology, Economics, Management, and Policy*. Oxford University Press, Oxford, pp. 42–55.

Pimentel, D., Zuniga, R. and Morrison, D. (2005) Update on the environmental and economic costs associated with alien-invasive species in the United States. *Ecological Economics* 52, 273–288.

Poland, T.M. and McCullough, D.G. (2006) Emerald ash borer: invasion of the urban forest and the threat to North America's ash resource. *The Journal of Forestry* 104, 118–124.

PPQ (Plant Protection and Quarantine) (2011) *Agricultural Quarantine Inspection Monitoring (AQIM) Handbook*. US Department of Agriculture, Marketing and Regulatory Programs and Animal and Plant Health Inspection Service, Plant Protection and Quarantine, Frederick, Maryland.

Rodrigue, J.-P., Comtois, C. and Slack, B. (2009) *The Geography of Transport Systems*, 2nd edn. Routledge, New York.

Southworth, F., Peterson, B.E., Hwang, H., Chin, S. and Davidson, D. (2011) *The Freight Analysis Framework version 3 (FAF3). A Description of the FAF3 Regional Database and How it is Constructed*. Oak Ridge National Laboratory, Oak Ridge, Tennessee. Available at: http://faf.ornl.gov/fafweb/Data/FAF3ODDoc611.pdf (accessed 10 December 2013).

Tukey, J.W. (1977) *Exploratory Data Analysis*. Addison-Wesley, Reading, Massachusetts.

USCB (United States Census Bureau) (2012) *2010 Census Summary File 1: Technical Documentation*. US Department of Commerce, Economics and Statistics Administration, US Census, Washington, DC. Available at: http://www.census.gov/prod/cen2010/doc/sf1.pdf (accessed 7 October 2014).

Westphal, M.I., Browne, M., MacKinnon, K. and Noble, I. (2008) The link between international trade and the global distribution of invasive alien species. *Biological Invasions* 10, 391–398.

4 Simulation Modelling of Long-distance Windborne Dispersal for Invasion Ecology

Hazel R. Parry,[1]* Debbie Eagles[2] and Darren J. Kriticos[3]

[1]*CSIRO Biosecurity Flagship, Brisbane, Queensland, Australia;*
[2]*CSIRO Australian Animal Health Laboratory, Geelong, Victoria, Australia;* [3]*CSIRO Biosecurity Flagship, Canberra, Australian Capital Territory, Australia*

Abstract

We present a method to simulate atmospheric dispersal events in invasion ecology using examples of two economically important and highly mobile insect species: the bird cherry-oat aphid, *Rhopalosiphum padi* (which vectors yellow dwarf viruses to maize, barley, oats and wheat); and the biting midge, *Culicoides imicola* (which vectors bluetongue virus to a wide range of domestic and wild ruminants). We demonstrate this method using two available atmospheric trajectory modelling tools: HYSPLIT and PMTRAJ.

Significance of Windborne Dispersal for Invasive Alien Species

In an increasingly interconnected world, many pests and pathogens can be moved through multiple dispersal pathways, not just natural, but human-mediated or human-enhanced (Wilson *et al*., 2009). However, biological transport processes through the atmosphere and the oceans have always been (Munoz *et al*., 2004), and still are (Kim and Beresford, 2008; Chapman *et al*., 2010; Hopkinson and Soroka, 2010), an important mechanism for long-distance dispersal for both motile organisms, such as birds and insects, and passive dispersers, such as plant seeds, fungal spores and aerosolized bacteria.

The concept of trajectory modelling for long-distance dispersal in ecology is not new (Hendrie *et al*., 1985; Aylor, 1986; Scott and Achtemeier, 1987). For approximately 25 years, researchers have developed trajectory models to simulate long-distance dispersal of both particles and organisms, including insect pests, through the atmosphere (Drake and Gatehouse, 1995). Recently, there has been increased interest in applying these methodologies to invasion ecology and biosecurity risk analysis (Russo and Isard, 2011). This interest is perhaps because the tools, resolution and accuracy of meteorological data, modelling techniques, computational power and knowledge about ecological behaviours have become sophisticated enough to establish realistic and robust simulations rapidly.

Aerial dispersal models are suitable for a range of economically important applications (Table 4.1), such as assessing biosecurity risks, pest alert systems, pest (including weeds) and disease management, and forensic investigations of pest incursions

---

* Corresponding author. E-mail: hazel.parry@csiro.au

© CAB International 2015. *Pest Risk Modelling and Mapping for Invasive Alien Species* (ed. R.C. Venette)

(Kim and Beresford, 2008; Chapman et al., 2010; Hopkinson and Soroka, 2010; Savage et al., 2010; Leskinen et al., 2011). Models are now progressing from simple analysis of air-parcel trajectories assuming passive organism transport to trajectory models that are able to incorporate simple organism behavioural features. In the future,

Table 4.1. Summary of some existing trajectory/dispersion models and their application in invasion ecology. An extended version of this table is available in the online supplement to Chapter 4 (Table S4.1) giving more details on each trajectory/dispersion model such as spatial and temporal resolution and availability.

Trajectory/dispersion model	Study subject	Application
SILAM (Leskinen et al., 2011)	Pest: Bird cherry-oat aphid (Rhopalosiphum padi); diamondback moth (Plutella xylostella)	Early warning of pest arrival via long-distance dispersal events
HYSPLIT (Garner et al., 2006; Zhu et al., 2006; Kim and Beresford, 2008; MacRae et al., 2011; Eagles et al., 2013; Otuka, 2013)	Pathogen: Wheat stripe rust (Puccinia striiformis); FMDV Pathogen/vector: Bluetongue/Culicoides Pest: Green peach aphid (Myzus persicae); rice planthoppers (Laodelphax striatellus, Sogatella furcifera and Nilaparvata lugens)	Identification of sites at which rust spores are likely to be deposited after transit from Australia to New Zealand Risk assessment of windborne spread of FMDV, to allocate activities like surveillance and vaccination on a risk basis Relate spring low-level jet streams to intensity of M. persicae flight activity and spread of PLRV and PVY Identify the migration source of rice planthoppers and Culicoides
PMTRAJ (Rochester et al., 1996; Deveson et al., 2005; Anderson et al., 2010; Parry et al., 2011; Eagles et al., 2012)	Pest: Bollworms (Helicoverpa punctigera and Helicoverpa armiger); Australian plague locust (Chortoicetes terminifera); planthopper (Eumetopina flavipes) Pest/vector: Aphid (R. padi) Pathogen/vector: Bluetongue/Culicoides	The basis of a system for forecasting moth migrations from inland habitat to coastal cropping regions The identification of dispersal mechanisms which facilitate particular biological invasions Tracing the source of locust pest outbreaks
TAPM (Hurley et al., 2005; Savage et al., 2010)	Fungal pathogen (generic)	Determine whether changes to the seasonal and circadian timing of propagule release can a have a significant effect on the area covered by resulting aerial dispersal
CMC (LRTAP) (Hopkinson and Soroka, 2010)	Pest: Diamondback moth (P. xylostella)	Use of both forward and back trajectories to identify likely sources of pest outbreaks
CALPUFF (Pfender et al., 2006)	Pathogen: Grass stem rust (Puccinia graminis)	Estimation of dispersal and deposition of grass stem rust at a landscape scale
MM5 (Hu et al., 2013)	Pest: Brown planthopper (N. lugens)	
NAME (Ågren et al., 2010; Chapman et al., 2010) (now NAME III)	Pest: Autographa gamma moths; air pollution Pathogen: Culicoides midges/bluetongue virus, FMDV	Trajectory analysis in combination with radar data showed that moth behaviours alter migration distances and directions of seasonal migration Estimation of likely source of bluetongue introduction to Sweden from Europe and likely points of introduction

FMDV, foot-and-mouth disease virus; PLRV, potato leafroll virus; PVY, potato virus Y.

trajectory models are likely to incorporate more advanced ecological parameters and be coupled with mechanistic, process-based simulations of flight initiation.

The aerial transport process (Fig. 4.1) has four key phases: (i) uplift; (ii) atmospheric transport; (iii) deposition; and (iv) redistribution (Hendrie et al., 1985; Isard and Irwin, 1993). For each phase of this process, three activities can increase our knowledge of pest risk from long-distance atmospheric dispersal events (Fig. 4.1): (i) data collection/surveillance; (ii) mapping/ data analysis; and (iii) simulation modelling. We address all three activities in this chapter, as they are often interdependent. We present data requirements, analysis and simulation modelling methods applicable in invasion ecology to the first three phases of the aerial transport process by two worked examples of insect pests (Table 4.2). We focus on the first three phases as the fourth phase of redistribution (i.e. further movement and spread in the destination area) warrants a full exploration in its own right due to its complexity, where multiple

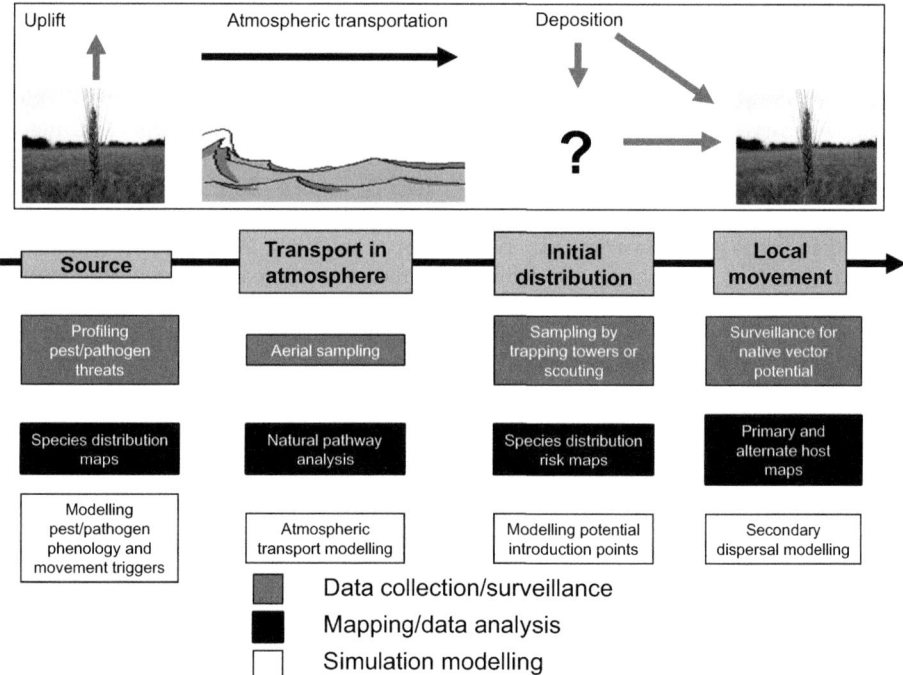

Fig. 4.1. Key phases in the aerial transport process for pests and pathogens (after Hendrie et al., 1985; Isard and Irwin, 1993). Integrated methods to study the processes are shown.

Table 4.2. Summary of long-distance windborne dispersal cases.

Species	Model aims	Means to estimate pest density at source	Software used
Rhopalosiphum padi (bird cherry-oat aphid)	Seasonal long-distance migration of insect pest into crop	Mechanistic/process-based model	PMTRAJ (part of GENSIM)
Culicoides imicola (biting midges)	Long-distance dispersal potential for pest entry	Arbitrary estimate	HYSPLIT

processes may operate at a range of spatial and temporal scales (Parry et al., 2013). Further data and analysis are required to calculate risk and model dispersal at the destination in this fourth phase. While these models may have many of the same characteristics as the long-distance dispersal models, they also have some important scale-dependent nuances and are, therefore, outside the scope of this chapter.

It should be noted, however, that the interpretation of phase 3 (i.e. deposition) in the aerial transport process is often affected by linkage with the events of phase 4 (i.e. redistribution). Identification of a deposition (i.e. arrival) point often occurs after some redistribution has occurred. Our ability to identify precisely these points can be limited. For example, the detection of new bluetongue serotypes is confounded by the difficulty of assignment of any particular possible arrival event/s or location as the entry point (Eagles et al., 2012, 2013).

We demonstrate the use of two atmospheric wind trajectory[1] modelling tools: HYSPLIT – Hybrid Single-Particle Lagrangian Integrated Trajectory model (Draxler and Hess, 1997, 1998; Draxler, 1999; Draxler and Rolph, 2013; Rolph, 2013) and PMTRAJ (Rochester et al., 1996; Rochester, 1999), both of which are freely available. We apply these tools to the dispersal of the bird cherry-oat aphid, *Rhopalosiphum padi*, and the biting midge, *Culicoides imicola*. Integrated simulation methods are now available to study the entire aerial transport process from the source, in the atmosphere, the initial distribution following transportation and through subsequent local movement and risk (Fig. 4.1). Such holistic approaches aim to advance an ecological understanding of the initiation of long-distance dispersal, the effects of atmospheric dynamics on an organism during dispersal and the consequences of such dispersal.

Aerial Dispersal of *R. padi* and *C. imicola*

Atmospheric pest and pathogen dispersal can be a regular event, although dispersal events may vary greatly by year and season, such as with locusts (Deveson et al., 2005) and aphids (Leskinen et al., 2011). Pest risks may arise as movement patterns shift with a changing global climate; for example, migration events to higher latitudes may become more common (Chapman et al., 2010). Significant threats from invasive alien pests also come from changing movement and host susceptibility patterns as crops, animals and other host species are translocated from their native environments. As a secondary consequence of new geographical risks, we may create additional 'stepping-stone' risks to new areas (Hopkinson and Soroka, 2010).

R. padi, the bird cherry-oat aphid, is the primary vector of yellow dwarf viruses in Australia (Parry et al., 2012). Worldwide, cereal aphids such as *R. padi* continue to be a major pest of cereal crops, largely due to their complex life cycle and dispersal behaviours making them difficult to suppress (Parry, 2013). Many studies have shown that cereal aphids are capable of long-distance movement; cereal aphids are known to have a distinct 'migratory' flight phase, in which they are often carried distances of tens, even hundreds, of kilometres from their source by wind trajectories. *R. padi* is a non-native species that has established in Australia, periodically 'invading' autumn crops from over-summer host sources such as ryegrass pasture, potentially many kilometres away from the crop (Parry et al., 2012). Therefore a better understanding and projection of aphid movement patterns at multiple spatial scales is important to build area-wide strategies for cereal aphid integrated pest management (Parry, 2013).

The genus *Culicoides* includes a number of species of biting midge known to vector a range of arboviruses, including new serotypes of bluetongue virus, worldwide. Such viruses are isolated occasionally from cattle and insects in northern Australia and are thought to be introduced via windborne dispersal of *Culicoides* from neighbouring land masses to the north (Eagles et al., 2013). Unlike cereal aphids, *Culicoides* are not regular long-distance dispersers.

However, due to their small size it is thought that on occasion they may become entrained in wind currents that enable them to move large distances. In view of this, trajectory modelling has been used to estimate what the potential may be for the long-distance windborne transport of *Culicoides* (Eagles *et al*., 2012, 2013) and thus the arrival of novel arboviruses into Australia: a major biosecurity threat.

Research Questions

In invasion ecology, simulations of long-distance wind dispersal have been applied to several spatially explicit research topics, for example: (i) exploring invasion risk and identifying likely sites of introduction (Pfender *et al*., 2006; Kim and Beresford, 2008; Anderson *et al*., 2010); (ii) developing early-warning systems for pest arrival (Rochester *et al*., 1996; Deveson and Hunter, 2002; Deveson *et al*., 2005; Leskinen *et al*., 2011); and (iii) identifying the potential source of known outbreaks (Deveson *et al*., 2005; Ågren *et al*., 2010; Hopkinson and Soroka, 2010). In addition, studies of aerial transport have been used to explore relationships between population processes at the source and resultant dispersal patterns (Savage *et al*., 2010); design surveillance and vaccination strategies for insect-vectored diseases of humans and animals (Garner *et al*., 2006); and explore in-transit behaviours of insects (Chapman *et al*., 2010; Drake and Wang, 2013).

Data and Software Needs

Modelling the aerial transport process can be data-intensive because data are required for each phase. For the first phase, data are needed to profile the initial pest migration threat (i.e. where might the species of concern be coming from). Methods to address this need range from simple, sometimes arbitrary, estimates of population size and distribution to dynamic population simulations coupled with empirical observations from the field for model initialization (i.e. biofixes) and verification. All aerial simulation methods require that a spatial location for the source be specified; however, this source may be specified with varying degrees of precision, from a point to a broad geographical area. Likewise, temporal dynamics at the source may be specified with differing degrees of complexity, from a single dispersive event to an ongoing migration threat. For the migration-trajectory model phase, the assumption of active or passive dispersal will dictate data requirements. The calculation of risk at the destination can be a risk-modelling-and-mapping exercise (e.g. identification of suitable climates and vulnerable hosts) or can involve dynamic simulation modelling methods to calculate immediate and longer-term risk (Parry *et al*., 2013). Both types of methods require data relating to the environment at the destination, the dispersal ecology of the organism and the assumed response of the organism to the new environment.

Profiling pest migration threat (Phase 1)

Three broad approaches are typically used to profile a pest or pathogen migration threat: (i) empirical field data are used to generate initial estimates of flight activity, simply by scaling up observations under certain environmental conditions (Deveson *et al*., 2005; Westbrook *et al*., 2011); (ii) life history data are used to develop mechanistic, process-based models that include factors that lead to migratory behaviour (Parry *et al*., 2011); and (iii) highly arbitrary migration estimates are used to consider a wide range of hypothetical dispersal scenarios (i.e. 'sensitivity' methods; Eagles *et al*., 2012, 2013).

In many cases, empirical data will not be available because they are costly to obtain. Nonetheless, field observations, if collected regularly with good spatial coverage, provide perhaps the best estimates of immediate migration threats. Empirical data may be from a variety of sources including traditional survey data, light-trap catches, suction traps, satellite imagery and direct observations from purpose-built insect-monitoring radar. Surveillance records may sometimes be available as used in locust migration modelling (Deveson *et al.*, 2005) or proxy information may also be useful such as habitat distribution.

In general, observational data are only likely to identify large populations because small, dispersed populations are much more difficult to detect. Therefore, an empirically driven profile of pest migration threat basically assumes that only easily detected, high-density populations will pose a migration threat. For many species of insects, this circumstance is likely to be true, particularly as some species (e.g. aphids) generate winged morphs (i.e. alates) at high population densities, although alates are produced in response to other factors such as a decline in food availability. The assumption that organisms must reach high densities before they are likely to disperse may be less applicable to plants or pathogens; for these cases, the start of migration periods might be forecast from phenological development models. In general, ground-level observations that confirm the presence of a species do not necessarily indicate a propensity for migration by that species. Other en-route aerial observation methods, such as suction trapping (Woiwod and Harrington, 1994) or radar (Nieminen *et al.*, 2000; Drake and Wang, 2013), are better suited to identify species movements but may leave little time to respond to these events when they occur.

If aerial or en-route sampling data are unavailable, ground-level empirical data are best supplemented by process-based models that include factors known to influence migratory behaviour (Parry *et al.*, 2006, 2011, 2013). Such models can include factors that affect population dynamics, propensity to produce migratory morphs (e.g. alate aphids) and the potential to alight. For example, an aphid model could be initialized with empirical observations on population structure and geographic distribution to identify when and where alate adults might appear. Alternatively, a spatially explicit population dynamics model could be used to infer the production of alate adults from suitable habitat and environmental drivers (Parry *et al.*, 2006, 2011, 2013). Data requirements for such process-based modelling vary considerably, depending on the structure of the model. The models must be interpreted with care because errors compound over simulated time. The models are most effective if they can be run with periodic biofixes to effectively re-calibrate the model to actual field conditions.

Most trajectory modelling software requires identification of a source point or region. The source may be characterized by a single point coordinate or multiple points of varying concentration (e.g. for use in HYSPLIT and PMTRAJ; software described in greater detail in the section 'Modelling migration trajectories (Phase 2)' below), or a region defined in a geographical information system and exported as an appropriate file type (e.g. PMTRAJ reads ASCII grid files and point locations). The grid cell values or points will contain an estimate of the number of migrants, which can range from a simple arbitrarily large value (i.e. the 'sensitivity' approach) to density values that fluctuate in space and time according to a population dynamics model. The model could include a sub-model to describe the propensity of individuals to migrate depending on a range of environmental and behavioural cues. Output then becomes input to initialize the redistribution trajectory model.

Examples of useful data for both profiling the propensity of species to migrate and estimating migration trajectories are given in Table 4.3. We were able to simulate flight initiation of *R. padi* due to the vast amount of information in the literature on the life history of this species in particular and aphids in general (Parry, 2013). This knowledge base allowed us to construct a mechanistic model of population dynamics for aphid populations at a source that considers factors which are likely to drive aphids to disperse. However, the biological literature is much more limited for *Culicoides*, so a 'sensitivity' method was used to give a less precise but still valuable indication of potential migration risk.

Table 4.3. Details of meteorological data used in Phase 1 (i.e. profiling pest migration threat) and Phase 2 (i.e. migration-trajectory model). (All data supplied by the Australian Bureau of Meteorology and the Air Resources Laboratory.)

Meteorological data	Spatial extent	Spatial resolution	Temporal scale	Format
Phase 1				
Temperatures[a]	Weather station	Point	Daily maximum and minimum	Text file
Day length[a]	Weather station	Point	Daily	Text file
Wind speed (ground)[a]	Weather station	Point	Mean daily	Text file
Wind direction (ground)[a]	Weather station	Point	Mean daily	Text file
Humidity[a]	Weather station	Point	Mean daily	Text file
Phase 2				
Wind speed and direction (multi-level): limited-area prediction system regional atmospheric circulation model[a]	Australia	0.75°	6 h	Hdf
NCEP/NCAR re-analysis[b]	Global	2.5°	6 h	Air Resources Laboratory format

[a]Data requirement for PMTRAJ (*Rhopalosiphum padi*) model only.
[b]NCEP/NCAR, National Center for Environmental Protection/National Center for Atmospheric Research (USA). Data requirement for HYSPLIT (*Culicoides imicola*) model only.

Modelling migration trajectories (Phase 2)

Aerial dispersal can be viewed generally as passive (i.e. where the movement of the organism is controlled primarily by movements of a medium such as the wind) or active (i.e. where organismal behaviours primarily dictate the speed and direction of movement). In reality a continuum between the two forms of dispersal may exist, but simplification of aerial dispersal into these two categories makes it easier to consider appropriate models. Plants, microbes and small insects are generally considered passive dispersers over long distances whereas large, winged insects usually exhibit some degree of active dispersal behaviour. Even weak flying insects may select conditions for migratory take-off, possibly related to photoperiod and seasonal conditions. For example, in China moths of the Old World bollworm, *Helicoverpa armigera*, actively fly northwards in southerly winds during spring and summer and return south on nights with northerly winds during autumn (Feng *et al.*, 2009). We use the term 'migrate' generally to describe long-distance active transport.

In order to parameterize an atmospheric circulation model with biological factors that are important to flight and simulate continuously across space and time, a 'front end' to most standard circulation models must be built to modify the estimated trajectories in relation to flight behaviours. Front-end models allow the user to create time-dependent, time-varying sources of dispersing biota. For passive dispersers, the front-end model can be very simple because there are very few data requirements.

HYSPLIT is a freely available model capable of computing simple air-parcel trajectories to running complex dispersion and deposition simulations (Draxler and Hess, 1997, 1998; Draxler, 1999; Draxler and Rolph, 2013; Rolph, 2013). HYSPLIT can be accessed via a web-based interface or the software can be downloaded and run locally. Meteorological data also can be downloaded with the software or the user can provide his/her own meteorological inputs, provided they are converted to a format that HYSPLIT can process. Sample programs for converting meteorological data into this format are available. HYSPLIT will not allow for complex organism behaviours (although recent development of a new web-based front-end 'TAPPAS' may change this; see Graham *et al.*, 2013), but will allow for multiple take-off locations, start times and dispersal durations. These input variables can be adjusted manually or by using the EMITIMES file.

GENSIM is the C++ program that runs models in the PMTRAJ package (Rochester *et al.*, 1996; Rochester, 1999). Most of the code for the program is in a C++ library. The ancillary C++ programs in the PMTRAJ package are also based on that library. For full information about GENSIM, see Rochester (1999). PMTRAJ is an extension of the back-trajectory analysis method of Scott and Achtemeier (1987). GENSIM/PMTRAJ is available as open source software from its developer (W. Rochester, Queensland, 2013, personal communication; wayne.rochester@csiro.au).

PMTRAJ can simulate passive, and some aspects of active, dispersal, with inputs as indicated in Table 4.4. For passive dispersal, specific parameters (and units) include:

1. Take-off location(s) (latitude and longitude).
2. Take-off time of day and duration of take-off within a single day (hour/hours).
3. Flight altitudes (metres).
4. Flight start date (dd–mm–yy).
5. Flight end date, i.e. the flight period (dd–mm–yy).
6. The size of the dispersing population – this may be the actual number of insects, an assumed number of insects or a sampled subset of individuals that are dispersing.

Table 4.4. Details of parameters and data used in migration-trajectory models for *Rhopalosiphum padi* and *Culicoides imicola*.

Parameter	Description	*R. padi* model	*Culicoides* model
Take-off			
Start time	Time of day when take-off begins	9 am local time	Dusk (~ 6 pm local time)
Duration	Length of time take-off continues	7 h	3 h
Latitude	Grid cell latitude at take-off	–33.13 decimal degrees (Wokalup, Western Australia)	Nine points in SE Indonesia, Timor-Leste and PNG
Longitude	Grid cell longitude at take-off	115.88 decimal degrees (Wokalup, Western Australia)	Nine points in SE Indonesia, Timor-Leste and PNG
Duration			
Minimum	Minimum flight duration	2 h	20 h
Maximum	Maximum flight duration	6.5 h	20 h
Height			
Minimum	Minimum height of flight	100 m	0 m
Maximum	Maximum height of flight	1000 m	1000 m
Flight speed			
Minimum	Minimum speed the organism is capable of flying unaided	0 km/h	NA
Maximum	Maximum speed the organism is capable of flying unaided	0 km/h	NA
Flight offset			
Minimum	Minimum flight bearing against the wind	0 degrees	NA
Maximum	Maximum flight bearing against the wind	0 degrees	NA
Number of flights	Number of individuals in simulation per time step	Determined by population dynamics/flight initiation sub-model at source	100 per hour

PNG, Papua New Guinea; NA, not applicable.

For active dispersal, PMTRAJ models can include additional front-end parameters:

1. Survival functions during flight.
2. The range of flight altitudes.
3. The range of flight durations.
4. The organisms' flight speed and flight direction (i.e. orientation relative to the wind).

Survival functions are frequently ignored because relevant data are unavailable. Some stochastic elements also can be incorporated into PMTRAJ simulations. For example, take-off time and duration within a flight period can be selected at random to allow for simulation of successive, intermittent flights.

Examples of data and data sources used to parameterize trajectory models in these ways are given for *R. padi* and *C. imicola* in Table 4.4. Where data, such as altitude or flight duration, have a range, values are randomly drawn for each individual (or aggregated individuals/sampled subset) in the dispersing population from the range. An alternative way to model the range of altitudes, particularly for very large populations, is to assign a fraction to each altitude based on a probability-density function within the range.

The front-end model requires multi-level wind data over an area to calculate the movement pathway. PMTRAJ was used for the *R. padi* model and was restricted to data derived from the Bureau of Meteorology Limited Area Prediction System (LAPS) data set for the Australasian region (Table 4.3). The Australian Plague Locust Commission (APLC) is currently developing a new wind trajectory model that utilizes the latest Bureau of Meteorology Australian Community Climate and Earth-System Simulator (ACCESS) data. HYSPLIT was used to model *C. imicola* dispersal, but several other, freely available trajectory-pathway modelling tools could have been used (Table 4.1). PMTRAJ or the new APLC software can be adapted to work with alternative weather data sets and HYSPLIT has the capability to use other user-specified data sets provided the data are converted to the correct format.

Calculation of flight termination (Phase 3)

For this phase of aerial transport, conditions that will end a flight must be described. In a simple model, these conditions simply may be that the estimated flight duration has elapsed. The model uses this information, the estimated flight speed (i.e. a combination of wind speed and the organism's flight velocity) and wind direction to estimate the deposition location and the distance travelled. Estimates of flight duration can be derived from field or laboratory sources; much flight data available in the literature has been obtained from tethered insects on a flight mill (Johnson, 1969). Flight mills may significantly overestimate the potential flight duration. In the field, flight duration is likely to depend on a range of factors (Drake and Gatehouse, 1995):

1. The individual's physiological state and age (Liquido and Irwin, 1986).
2. Energetics, i.e. body fat and weight affect flight capacity (Cockbain, 1961).
3. Physiological stress experienced during flight (Ward *et al.*, 1998).
4. Inherent behavioural cues to switch from migratory behaviour to landing behaviour, e.g. visual responses to plant-related wavelengths (Kennedy *et al.*, 1961; Rautapää, 1980).
5. Local weather conditions, e.g. precipitation (Hendrie *et al.*, 1985), wind (Symmons and Luard, 1982), temperature and updrafts (Achtemeier, 1992).

These factors are usually ignored in dispersal models (but see Hu *et al.*, 2013). We demonstrate only the simplest flight termination method here but wish to highlight the potential importance of a deeper consideration of biological landing cues (Parry, 2013).

Flight termination is presently incorporated into the *R. padi* and *C. imicola* trajectory models as estimates of flight duration based on empirical data from the laboratory and field. While the HYSPLIT model can account for insects that settle out of the atmosphere by gravity (i.e. dry deposition parameters) or precipitation (i.e. wet deposition parameters), these features were not used. More sophisticated models of flight termination could be coupled with the trajectory models to incorporate factors that can alter duration of flight and better estimate where organisms might be deposited.

Once the arrival locations have been determined, spatial analysis software can be used to visualize results from single dispersal events or multiple events over time. Combined arrival maps can give a spatial probability-density estimate for deposition in a given time period. Further calculation of risk at the destination as the individuals redistribute (i.e. Phase 4) can be achieved via risk mapping or a simulation modelling approach to calculate both immediate and longer-term risk; several methods and software packages can be employed in this phase (Parry et al., 2013).

Analyses

In this section, we bring the data and software together to estimate the likelihood that a species might be uplifted into the atmosphere from a source location, transported long distances and deposited in an area of concern.

One simple method for estimating pest threat is by using sensitivity scenario testing to demonstrate the possibility of migration from an origin to a destination and to assign a relative probability of arrival in terms of how frequently migration might occur. For this method, a point source or region is used as the origin and an arbitrarily large number of migrants is assumed to originate from this source at each time step (i.e. on the order of 10,000 individuals; cf. Eagles et al., 2012). The analysis can identify seasonal risk patterns, with the underlying assumption that the availability of migrants does not vary over time. However, there are clear limitations to this method. It does not account for population dynamics at the source location or the propensity of individuals to migrate. Consequently, such a method cannot estimate an actual probability of migration. Field observations and population models for the source location to estimate the probability of migration events occurring in space and time could help overcome this limitation, particularly if in-transit survival is also modelled accurately.

For PMTRAJ and HYSPLIT, there are four key steps to running a simulation:

1. Prepare the input data for the source location(s), either individual points or a raster data layer. For each point or grid cell, an estimate of density is needed. In our examples, we characterized source locations for R. padi with outputs from a population dynamics simulation model and for C. imicola with location information from published studies and source numbers chosen arbitrarily.
2. Prepare the parameter file, as described in the section 'Modelling migration trajectories (Phase 2)'.
3. Run the model.
4. Display or process the output data (e.g. plot trajectory lines from a single point-source trajectory simulation or calculate densities in grids for raster output).

To simulate R. padi aphid population movement, PMTRAJ (Rochester et al., 1996) is linked to LAPS data to parameterize and perform the wind trajectory simulation. Movement is simulated from potential source habitats of irrigated pasture to cereal-cropping regions in Western Australia. PMTRAJ simulates change in the distribution of the aphid population that results from a period of migration. The model acts as an individual-based model when distributing the aphids in space. Movement of aphids is simulated along a wind trajectory that is determined by the wind velocities around the aphids. The state of the aphids upon termination relates to the environmental conditions that they experience during flight. Movement and condition dictate the final distribution in the area of concern. PMTRAJ requires several parameters to initialize the model, from which are derived stochastic flight attributes (Table 4.4). The sub-model also requires multi-level wind speeds and directions (Table 4.3). PMTRAJ calculates the wind trajectory during a 24 h period by representing movement pathways as an aggregation of eight 3-h straight-line segments at multiple elevations. PMTRAJ parameters can be set to interpolate smaller segments (e.g. 1 h).

To simulate the atmospheric dispersal of *C. imicola* midges from putative sources across Indonesia, Timor-Leste and Papua New Guinea, the HYSPLIT_4 model was used. The particle-dispersal-simulation mode of the model calculated the transport of particles based on mean wind speed and a random component to account for turbulence (Garner *et al.*, 2006). Particle dispersal simulation also takes into account extra factors such as particle diameter, density and shape. The particle source is simulated by the release of a user-defined number of particles (i.e. midges, in this case) over a specified period. As there were nine source sites in this study, the specified number of particles was evenly distributed among source sites (Eagles *et al.*, 2013).

In the examples presented here, we have not considered the potential complexity of flight termination. We assume flights terminate with 100% survival of the transported insects. The resultant distribution pattern of aphid densities as a result of wind dispersal, calculated from the PMTRAJ model for April and May 1999 from an irrigated pasture in Western Australia, shows that aphids can be dispersed in any direction, and frequently they may be blown out to sea (Fig. 4.2). The model suggests that most aphids land within 100 km of the pasture; however, they can be blown >300 km. This result means that the aphids from the irrigated pasture source may arrive at wheat-farming areas around Avondale and Mount Barker (Fig. 4.2), illustrating the potential importance of a regional approach to the prediction and management of aphid outbreaks.

Based on the simulation model, dispersal of *C. imicola* into Australia is

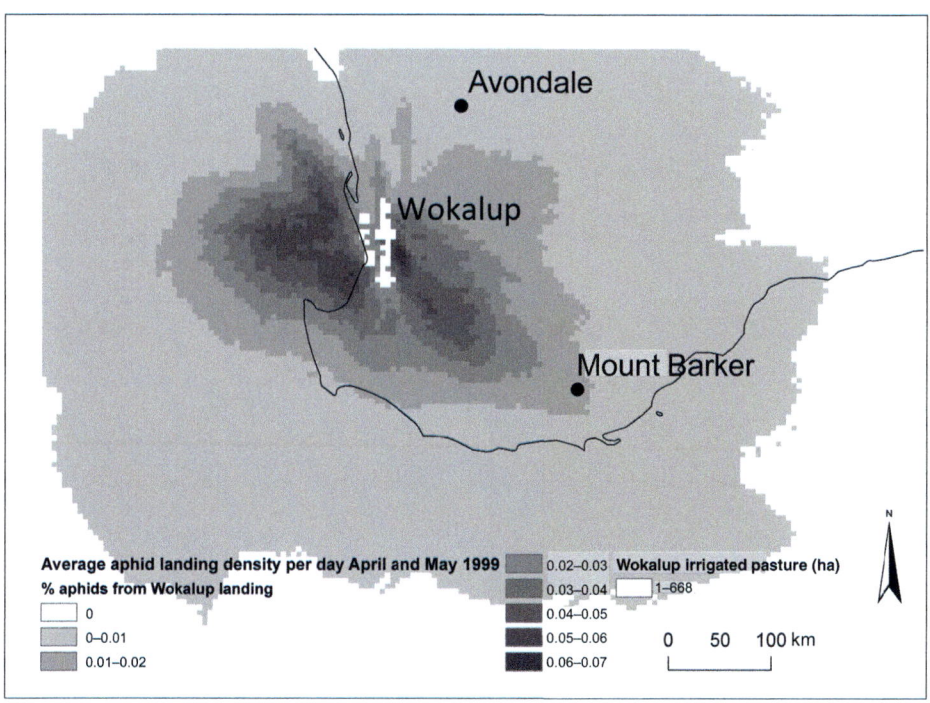

Fig. 4.2. Summarized regional wind dispersal pattern of the percentage of aphids (per 5 km × 5 km grid cell) for all days in April and May 1999 from PMTRAJ model simulation runs. Output from PMTRAJ as an ASCII grid file was read into ArcGIS to create this map.

possible from at least as far west as Lombok (Indonesia) and as far east as Fly (Western Province) in Papua New Guinea (Fig. 4.3). While the temporal and spatial dispersal patterns are source- and site-dependent, the greatest potential for arrival appears to be in December–March, coinciding with the monsoon season in this region. All three states comprising the northern coastline of Australia (Western Australia, Northern Territory and Queensland) appear to be at risk from *C. imicola* incursions from neighbouring regions. The patterns of dispersal postulated by models may be helpful for surveillance of diseases for which *C. imicola* and other *Culicoides* spp. are vectors, such as the economically important bluetongue virus of ruminant livestock.

Discussion

The methods presented here allow researchers to perform analyses of windborne transport of pests. The models may not yet be adequate to reliably simulate a movement pathway for a particular event. In general, the accuracy of windborne transport simulations is limited by data availability and ecological knowledge about flight initiation, in-flight behaviours, flight termination and survival after transport. Their general use in invasion ecology is, therefore, perhaps best restricted to estimations of incursion risks, possible introduction sites or potential sources of known outbreaks. These forecasts may be especially useful for early-warning systems.

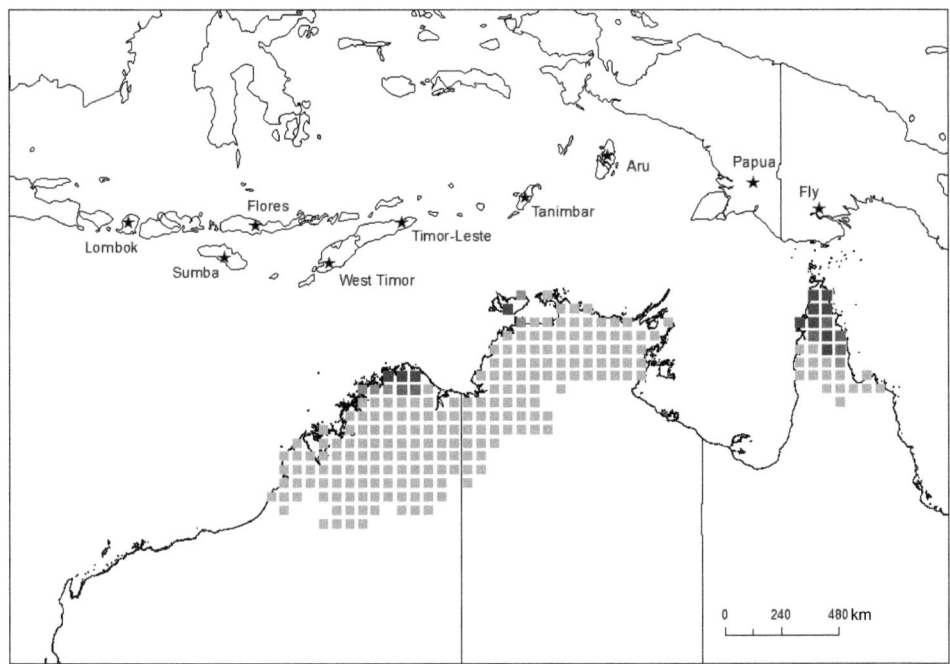

Fig. 4.3. Summarized regional wind dispersal pattern of *Culicoides* from nine putative source sites (stars) across East Indonesia, Timor-Leste and Papua New Guniea into northern Australia. Dispersal was assessed retrospectively for each December–March over a 15-year period (1995–2010). Each 0.5° × 0.5° grid cell contains cumulative deposition from all sites across the 60 months (= 4 months/year × 15 years) assessed. The darker the square, the greater the relative deposition into that site. Output from HYSPLIT was read into ArcGIS to create this map.

Relatively recent trends in ecological modelling have seen an increase in mechanistic models, particularly individual-based models (Grimm and Railsback, 2005). Such models have the advantage of incorporating behaviours and physical dynamics that can help to simulate the complex range of phenomena that may lead to initiation of long-distance dispersal (Jongejans et al., 2008). Such models can also contribute to an understanding of the effects of atmospheric conditions on the organism during dispersal (Shamoun-Baranes et al., 2010), an emerging branch of ecology that has been labelled 'aeroecology' (Kunz et al., 2008). By incorporating more sophisticated demographic simulations into mechanistic spatial dispersal models, scientists should better understand the initiation of migratory and long-distance dispersal events, the spatial extent, biological cost, speed and direction of dispersal, and the nature of possible feedbacks (e.g. between the deposition and redistribution phase). These models are likely to have greater realism and predictive power at multiple scales, as well as the ability to increase understanding of likely sources of historic migration events.

Acknowledgements

This work was completed primarily during the course of a post-doctoral fellowship at CSIRO, supported by the Cooperative Research Centre for Plant Biosecurity (H.R.P., D.J.K.). The *Culicoides* work was carried out as part of a PhD project funded by the Australian Biosecurity Cooperative Research Centre (D.E.). We would like to thank the International Pest Risk Modelling Workgroup (http://www.pestrisk.org/) for their support and helpful discussions that have led to the development of this chapter.

Notes

[1] 'The trajectory or path of an air parcel is a curve denoting successive three-dimensional positions in time of the air parcel' (Hopkinson and Soroka, 2010, p. 2).

This chapter is enhanced with supplementary resources.

To access the material please visit:

www.cabi.org/openresources/43946/

References

Achtemeier, G.L. (1992) Grasshopper response to rapid vertical displacements within a 'clear air' boundary layer as observed by Doppler radar. *Environmental Entomology* 21, 921–938.

Ågren, E.C.C., Burgin, L., Lewerin, S.S., Gloster, J. and Elvander, M. (2010) Possible means of introduction of bluetongue virus serotype 8 (BTV-8) to Sweden in August 2008: comparison of results from two models for atmospheric transport of the *Culicoides* vector. *Veterinary Record* 167, 484–488.

Anderson, K.L., Deveson, T.E., Sallam, N. and Congdon, B.C. (2010) Wind-assisted migration potential of the island sugarcane planthopper *Eumetopina flavipes* (Hemiptera: Delphacidae): implications for managing incursions across an Australian quarantine frontline. *Journal of Applied Ecology* 47, 1310–1319.

Aylor, D.E. (1986) A framework for examining inter-regional aerial transport of fungal spores. *Agricultural and Forest Meteorology* 38, 263–288.

Chapman, J.W., Nesbit, R.L., Burgin, L.E., Reynolds, D.R., Smith, A.D., Middleton, D.R. and Hill, J.K. (2010) Flight orientation behaviours promote optimal migration trajectories in high-flying insects. *Science* 327, 682–685.

Cockbain, A.J. (1961) Fuel utilization and duration of tethered flight in *Aphis fabae* Scop. *Journal of Experimental Biology* 38, 163–174.

Deveson, E.D. and Hunter, D.M. (2002) The operation of a GIS-based decision support system for Australian locust management. *Entomologia Sinica* 9, 1–12.

Deveson, E.D., Drake, V.A., Hunter, D.M., Walker, P.W. and Wang, H.K. (2005) Evidence from traditional and new technologies for northward migrations of Australian plague locusts (*Chortoicetes terminifera*) (Walker) (Orthoptera: Acrididae) to western Queensland. *Austral Ecology* 30, 928–943.

Drake, V.A. and Gatehouse, A.G. (1995) *Insect Migration: Tracking Resources through Space and Time*. Cambridge University Press, Cambridge.

Drake, V.A. and Wang, H. (2013) Recognition and characterization of migratory movements of Australian plague locusts, *Chortoicetes terminifera*, with an insect monitoring radar. *Journal of Applied Remote Sensing* 7, 075095.

Draxler, R.R. (1999) *HYSPLIT4 User's Guide. NOAA Technical Memorandum ERL ARL-230*. National Oceanic and Atmospheric Administration Air Resources Laboratory, Silver Spring, Maryland.

Draxler, R.R., and Hess, G.D. (1997) *Description of the HYSPLIT_4 Modelling System. NOAA Technical Memorandum ERL ARL-224*. National Oceanic and Atmospheric Administration Air Resources Laboratory, Silver Spring, Maryland.

Draxler, R.R. and Hess, G.D. (1998) An overview of the HYSPLIT_4 modelling system for trajectories, dispersion and deposition. *Australian Meteorological Magazine* 47, 295–308.

Draxler, R.R. and Rolph, G.D. (2013) *HYSPLIT (HYbrid Single-Particle Lagrangian Integrated Trajectory)*. National Oceanic and Atmospheric Administration Air Resources Laboratory, Silver Spring, Maryland. Available at: http://ready.arl.noaa.gov/HYSPLIT.php (accessed 17 December 2014).

Eagles, D., Deveson, T., Walker, P.J., Zalucki, M.P. and Durr, P. (2012) Evaluation of long-distance dispersal of *Culicoides* midges into northern Australia using a migration model. *Medical and Veterinary Entomology* 26, 334–340.

Eagles, D., Walker, P.J., Zalucki, M.P. and Durr, P.A. (2013) Modelling spatio-temporal patterns of long-distance *Culicoides* dispersal into northern Australia. *Preventive Veterinary Medicine* 110, 312–322.

Feng, H., Wu, X., Wu, B. and Wu, K. (2009) Seasonal migration of *Helicoverpa armigera* (Lepidoptera: Noctuidae) over the Bohai sea. *Journal of Economic Entomology* 102, 95–104.

Garner, M.G., Hess, G.D. and Yang, X.B. (2006) An integrated modelling approach to assess the risk of wind-borne spread of foot-and-mouth disease virus from infected premises. *Environmental Modelling and Assessment* 11, 195–207.

Graham, K., Durr, P.A. and van Klinken, R. (2013) Modelling the risk of long-distance wind-borne spread of pests and pathogens: the *TAPPAS* application – taking GIS-based research into operational use. Poster presented at the GeoVet 2013 Conference, 21–23 August 2013, London.

Grimm, V. and Railsback, S.F. (2005) *Individual-Based Modelling and Ecology*. Princeton University Press, Princeton, New Jersey.

Hendrie, L.K., Irwin, M.E., Liquido, N.J., Ruesin, W.G., Mueller, E.A., Voegtlin, D.J., Achtemeier, G.L., Steiner, W.M. and Scott, R.W. (1985) Conceptual approach to modelling aphid migration. In: MacKenzie, D.R., Barfield, C.S., Kennedy, G.C. and Smith, J.C. (eds) *The Movement and Dispersal of Agriculturally Important Biotic Agents*. Claitor's Publishing Division, Baton Rouge, Louisiana, pp. 541–582.

Hopkinson, R.F. and Soroka, J.J. (2010) Air trajectory model applied to an in-depth diagnosis of potential diamondback moth infestations on the Canadian Prairies. *Agricultural and Forest Meteorology* 150, 1–11.

Hu, G., Lu, F., Lu, M.-H., Liu, W.-C., Xu, W.-G., Jiang, X.-H. and Zhai, B.-P. (2013) The influence of Typhoon Khanun on the return migration of *Nilaparvata lugens* (Stål) in Eastern China. *PLoS ONE* 8, e57277.

Hurley, P.J., Physick, W.L. and Luhar, A.K. (2005) TAPM: a practical approach to prognostic meteorological and air pollution modelling. *Environmental Modelling and Software* 20, 737–752.

Isard S.A. and Irwin, M.E. (1993) A strategy for studying the long-distance aerial movement of insects. *Journal of Agricultural Entomology* 10, 283–297.

Johnson, C.G. (1969) The air speed and the intrinsic duration and range of flight. In: Johnson, C.G. (ed.) *Migration and Dispersal of Insects by Flight*. Methuen & Co. Ltd, London, pp. 85–129.

Jongejans, E., Skarpaas, O. and Shea, K. (2008) Dispersal, demography and spatial population models for conservation and control management. *Perspectives in Plant Ecology, Evolution and Systematics* 9, 153–170.

Kennedy, J.S., Booth, C.O. and Kershaw, W.J.S. (1961) Host finding by aphids in the field III. Visual attraction. *Annals of Applied Biology* 49, 1–21.

Kim, K.S. and Beresford, R.M. (2008) Use of a spectrum model and satellite cloud data in the simulation of wheat stripe rust (*Puccinia striiformis*) dispersal across the Tasman Sea in 1980. *Agricultural and Forest Meteorology* 148, 1374–1382.

Kunz, T.H., Gauthreaux, S.A., Hristov, N.I., Horn, J.W., Jones, G., Kalko, E.K.V., Larkin, R.P., McCracken, G.F., Swartz, S.M., Srygley, R.B., Dudley, R., Westbrook, J.K. and Wikelski, M. (2008) Aeroecology: probing and modelling the aerosphere. *Integrative and Comparative Biology* 48, 1–11.

Leskinen, M., Markkula, I., Kostinen, J., Pylkko, P., Ooperi, S., Siljamo, P., Ojanen, H., Raiskio, S. and Tiilikkala, K. (2011) Pest insect immigration warning by an atmospheric dispersion model, weather radars and traps. *Journal of Applied Entomology* 135, 55–67.

Liquido, N.J. and Irwin, M.E. (1986) Longevity, fecundity, change in degree of gravidity and lipid content with adult age, and lipid utilisation during tethered flight of alates of the corn leaf aphid, *Rhopalosiphum maidis*. *Annals of Applied Biology* 108, 449–459.

MacRae, I., Carroll, M. and Zhu, M. (2011) Site-specific management of green peach aphid, *Myzus persicae* (Sulzer). In: Clay, S.A. (ed.) *GIS Applications in Agriculture. Volume 3: Invasive Species.* CRC Press, Boca Raton, Florida, pp. 167–190.

Munoz, J., Felicisimo, A.M., Cabezas, F., Burgaz, A.R. and Martinez, I. (2004) Wind as a long-distance dispersal vehicle in the Southern Hemisphere. *Science* 304, 1144–1147.

Nieminen, M., Leskinen, M. and Helenius, J. (2000) Doppler radar detection of exceptional mass-migration of aphids into Finland. *International Journal of Biometeorology* 44, 172–181.

Otuka, A. (2013) Migration of rice planthoppers and their vectored re-emerging and novel rice viruses in East Asia. *Frontiers in Microbiology* 4, 1–11.

Parry, H. (2013) Cereal aphid movement: general principles and simulation modelling. *Movement Ecology* 1, 14.

Parry, H., Evans, A. and Morgan, D. (2006) Aphid population response to agricultural landscape change: a spatially explicit, individual-based model. *Ecological Modelling* 199, 451–463.

Parry, H., Aurambout, J.-P. and Kriticos, D. (2011) Having your cake and eating it: a modelling framework to combine process-based population dynamics and dispersal simulation. In: Chan, F., Marinova, D. and Anderssen, R. (eds) *MODSIM2011, 19th International Congress on Modelling and Simulation.* Modelling and Simulation Society of Australia and New Zealand, Perth, Australia, pp. 12–16.

Parry, H., Macfadyen, S. and Kriticos, D. (2012) The geographical distribution of Yellow dwarf viruses and their aphid vectors in Australian grasslands and wheat. *Australasian Plant Pathology* 41, 375–387.

Parry, H., Sadler, R. and Kriticos, D. (2013) Practical guidelines for modelling post-entry pest spread in invasion ecology. *NeoBiota* 18, 41–66.

Pfender, W., Graw, R., Bradley, W., Carney, M. and Maxwell, L. (2006) Use of a complex air pollution model to estimate dispersal and deposition of grass stem rust urediniospores at landscape scale. *Agricultural and Forest Meteorology* 139, 138–153.

Rautapää, J. (1980) Light reactions of cereal aphids (Homoptera, Aphididae). *Annales Agriculturae Fenniae* 46, 1–12.

Rochester, W.A. (1999) The migration systems of *Helicoverpa punctigera* (Wallengren) and *Helicoverpa armigera* (Hübner) (Lepidoptera: Noctuidae) in Australia. PhD thesis, The University of Queensland, Brisbane, Australia.

Rochester, W.A., Dillon, M.L., Fitt, G.P. and Zalucki, M.P. (1996) A simulation model of the long-distance migration of *Helicoverpa* spp. moths. *Ecological Modelling* 86, 151–156.

Rolph, G.D. (2013) *Real-time Environmental Applications and Display sYstem (READY).* National Oceanic and Atmospheric Administration Air Resources Laboratory, Silver Spring, Maryland. Available: at http://ready.arl.noaa.gov (accessed 17 December 2014).

Russo, J.M. and Isard, S.A. (2011) Online aerobiology process model. In: Clay, S.A. (ed.) *GIS Applications in Agriculture. Volume 3: Invasive Species.* CRC Press, Boca Raton, Florida, pp. 159–166.

Savage, D., Barbetti, M.J., MacLeod, W.J., Salam, M.U. and Renton, M. (2010) Timing of propagule release significantly alters the deposition area of resulting aerial dispersal. *Diversity and Distributions* 16, 288–299.

Scott, R.W. and Achtemeier, G.L. (1987) Estimating pathways of migrating insects carried in atmospheric winds. *Environmental Entomology* 16, 1244–1254.

Shamoun-Baranes, J., Bouten, W. and van Loon, E. (2010) Integrating meteorology into research on migration. *Integrative and Comparative Biology* 50, 280–292.

Symmons, P.M. and Luard, E.J. (1982) The simulated distribution of night-flying insects in a wind convergence. *Australian Journal of Zoology* 30, 187–198.

Ward, S.A., Leather, S.R., Pickup, J., Harrington, R. (1998) Mortality during dispersal and the cost of host specificity in parasites: how many aphids find hosts? *Journal of Animal Ecology* 67, 763–773.

Westbrook, J.K., Eyster, R.S. and Allen, C.T. (2011) A model for long-distance dispersal of boll weevils (Coleoptera: Curculionidae). *International Journal of Biometeorology* 55, 585–593.

Wilson, J.R.U., Dormontt, E.E., Prentis, P.J., Lowe, A.J. and Richardson, D.M. (2009) Something in the way you move: dispersal pathways affect invasion success. *Trends in Ecology & Evolution* 24, 136–144.

Woiwod, I.P. and Harrington, R. (1994) Flying in the face of change: The Rothamsted Insect Survey. In: Leigh, R.A. and Johnson, A.E. (eds) *Long-Term Experiments in Agricultural and Ecological Sciences*. CAB International, Wallingford, UK. pp. 321–342.

Zhu, M., Radcliffe, E.B., Ragsdale, D.W., MacRae, I.V. and Seely, M.W. (2006) Low-level jet streams associated with spring aphid migration and current season spread of potato viruses in the US northern Great Plains. *Agricultural and Forest Meteorology* 138, 192–202.

5 Using the MAXENT Program for Species Distribution Modelling to Assess Invasion Risk

Catherine S. Jarnevich[1]* and Nicholas Young[2]

[1]*US Geological Survey, Fort Collins Science Center, Fort Collins, Colorado, USA;* [2]*Natural Resource Ecology Laboratory, Colorado State University, Fort Collins, Colorado, USA*

Abstract

MAXENT is a software package used to relate known species occurrences to information describing the environment, such as climate, topography, anthropogenic features or soil data, and forecast the presence or absence of a species at unsampled locations. This particular method is one of the most popular species distribution modelling techniques because of its consistent strong predictive performance and its ease to implement. This chapter discusses the decisions and techniques needed to prepare a correlative climate matching model for the native range of an invasive alien species and use this model to predict the potential distribution of this species in a potentially invaded range (i.e. a novel environment) by using MAXENT for the Burmese python (*Python molurus bivittatus*) as a case study. The chapter discusses and demonstrates the challenges that are associated with this approach and examines the inherent limitations that come with using MAXENT to forecast distributions of invasive alien species.

Intended Uses of MAXENT

MAXENT is part of a suite of models termed alternatively species distribution models, habitat suitability models, niche models, environmental niche models, bioclimatic models and climate envelopes, among others. This group of techniques determines relationships between sampled locations (i.e. occurrence or abundance information) for a species or group of species and associated environmental variables (i.e. covariates) at the locations (e.g. climate, topography, vegetation or land-use features) that are used to estimate some function of occurrence at unsampled locations. These types of models have been reviewed previously in the scientific literature (Guisan and Zimmermann, 2000; Elith and Leathwick, 2009) and books (Franklin, 2009; Peterson *et al.*, 2011). Several insightful reviews exist that are specific to invasive alien species and the challenges of using correlative models to assess invasion risk (Jiménez-Valverde *et al.*, 2011; Elith, 2014).

MAXENT is a correlative model in that the algorithm determines relationships between locations and environmental information. It is a generative approach because relationships are determined by comparing environmental characteristics at known locations of a species to the available environment as characterized by a set of background locations. Alternatively, discriminative approaches try to distinguish differences between where a species is found (i.e. presence) and where it is not found (i.e.

* Corresponding author. E-mail: jarnevichc@usgs.gov

absence). The method is similar to the concept of use–available models (i.e. resource selection functions) in the wildlife literature (Warton and Aarts, 2013). The primary MAXENT output is a ranking of locations across the area of interest based upon their relative suitability for the species of interest rather than actual probabilities of occurrence. Information on species prevalence, which requires presence and absence data, is necessary for true occurrence probabilities (Phillips and Elith, 2013). MAXENT generates predictions by finding the probability distribution of maximum entropy for the species being modelled constrained by the predictor variables in relation to the known presence locations (Phillips et al., 2006). Six relationships (i.e. 'feature types') can be fit in MAXENT: linear, quadratic, product (i.e. interaction between two predictors), categorical, hinge and threshold (Phillips and Dudik, 2008).

MAXENT was created to be used with presence-only data, such as location data available from herbarium and museum collections, for a species at equilibrium with its environment. 'Equilibrium' conditions imply that a species has been present in the region with enough time to spread to an available suitable habitat. When this condition is met, a systematic study producing presence-and-absence or abundance data should be used with an appropriate method such as a regression technique that can take advantage of the rich data set. However, invasive alien species present a unique set of issues for correlative models. In this case, within the native range a species may approach the assumed equilibrium conditions. In the invaded range, however, this condition is violated unless the species has been present for enough time to spread to all possible suitable habitats.

Models created using native range data may be applied to the invaded range; the ability of a model to predict accurately distributions in an unsampled region such as a potentially invaded range is commonly termed transferability. The distribution of a species in its native range may be controlled by environmental factors such as climate, but other factors such as geographic barriers, competition and predation may limit the distribution. These other limiting factors may act differently, be absent entirely or may be replaced by novel limiting factors in the invaded range compared with the native range. These differences can lead to underestimates of the potential geographic range size when models are developed on an invasive alien species' native range and are transferred to its invaded range (Jiménez-Valverde et al., 2011).

Presence-only techniques can be highly sensitive to sampling bias, i.e. the non-random placement of sampling locations (Pearce and Boyce, 2006). For methods that use presence-and-absence data, sampling bias is largely accounted for because generally the presence data have the same bias as the absence data and, therefore, the bias cancels out (Phillips et al., 2009). Presence-only techniques are limited in this regard because unless models adjust the selection of the random background locations, the model provides forecasts of the species' suitable habitat and any bias in the presence data, producing a forecast of habitat suitability and sampling intensity rather than only habitat suitability. Fortunately, strategies exist to account for sampling bias in presence-only modelling (e.g. Phillips et al., 2009) and will be discussed in more detail below.

Context for a Case Study: Burmese Python

We used the Burmese python (*Python molurus bivittatus*), native to South-East Asia and currently established in Florida, USA, to explore climate matching between a native range and an invaded range with MAXENT. The Burmese python was introduced to Florida, most likely through the pet trade, from accidental and intentional releases by their owners. These constrictors prey on a wide variety of native fauna. Identifying the potential distribution of this invader is important for natural resource managers and conservation scientists. The optimal method for performing this analysis is uncertain, and important methodological issues related to background location selection and assessment of model complexity need to be addressed.

MAXENT is a user-friendly application, providing a simple graphical user interface and requiring minimal inputs. Location data for the Burmese python and other species can readily be downloaded from an online source such as Global Biodiversity Information Facility (http://www.gbif.org) and raster layers of global bioclimatic data are freely available from WorldClim (http://www.worldclim.org/). The ease of producing a result (i.e. a map) with MAXENT and its generally good performance relative to other species distribution models (Elith et al., 2006) have contributed to the proliferation of papers using this technique. Ease of use also has led to misuse of MAXENT. Some users have not evaluated the implications of running MAXENT with just the default settings or considered whether this tool is appropriate for particular research questions.

The example within this chapter stems from our previously published assessment of MAXENT and its utility to forecast climatically suitable areas for invasive alien pythons (Rodda et al., 2011). For that case, native range data from Asia were used to forecast potential distribution in an invaded range (i.e. southern North America and northern South America). We argued that MAXENT was not an appropriate tool for this purpose, although it may be useful in other contexts. This chapter describes the process by which we created a MAXENT model. To show the importance each decision can make, we begin with the default settings and slowly change parameters to end with what we felt was an appropriate implementation of MAXENT given the data we have. This approach to MAXENT model construction for invasive alien species should be broadly applicable and some of the insights gained from our model for pythons may be universal.

Resources for MAXENT Models

This section describes where to obtain the resources required to construct a MAXENT model.

MAXENT software and support

MAXENT software is freely available for download via the internet (http://www.cs.princeton.edu/~schapire/maxent/) and a well-established MAXENT Google Group exists to facilitate discussion within the user community. Also available is a MAXENT tutorial (http://www.cs.princeton.edu/~schapire/maxent/tutorial/tutorial.doc) that provides an introduction to the program and describes an example application. Several software packages exist to assist in using MAXENT, especially to pre-process data and facilitate analyses of model output. These packages include, but are not limited to, ENMTOOLS (Warren et al., 2010), DISMO R package (Hijmans et al., 2011), BIOMOD (Thuiller et al., 2009) and software for assisted habitat modelling (SAHM; Morisette et al., 2013).

MAXENT data files

MAXENT requires species occurrence data (i.e. x, y coordinates such as longitude and latitude of where a species has been observed) and relevant spatial predictor variables, termed environmental variables, in ASCII raster format. Gathering and pre-processing these data can often be the most time-consuming part of creating a MAXENT model, although the software packages mentioned above can simplify these efforts and greatly reduce pre-processing time.

The occurrence data should adequately represent the species' geographic distribution (i.e. not be spatially biased) and the environmental gradients occupied by the species (e.g. cover the full range of environments available to the species). The definition of 'adequate', however, is ambiguous. Smaller sets of occurrence data (i.e. fewer than ten records) have been used successfully to generate models of potentially suitable habitat (not potential distribution) and locate new populations of rare species (Pearson et al., 2007). Wisz et al. (2008) recommended >30 occurrence records, but there is no agreement on a requisite number of occurrence records for MAXENT. Occurrence data can be compiled from the literature, museum records and other sources, such as field surveys or reviews. The Global Biodiversity Information Facility (http://www.gbif.org) can be a good source of aggregated museum records. Once compiled, these data need to be examined carefully for coordinate or identification errors. Errors in georeferencing are most easily identified after the data are mapped. Errors may include terrestrial species appearing in oceans, gridded data that may be coarser than the desired resolution and point locations lying outside the bounds of the country in which they were reported to occur. The DISMO R package (Jiménez-Valverde et al., 2013) can be particularly useful in finding and repairing errors in occurrence data.

Species occurrence data must be in the form of a comma-separated value (CSV) format file with a header row, followed by rows representing a species' occurrence. There should be three columns that include the species' name in column one, x-coordinates (i.e. longitude) in column two and the y-coordinates (i.e. latitude) in column three. MAXENT does not require a particular coordinate system, but the x- and y-coordinates must match the coordinate system of any raster layers provided as environmental layers.

Users can provide environmental data to MAXENT in one of two ways. The first option is to supply a path to the folder containing rasters of environmental data in ASCII (.asc) format. With this option, MAXENT will randomly generate background samples for model fitting and can produce maps for the area of interest. The second option is termed 'Samples with Data' or SWD format and entails supplying a CSV file of background locations with the environmental variable data. The first three columns match those described for occurrence data, but now the first column should be blank, the second should contain x coordinates and the third y coordinates. The following columns will have the name of each environmental variable to be considered in the model as the column headings and with that environmental variable's value extracted for each location (i.e. in each row). If SWD format is used, the occurrence data file must also include the environmental variable values for each point. The task of creating an SWD file can be done in any number of ways including the DISMO R package, DIVA-GIS, SAHM or ArcGIS; the first three are freely available online.

The availability and quality of spatial data have increased rapidly in the past decade, so modellers can consider a greater number and variety of environmental variables for species distribution modelling. Which environmental variables should be used to develop distribution models depends on the scale, the species, the research question and the available data. Climate data are most often used as environmental variables in developing native range models that are then transferred to the potentially invaded range. In this chapter, we present models that only use climate environmental variables, but many other variables could be used including topography, anthropogenic features, land use, remotely sensed variables or a combination of these. However, when the intent of the model is to transfer it in space or time, predictors should be causally, not indirectly, related to distribution (Jiménez-Valverde et al., 2011). Predictors that are contingent on spatial locations such as dispersal barriers or biotic interactions should also not be included.

All environmental data used for MAXENT modelling must be the same coordinate reference system (i.e. projection and datum), extent (i.e. scale) and grain (i.e. resolution). They must also be in ASCII raster format and match the coordinate reference system of the species occurrence data. Many software applications exist that can convert from one raster format to another (e.g. DIVA-GIS, SAHM, R-CRAN 'raster' package or ArcGIS).

MAXENT Analyses

For our analysis of climatic matching between the Burmese python's native range and its invaded range, we used 86 points that had been collated by Pyron et al. (2008) from georeferenced museum voucher specimens, the literature and wildlife agencies to represent the native range. We removed four of the original 90 data points from Pyron et al. (2008) through a data review and cleaning process, after we had determined that four locations taken from Nabhitabhata and Chan-Ard (2005) were for a different species, the blood python (*Python brongersmai*).

For environmental variables, we used the 19 bioclimatic variables available from WorldClim at a 2.5 arc-minutes resolution, following Pyron et al. (2008). (Later in this section, we discuss the importance of limiting this highly inter-correlated set of environmental variables before running models.) As mentioned above, MAXENT requires all raster layers to have the same extent and cell size and for the grids to line up perfectly. Often this can be one of the hardest tasks. However, as all our raster layers came from the same source, they already met this requirement.

While MAXENT provides a user-friendly interface to run models, it can be executed via command line. In the analyses that follow, we provide the exact command line code used to run MAXENT. Running MAXENT using this technique will still launch the user interface with all the settings defined based on the specific commands listed.

Analysis with default settings

We started with the default settings for MAXENT but later ran 25 replicates while withholding a different 10% of the presence locations to use as a test data set. We were interested in variable contribution, so we turned on creation of response curves and jackknife measure of variable importance. For comparison purposes, we wanted to use the same set of background points across all runs focused on the same extent, so we turned on the *writebackgroundpredictions* option to obtain the coordinates of the background points generated by MAXENT. We then could create an SWD format file including these same background locations to use in further model runs.

The following text was entered (without line breaks) into the command line to run the first model:

java -jar maxent.jar -E Python_molurus responsecurves jackknife writebackgroundpredictions outputdirectory=J:\Projects\python\maxent_default_90_global samplesfile=J:\Projects\python\90locations.csv environmentallayers=J:\Projects\python\Global_extent randomseed nowriteclampgrid nowritemess randomtestpoints=10 replicates=25 replicatetype=subsample nooutputgrids

In the first part of this command, '-*E Python_molurus*' selects the species for which we want to run the model. In our samples file, the first column, species' name, contains the text '*Python_molurus*' which represents our species of interest. If this column had contained records for multiple species, identified by their unique names, the other species would have been ignored. The next three commands tell MAXENT to create the response curves, to conduct a jackknife of variable importance and to produce a CSV file that contains the coordinates for the background points MAXENT generates (i.e. *responsecurves*, *jackknife* and *writebackgroundpredictions*, respectively).

In the next part of the command, we specified the folders and the files to use. The output directory is the folder we created to store the MAXENT output. The *samplesfile* command provides the path to the occurrence data CSV file. The *environmentallayers* command gives the location of the folder containing the raster environmental layers in ASCII format at the global extent. MAXENT is set by default to select 10,000 background points within the extent of these rasters and will use these environmental variables to extract values at presence and background locations to use in model fitting and to create the final suitability maps.

The *randomseed* command tells MAXENT how to initiate its random number generator. If we had supplied a number for the seed, the same random set of numbers would have been selected each time the model was run. With the *randomseed* command, we tell MAXENT to draw a random number and initiate its random number generator with that value. This approach will generate a different set of random numbers each time the model is run. As a result, MAXENT will select a different subset of points for testing and to generate background points each time the model is run. We have MAXENT write out the coordinates for the background locations (i.e. by using *writebackgroundpredictions* in the first line) so we can use these same background points in future runs. *Nowriteclampgrid* turns off the creation of output grids including where clamping occurs. 'Clamping' is the treatment of variables outside the training range as if they were at the edges of the training range. *Nowritemess* turns off the creation of a multivariate environmental similarity surface which describes how similar the environment at each location within the surface is to the environmental range captured by the training data set. The commands *randomtestpoints=10*, *replicates=25* and *replicatetype=subsample* were used to explore how much impact specific data points might have on the results and to calculate evaluation metrics on a subset of data not used to train the model, i.e. the test data. Training data are defined as the presence locations used by MAXENT to fit a model, while test data include presence locations withheld from model fitting that can be used to evaluate a model. Specifically, we set the replicate type to subsample, the number of replicates to 25 and the random test percentage to 10% which will result in 25 models, each created using a different random subsample of 90% of the presence data to train the model with the remaining randomly chosen 10% used to test the model. We chose subsample over the other options of cross-validation or jackknife so we could specify the test percentage and because we had a moderate number of presence records.

The choice of replicate type should depend on the number of presence records. Bootstrapping or jackknife (i.e. repeating an analysis while leaving out a record) works well with small data sets (i.e. approximately <30 presence records) when most of the records are needed to fit the model. For large data sets (i.e. approximately >100 presence records), cross-validation, where the data are split into *k* folds of equal size and the model is run *k* times with each fold being withheld once for testing, is appropriate for testing. The number of folds will dictate the number of replicates and the size of the test data set. Subsampling is the most flexible and allows the user to specify the test percentage and number of replicates. The replicates are not independent.

Finally, to reduce processing time and save storage space, we chose to have MAXENT produce only summary grids, rather than grids for each replicate run, by issuing the *nooutputgrids* command and turning off the option to write output grids. The maps resulting from this set of code appear in Fig. 5.1a and b (see colour plate section).

Analysis with default settings and corrected presence data

We examined the effect of using the corrected presence location data (i.e. the set of locations excluding those identified as having a different species) on the model. We made only one change to the first command by removing the *writebackgroundpredictions* command. Other command changes, in bold below, included: updating the output directory; changing the samples file to the taxonomically corrected list in SWD format; changing the environmental variables path to point to the new SWD format rather than ASCII files; and adding a projection-layers directory pointing to the folder containing the global ASCII grids so that we could still make maps. SWD format files do not include information about where the rasters are located and therefore maps of the study area cannot be produced without using a projection-layers directory. The maps resulting from this set of code appear in Fig. 5.1c and d (see colour plate section).

The new command lines were thus:

*java -jar maxent.jar -E Python_molurus responsecurves jackknife outputdirectory=J:\ Projects\python***maxent_default_86_ global projectionlayers=J:\Projects\ python\Global_extent*** samplesfile=J:\ Projects\python***86locations_SWD.csv*** *environmentallayers=J:\Projects\ python***86locations_SWD.csv*** *randomseed nowriteclampgrid nowritemess randomtestpoints=10 replicates=25 replicatetype=subsample nooutputgrids*

Accounting for accessible areas

By using data layers at a global extent for the environmental variables, we have MAXENT randomly generate background locations across the entire globe. When creating correlative models using presence and background or pseudo-absence locations, it is important to restrict background locations to the area accessible to a species (VanDerWal et al., 2009; Barve et al., 2011). A model to predict potential habitat based on locations in a native range, such as Burmese python habitat in the USA based on locations in Asia, should be modelled with background locations taken only from the native range (Phillips, 2008). One method to limit background point selection is to generate a minimum convex polygon (MCP) around the presence locations. There are many tools to generate an MCP, but for this analysis, we used Hawth's Tools (Beyer, 2004) within ArcGIS 9.3 software. However, with the release of ArcGIS 10.0, Hawth's Tools has been discontinued and replaced with Geospatial Modelling Environment, which is freely available at http://www.spatialecology.com/gme/index.htm to ArcGIS 10.x users. The specific command to use within Geospatial Modelling Environment is:

genmcp(in="C:/python/pointlocations.shp", out="C:/python/MCP_pointlocations.shp");

where the *in* data set is the path and name of the shapefile with point locations and the *out* data set specifies the path and name for the MCP shapefile that is being generated. To ensure inclusion of all point locations where Burmese python has been observed, we added a one-pixel (2.5 arc-minute) buffer around our MCP by using the buffering tool in ArcGIS 10.

Other ways to limit the area from which MAXENT might draw background locations have been suggested in the literature. One way is to create a bias surface that reflects what is known about sampling bias in the presence locations. MAXENT then uses the bias surface to constrain background locations to reflect a similar bias (e.g. if all samples are within 1 km of roadsides, then the bias surface would have high values near roadsides and zeros away from roadsides so that background data would be constrained to within 1 km of roadsides). Other methods include selecting background locations within similar climatic zones (Webber et al., 2011), limiting the selection to within a certain distance of presence sample locations (VanDerWal et al., 2009) and applying a kernel density approach that assumes higher-density locations are where the species has been present longer when the goal is to develop a model for a spreading invasive alien species (Elith et al., 2010).

To force MAXENT to restrict the possibilities for selecting background data, we clipped our input raster layers to the buffered MCP using the ArcGIS Spatial Analyst extract-by-mask function with each environmental variable as the input raster and the buffered MCP shapefile as the feature mask. After this process, we had two folders, each with 19 ASCII rasters with the same names: one at the global extent and another at the MCP extent.

As described above, MAXENT, as with any presence–background/pseudo-absence algorithm, is highly sensitive to sampling bias (Phillips et al., 2009). One recommended way to control for unknown bias is to use a target background approach where background points are chosen that may have a similar bias (Phillips, 2008) and generally this method is a good approach. In our case, we had reservations about museum records for large constrictor species, so we did not feel this method was appropriate. Rodda et al. (2011) discussed these reservations in detail. Georeferenced positions in museum records often lack precision, especially when collection localities are reported only to the nearest degree of longitude and latitude. Occasionally, collection localities may be reported just as the nearest market or town, not the actual location where a specimen was obtained. In addition, locations likely exist where a species is present but has gone undetected. The goal of Rodda et al. (2011) was not to produce the best MAXENT model, but to evaluate the sensitivity of MAXENT to different options. Had the goal been different, the approach to target background locations within the MCP might have been useful. Given the lack of information about sampling bias, we allowed MAXENT to generate background points randomly within the MCP.

Our next run in MAXENT was the same as the previous, but we changed the extent from which background points were drawn from the globe to the MCP. Once again, we wanted a record of the selected background points so that we could use the same background locations in future model iterations.

Commands for the next run were:

java -jar maxent.jar -E Python_molurus responsecurves jackknife **writebackgroundpredictions** *outputdirectory=J:\Projects\python* **maxent_86_MCP** *projectionlayers=J:\Projects\python\Global_extent samplesfile=J:\Projects\python\86locations_SWD.csv environmentallayers=J:\Projects\python***MCP_bckgrnd** *randomseed nowriteclampgrid nowritemess randomtestpoints=10 replicates=25 replicatetype=subsample nooutputgrids*

Text in bold represents a change from the previous command submission ('Analysis with default settings and corrected presence data' subsection). Maps resulting from this run appear in Fig. 5.1e and f (see colour plate section).

Extrapolation assessment

Now that we had projected the model to a broader extent than our samples (background locations within MCP, projected to globe), we wanted to run the multivariate environmental similarity surface (MESS) analysis in MAXENT (Elith *et al.*, 2010). This analysis assesses the amount of extrapolation that might be required to project a model to new areas by comparing, in a univariate sense, the range of values for each environmental layer captured by the training and background data with the values for that layer at each location to which the model will be applied. A location is assigned an increasing negative value as the value for the environmental variable at a location is increasingly outside the range of values for that variable in the sample and background locations. Areas with negative values have been termed 'novel areas'. However, for initial runs in the code above, the *nowritemess* option turned off the MESS analysis to save processing time and storage space, and the *outputgrids* needed to be turned on to produce MESS maps. We can obtain the same MESS map by running the code from the subsection 'Accounting for accessible areas' for a single iteration, changing the items highlighted in bold and removing everything after *nowriteclampgrid* as follows:

*java -jar maxent.jar -E Python_molurus responsecurves jackknife outputdirectory=J:\Projects\python***maxent_86_MCP_MESS** *projectionlayers=J:\Projects\python\Global_extent samplesfile=J:\Projects\python\86locations_SWD.csv environmentallayers=J:\Projects\python***MCP_bckgrnd.csv** *randomseed nowriteclampgrid*

Text in bold represents a change from the previous command submission ('Accounting for accessible areas' subsection).

Controlling for overfitting

Another potential concern with highly parameterized models is overfitting. Overfitting occurs when a model is overly complex, sometimes describing random noise or error in data rather than actual underlying relationships. This problem can result in the underestimation of suitable conditions in novel locations or climates.

Two methods are commonly used to determine if a model might be overfit. The first approach is to examine differences in the area under the receiver-operating characteristic curve (AUC) between the test and training area. AUC is a measure of the ability of a model to discriminate presences from absences (Fielding and Bell, 1997) and is explained in more detail in the next subsection ('Model assessment'). A much larger AUC calculated from the model-training location data (i.e. presence and background) compared with the AUC calculated from test location data can indicate overfitting. The second approach to assess the potential for overfitting is to examine the complexity of the response curves. Highly jagged response curves indicate overfitting.

One approach to reduce the likelihood of overfitting is for the user to turn on and off the feature types in MAXENT. The alternative is to let MAXENT control for overfitting automatically.

MAXENT controls for overfitting by using a 'regularization' parameter, a statistical penalty which balances model fit and model complexity. Only coefficients that improve the model enough to outweigh a regularization penalty are retained. Many coefficients near zero are removed from the model. The regularization parameter, known specifically as the 'betamultiplier' in MAXENT, has a default value based on an optimization by Phillips and Dudik (2008) for 12 species with 11–13 climate predictors, but users should not assume the default is appropriate in all cases.

Warren and Seifert (2011) created a tool within the ENMTOOLS program (available for download at http://enmtools.blogspot.com) to optimize the regularization value using the Akaike Information Criterion for small sample sizes (AICc). This tool requires the input locations that were submitted to MAXENT, the MAXENT ASCII output raster in raw format (as opposed to the logistic format which is the default) and the MAXENT lambdas file. The lambdas file (.lambdas) is one of the outputs automatically generated by MAXENT with each model run, found in the output folder that describes the model fit, covariates and coefficient values for covariates. To reduce running time, turn other features in MAXENT off (*nopictures, nowriteclampgrid, nowritemess, noplots* and remove *responsecurves* and *jackknife*). The resulting commands are as follows:

> java -jar maxent.jar -E Python_molurus nopictures outputformat=raw outputdirectory=J:\Projects\python\reg_test\maxent_90_MCP_Reg1 projectionlayers=J:\Projects\python\Global_extent samplesfile=J:\Projects\python\90locations.csv

> environmentallayers=J:\Projects\python\MCP_bckgrnd.csv randomseed nowarnings noaskoverwrite nowriteclampgrid nowritemess randomtestpoints=10 betamultiplier=1.0 replicates=25 replicatetype=subsample noplots noprefixes

The models, and thus the number of parameters and AICc values, change each time the set of input presence-points changes, so we performed this assessment multiple times on the 25 subsamples to mimic the method we would use in the final run. Each assessment progressively increased the betamultiplier from 1 to 10 by intervals of 1. An increase in the betamultiplier lowers the complexity of the model (e.g. the response curve for annual mean temperature in Fig. 5.2a from a betamultiplier of 1 is more complex than Fig. 5.2b from a betamultiplier of 4). We used the same process for our data sets with the 86 points and the 90 points (see section 'MAXENT Analyses'), then created CSV format files without a header row with each column containing the path to samples file, the path to raw ASCII output and the path to the lambdas file. Within ENMTOOLS, we opened the model selection tool under the measurements and tools heading. The output file from this tool contains a log-likelihood estimate, the number of parameters calculated from the lambdas file, the number of presence locations, and AIC, AICc and BIC (Bayesian Information Criterion) metric values. For each model replicate (1 to 25), we selected the regularization value with the smallest AICc value and calculated the mean of these values across the 25 runs. The mean became the regularization value for all subsequent runs using that particular set of presence and background locations.

To run the model using the optimized regularization value, we started with the commands from the subsection 'Extrapolation assessment' and added the command *betamultiplier*. To evaluate the 90-location data set, the regularization value was 3 and for the 86-location data set, the regularization value was 4. Commands for the 90-location data set assessment were as follows:

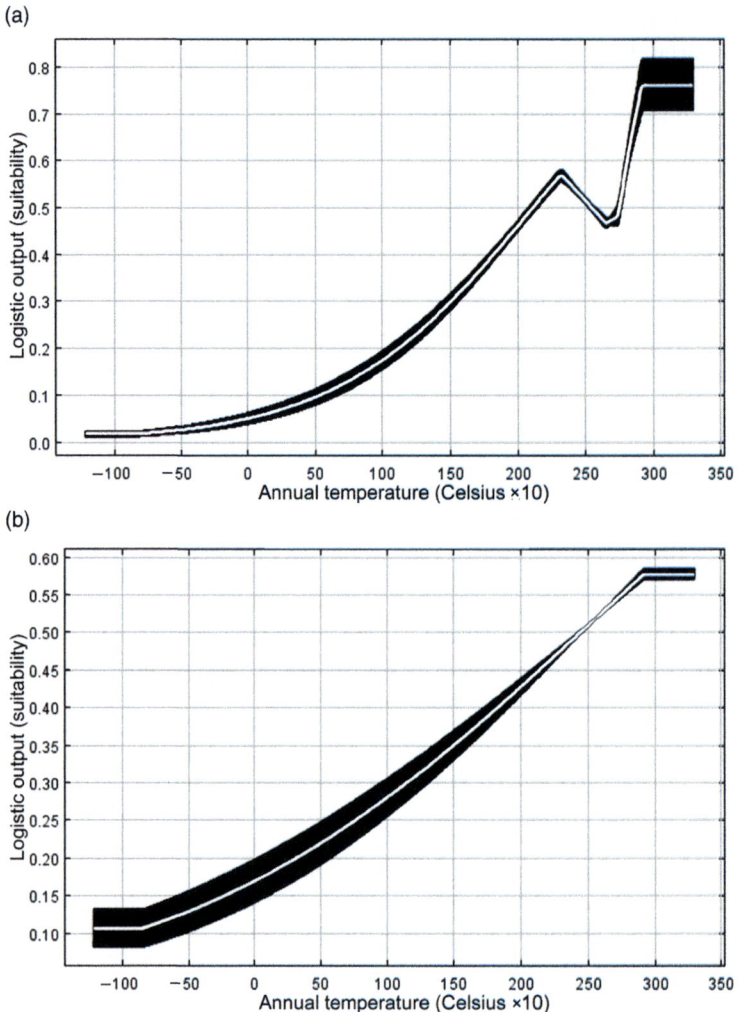

Fig. 5.2. Response curve for 'bio 1', annual mean temperature, for the 'MCP 86' model with: (a) the default regularization value of 1; and (b) the optimized regularization value of 4. The white line represents the average response while the black surrounding the white line represents the variation across 25 model runs. Model 'MCP 86' has background data restricted to a minimum convex polygon around corrected presence points.

*java -jar maxent.jar -E Python_molurus responsecurves jackknife outputdirectory=J:\Projects\python***maxent_90_MCP_reg3*** projectionlayers=J:\Projects\python\Global_extent samplesfile=J:\Projects\python***90locations_SWD.csv*** environmentallayers=J:\Projects\python\MCP_bckgrnd.csv randomseed nowriteclampgrid* **nowritemess randomtestpoints=10 betamultiplier=3.0 replicates=25 replicatetype=subsample nooutputgrids**

Text in bold reflects a change from the previous command submission (subsection 'Extrapolation assessment'). The above command lines can be used to run the 86 locations by changing the values in the output directory, the samples file and the regularization value (*betamultiplier*). To produce a MESS map for both of these runs, we followed the modifications described in subsection 'Extrapolation assessment'. Maps from models using all 90 locations are presented in Fig. 5.1g and h and maps using the cleaned 86 locations are presented in Fig. 5.1i and j (see colour plate section).

Model assessment

Examining the output maps (Fig. 5.1; see colour plate section) that resulted from each of the models provides some insight into the importance of the different MAXENT settings. The only difference in the models shown in Fig. 5.1a and b (from subsection 'Analysis with default settings') and Fig. 5.1c and d (from subsection 'Analysis with default settings and corrected presence data') is the Burmese python location points used to train the model; Fig. 5.1c and d exclude the four locations with incorrect taxonomy. This small change in input data creates relatively large differences in the maps, with a much greater area of suitability with the restricted presence locations. With the removal of the four points, the predictor variables and response curves change. Changes in the underlying model alter the predicted values for the presence locations. The second model had a much lower minimum training presence (MTP; the smallest value predicted by the model for any of the training points) value, which led to a greater area of suitable habitat.

The models behind Fig. 5.1e–j are based on background points that came from the MCP. With each model, we had MAXENT generate a layer indicating where extrapolation occurred (i.e. negative values in the novel layer produced by the MESS analysis). Many environments around the world are substantially different from the presence and background locations for Burmese pythons and could be considered novel, including a large portion of the USA, the area of interest for this assessment (see hash marks in Fig. 5.1e–j). Forecasts where the model is extrapolating should be treated with extra caution.

Figure 5.1e and f (from subsection 'Accounting for accessible areas') differ from Fig. 5.1c and d only in that background points are restricted to the MCP around the presence locations rather than drawn from across the globe. Figure 5.1g and h and Fig. 5.1i and j (both from subsection 'Controlling for overfitting') differ only in the location data used to generate the models as did Fig. 5.1a and b and Fig. 5.1c and d; in this case, however, we restricted MAXENT selection of background points to the MCP and altered the betamultiplier to control for overfitting. With those two alterations from the default settings, the removal of four presence locations has a minimal effect on the model predictions, and the resultant maps are much more similar (i.e. Fig. 5.1g and h compared with Fig. 5.1i and j) than those created with the default settings (i.e. Fig. 5.1a and b compared with Fig. 5.1c and d). Once we restricted the background data and model complexity, the models became more robust to the locations used.

There are several ways to assess model performance. MAXENT will generate and report standard statistical values in several files. When running replicates, MAXENT produces files for each replicate along with a summary file that contains means and standard deviations of model results (assessment metrics, variable importance, response curves, maps, etc.) across replicate model runs. The file MAXENTResults.csv includes output for each run along with a summary across all runs. Table 5.1 shows a variety of assessment metrics for each model generated.

One step in the model assessment process is to identify any model overfitting. As stated above, signs of overfitting include a large difference between the test and train AUC values (e.g. model MCP 86 in Table 5.1); jagged response curves showing a response to noise (e.g. Fig. 5.2a compared with Fig. 5.2b); a very spotty, non-contiguous distribution (e.g. Fig. 5.1a – South America); and high standard deviations in response curves and predictions across runs. Another method is to compare the number of parameters calculated from the lambdas file (in the example for this chapter, by using ENMTOOLS) with the number warranted given the input data. The first three models had more than four times the number of parameters than the last two that had optimized complexity (Table 5.1). Simpler

Table 5.1. Assessment metrics for five MAXENT models.

Model run name[a]	Regularization	Parameters, $n \pm$ SE	Mean MTP[b]	AUC for test (train)[c]
Default 90	1	62 ± 5	0.092	0.971 (0.984)
Default 86	1	59 ± 4	0.013	0.973 (0.982)
MCP 86	1	56 ± 5	0.159	0.702 (0.816)
MCP 90 reg3	3	13 ± 2	0.237	0.711 (0.747)
MCP 86 reg4	4	10 ± 2	0.222	0.624 (0.718)

Values are for 25 replicates.
[a]Names: 'Default 90', run with default settings and 90 presence points; 'Default 86', run with default settings and corrected presence points; 'MCP 86', background data restricted to a minimum convex polygon around corrected presence points; 'MCP 90 reg3', background data restricted to a minimum convex polygon around 90 presence points and the regularization value optimized; 'MCP 86 reg4', background data restricted to a minimum convex polygon around 86 presence points and regularization value optimized.
[b]MTP, minimum training presence. The lowest predicted value from the underlying model that correctly classifies all training presences as 'present'.
[c]AUC, area under the receiver-operating characteristic curve for test and training data sets. AUC measures the capacity of a model to distinguish presence from absence points. Values generally range from 0.5 (i.e. no better than chance) to 1.0 (i.e. perfect classification).

models may be better, especially when transferring a model to new space or time, as we are to the USA. More complex models may capture relationships specific to the location or temporal span of the model training data and thus may be more suited to interpolation within the extent and time of model training. Simpler models can also be distinguished by the complexity of response curves.

Before we evaluate how well a model performs in the invaded range, it is important to ensure that a model performs well in the known native range. MAXENT calculates the AUC score, a threshold-independent metric that measures how well a model discriminates between presence and background locations. AUC generally has values between 0.5 (i.e. no better than chance) and 1.0 (i.e. a perfect fit). A general guideline for evaluating AUC scores is that values between 0.5 and 0.7 represent models with low accuracies; values between 0.7 and 0.9 indicate useful models; and values above 0.9 represent models with high accuracy (Swets, 1988; Fielding and Bell, 1997).

Because AUC is measuring discrimination, we typically get very high AUC scores for models that use background locations from around the globe. Drawing background data points (i.e. 'pseudo-absence' areas) from the larger region increases the probability that these points will be environmentally distant from the presence locations. As a result, the rate at which these background locations are correctly identified will be artificially high; thereby elevating the AUC scores (Lobo et al., 2008). When background points are restricted to the area from which presence points were obtained (e.g. the area within the MCP in the example in this chapter), the AUC scores drop greatly (e.g. compare AUC scores in Table 5.1 from default MAXENT runs to AUC scores when MCP was used). Looking at AUC scores without understanding the limitations of the metric for MAXENT models might lead a user to conclude that the model from the default MAXENT run was great. Given the low test AUC value for the MCP 86 reg4 run (Table 5.1), it would probably be inappropriate to transfer the model to the USA, even though this model is very reasonable, as judged by the settings for this run.

Differences in AUC scores (Table 5.1) illustrate one of the problems in relying on a single metric to evaluate model performance. To obtain a more rigorous assessment, the statistical software R can be used easily to calculate other calibration metrics (e.g. presence-only calibration plot; Phillips and Elith, 2010) and other discrimination metrics (e.g. the true skill statistic; Allouche et al., 2006).

Experts on the species can also provide valuable assessments of the model by simply inspecting maps and response curves for the native range. In our example, two of the five models predicted low suitability for large regions of interior India known to have Burmese pythons (Fig. 5.1a and e). These regions could be sink habitats (i.e. areas where Burmese pythons are present, but are incapable of sustaining populations), but this result seems unlikely, given the way specimen records were collected for this study. Sink habitats within the native range of most reptiles and amphibians are unusual when considering climate variables alone, the scale of climate variation and the limited dispersal capacity of these species.

MAXENT models must also be assessed on their ability to meet the needs of users. Users, such as managers and policy makers, frequently want a binary map of suitable and unsuitable habitat. To make this product, the modeller must select a threshold to classify the continuous suitability values into presence or absence. Most threshold metrics (Liu et al., 2005; Freeman and Moisen, 2008) have been evaluated with presence–absence data, not simulated values such as those from MAXENT. MAXENT automatically calculates some threshold metrics and includes them in the Results.csv file and the HTML file for each run. Some threshold selection tools are from adaptations of common presence–absence methods (e.g. equal test or train sensitivity and specificity) and some are particular to presence-only techniques, such as MTP and 10-percentile training presence. MTP and 10-percentile training presence involve extracting predicted values for all training locations, ordering them from least to greatest, and selecting the lowest value for MTP or selecting the value for the ordered location that separates the lowest 10% of values from the highest 90% (e.g. the value for the location in the tenth position if there were 100 locations). We used MTP in this analysis as we did not feel the adapted methods were suited to this data set and the somewhat arbitrary omission of some localities (i.e. 10%) was also not appropriate for this species. For invasive species risk assessments, Jiménez-Valverde et al. (2011) recommend focusing on presence data, as we do by using the MTP, for applications of species distribution data that have the goal of identifying environmentally suitable localities.

MAXENT also provides sets of univariate and multivariate response curves and several ways to examine the importance of different predictors. MAXENT response curves are created between the minimum and maximum values from the combined occurrence and background locations for each environmental variable, covering the range of interpolation for the model. For areas outside the model's training range, response curves are continued to the left with a minimum value and to the right with the maximum value. Areas where this occurs are captured with the MESS analysis.

Our example included several highly correlated variables so most of these assessments are difficult to interpret and should be used cautiously, as they will be influenced by the covariance structure. Had the purpose of the exercise in this chapter been to create the best MAXENT model possible, rather than to evaluate the challenges in using MAXENT for this type of application, we would have examined a correlation matrix of the 19 bioclimatic variables measured at the presence and background points for each model. From this matrix, we would have selected a subset of uncorrelated predictors ($|r| < 0.7$ is most common; Dormann et al., 2013) that was relevant to the known biology of the species. This analysis can be performed in several software packages, though we prefer the SAHM CovariateCorrelationAndSelection module. Using fewer predictors that are not highly correlated is desirable when the objective is to transfer a model in space or time because the influence of changes in the correlation structure between the native range and the region of interest may be somewhat minimized.

Discussion

Our motivation for this chapter was to describe how to develop a correlative statistical model to assess pest risk and to highlight the challenges associated with this process. We focused on MAXENT for climate matching, but as with any modelling technique, having an understanding of the assumptions, data requirements, parameters and inherent caveats is critical for appropriate model application and interpretation. Specifically, the default settings of MAXENT are likely to be inappropriate for generating predictions about where invasive alien species might establish outside their native range. The processes by which background data are selected and requisite parameters are adjusted likely will vary from case to case and should be adapted to the species, the region of interest and the specific question at hand.

Even when great care is taken to select informed and tailored settings, MAXENT may not be the best choice for climate matching between the native range and the invaded range of an invasive alien species. MAXENT may not have been the best tool for Burmese python for two reasons. First is an issue related to any correlative modelling technique, in that our interest lies in characterizing potential habitat rather than the climate space currently occupied by the species. Second relates to the location data used to develop the model and the assumption that the available museum localities adequately represent the climate range inhabited by Burmese pythons. A more in-depth discussion of these concerns is found in the 'Conceptual issues' section of Rodda et al. (2011).

Other circumstances may exist where correlative techniques such as MAXENT are appropriate for invasive alien species. Crall et al. (2013) developed MAXENT models for invasive alien weed species in Wisconsin, USA, to guide future sampling efforts. In this case, sampling guided by MAXENT models located more unknown populations of these weeds than methods currently used by managers.

Many other approaches exist to predict and assess pest risk for a species in its current or potential invaded range other than the correlative climate matching model described in this chapter. For example, using expert and empirical information on the species, an ecological climate matching model can be developed and applied to the region of interest. This approach would take into account known precipitation and thermal thresholds that limit the Burmese python distribution in its native range and identify similar climatic zones in an invaded range (Rodda et al., 2009). Furthermore, others have suggested incorporating a mechanistic model or use of known locations of the pest in the invaded range to train the model to inform risk predictions (Elith et al., 2010). This has been recommended because often the potential range of the introduced species encompasses a wider climatic range than that found in the native range due to the absence of other limiting, non-climatic variables.

When calculating a model based on the native range, location data should cover the range of environments where the species occurs. The northern and western fringes of the range were missing from the presence points used in the MAXENT models. Rodda et al. (2011) were able to identify 14 additional locations from the literature and the California Academy of Sciences. Again, if our intent had been to create the best MAXENT model possible, using this broader suite of locations would be important and could potentially alter the predictions by expanding the environmental envelope captured by the presence locations.

The amount of extrapolation should also be assessed when a model is applied to a new area or time; it is also important to understand the way a model extrapolates. While a simple linear regression can extend the trend beyond the bounds of the data, MAXENT extrapolates by taking the predicted value at a minimum and applying that predicted value to any smaller values (Fig. 5.2); it does the same for the maximum value for a predictor and those beyond it. So, while the trend in Fig. 5.2 is increasing suitability with increasing temperature, at a maximum temperature of about 27.5°C the trend is cut off, and the suitability value at 27.5°C is substituted for higher temperatures. While the MESS analysis provides some information on areas of extrapolation, it is limited to univariate analyses. A new technique provides a mechanism to examine novel combinations of predictors that are within the sampled range of each predictor (Zurell et al., 2012).

As mentioned previously, which environmental variables are included and how those that are highly correlated are removed can have significant impacts on model prediction, accuracy and interpretation. Ideally, one of each pair of highly correlated environmental variables would be removed before generating a model. Furthermore, when projecting/transferring correlative models from a native range to an invaded range of interest, it is important to look at the correlation structure among environmental variables (i.e. the statistical relationship between pairs of variables) in both ranges to identify any differences. When these differ, this is violating one of the major assumptions of correlative presence-based statistical models such as MAXENT and predictions should be interpreted with caution.

As the prevalence of invasive alien taxa continues to increase, land managers and policy makers will rely on pest risk models to inform their decisions and actions. Correlative species distribution models that use species–climate relationships in a native range to predict invasion potential elsewhere can be important tools to develop pest risk predictions. There is always a need to assess the data, the question being asked, and the available tools with their unique limitations and assumptions to develop the best possible assessment given available data.

Acknowledgements

We would like to thank the US Geological Survey Fort Collins Science Center and the Natural Resource Ecology Laboratory at Colorado State University for logistical support. We also thank the anonymous reviewers for their constructive comments on earlier drafts of the chapter. Any use of trade, firm or product names is for descriptive purposes only and does not imply endorsement by the US Government.

References

Allouche, O., Tsoar, A. and Kadmon, R. (2006) Assessing the accuracy of species distribution models: prevalence, kappa and the true skill statistic (TSS). *Journal of Applied Ecology* 43, 1223–1232.

Barve, N., Barve, V., Jiménez-Valverde, A., Lira-Noriega, A., Maher, S.P., Peterson, A.T., Soberón, J. and Villalobos, F. (2011) The crucial role of the accessible area in ecological niche modeling and species distribution modeling. *Ecological Modelling* 222, 1810–1819.

Beyer, H.L. (2004) Hawth's Analysis Tools for ArcGIS. Available at: http://spatialecology.com/htools (accessed 19 December 2014).

Crall, A.W., Jarnevich, C.S., Panke, B., Young, N., Renz, M. and Morisette, J. (2013) Using habitat suitability models to target invasive plant species surveys. *Ecological Applications* 23, 60–72.

Dormann, C.F., Elith, J., Bacher, S., Buchmann, C., Carl, G., Carré, G., Marquéz, J.R.G., Gruber, B., Lafourcade, B., Leitão, P.J., Münkemüller, T., McClean, C., Osborne, P.E., Reineking, B., Schröder, B., Skidmore, A.K., Zurell, D. and Lautenbach, S. (2013) Collinearity: a review of methods to deal with it and a simulation study evaluating their performance. *Ecography* 36, 27–46.

Elith, J. (2014, forthcoming) Predicting distributions of invasive species. In: Walshe, T.R., Robinson, A., Nunn, M. and Burgman, M.A. (eds) *Invasive Species: Risk Assessment and Management*. Cambridge University Press, Cambridge, Chapter 6. Chapter now available at: http://arxiv.org/abs/1312.0851 and citable as an arXiv document.

Elith, J. and Leathwick, J.R. (2009) Species distribution models: ecological explanation and prediction across space and time. *Annual Review of Ecology, Evolution, and Systematics* 40, 677–697.

Elith, J., Graham, C.H., Anderson, R.P., Dudik, M., Ferrier, S., Guisan, A., Hijmans, R.J., Huettmann, F., Leathwick, J.R., Lehmann, A., Li, J., Lohmann, L.G., Loiselle, B.A., Manion, G., Moritz, C., Nakamura, M., Nakazawa, Y., Overton, J.M., Peterson, A.T., Phillips, S.J., Richardson, K., Scachetti-Pereira, R., Schapire, R.E., Soberon, J., Williams, S., Wisz, M.S. and Zimmermann, N.E. (2006) Novel methods improve prediction of species' distributions from occurrence data. *Ecography* 29, 129–151.

Elith, J., Kearney, M. and Phillips, S. (2010) The art of modelling range-shifting species. *Methods in Ecology and Evolution* 1, 330–342.

Fielding, A.H. and Bell, J.F. (1997) A review of methods for the assessment of prediction errors in conservation presence/absence models. *Environmental Conservation* 24, 38–49.

Franklin, J. (2009) *Mapping Species Distributions: Spatial Inference and Prediction.* Cambridge University Press, New York.

Freeman, E.A. and Moisen, G.G. (2008) A comparison of the performance of threshold criteria for binary classification in terms of predicted prevalence and kappa. *Ecological Modelling* 217, 48–58.

Guisan, A. and Zimmermann, N.E. (2000) Predictive habitat distribution models in ecology. *Ecological Modelling* 135, 147–186.

Hijmans, R.J., Phillips, S., Leathwick, J. and Elith, J. (2011) *Package 'dismo'.* R Foundation for Statistical Computing, Vienna. Available at: http://cran.r-project.org/web/packages/dismo/index.html (accessed 19 December 2013).

Jiménez-Valverde, A., Peterson, A.T., Soberón, J., Overton, J.M., Aragón, P. and Lobo, J.M. (2011) Use of niche models in invasive species risk assessments. *Biological Invasions* 13, 2785–2797.

Jiménez-Valverde, A., Acevedo, P., Barbosa, A.M., Lobo, J.M. and Real, R. (2013) Discrimination capacity in species distribution models depends on the representativeness of the environmental domain. *Global Ecology and Biogeography* 22, 508–516.

Liu, C.R., Berry, P.M., Dawson, T.P. and Pearson, R.G. (2005) Selecting thresholds of occurrence in the prediction of species distributions. *Ecography* 28, 385–393.

Lobo, J.M., Jimenez-Valverde, A. and Real, R. (2008) AUC: a misleading measure of the performance of predictive distribution models. *Global Ecology and Biogeography* 17, 145–151.

Morisette, J.T., Jarnevich, C.S., Holcombe, T.R., Talbert, C.B., Ignizio, D., Talbert, M.K., Silva, C., Koop, D., Swanson, A. and Young, N.E. (2013) VisTrails SAHM: visualization and workflow management for species habitat modeling. *Ecography* 36, 129–135.

Nabhitabhata, J. and Chan-Ard, T. (2005) *Thailand Red Data: Mammals, Reptiles and Amphibians.* Office of Natural Resources and Environmental Policy and Planning, Bangkok.

Pearce, J.L. and Boyce, M.S. (2006) Modelling distribution and abundance with presence-only data. *Journal of Applied Ecology* 43, 405–412.

Pearson, R.G., Raxworthy, C.J., Nakamura, M. and Peterson, A.T. (2007) Predicting species distributions from small numbers of occurrence records: a test case using cryptic geckos in Madagascar. *Journal of Biogeography* 34, 102–117.

Peterson, N., Soberon, J., Pearson, R.G., Anderson, R.P., Martinez-Meyer, E., Nakamura, M. and Araujo, M.B. (2011) *Ecological Niches and Geographic Distributions.* Princeton University Press, Princeton, New Jersey.

Phillips, S.J. (2008) Transferability, sample selection bias and background data in presence-only modelling: a response to Peterson et al. (2007). *Ecography* 31, 272–278.

Phillips, S.J. and Dudik, M. (2008) Modeling of species distributions with MAXENT: new extensions and a comprehensive evaluation. *Ecography* 31, 161–175.

Phillips, S.J. and Elith, J. (2010) POC plots: calibrating species distribution models with presence-only data. *Ecology* 91, 2476–2481.

Phillips, S.J. and Elith, J. (2013) On estimating probability of presence from use–availability or presence–background data. *Ecology* 94, 1409–1419.

Phillips, S.J., Anderson, R.P. and Schapire, R.E. (2006) Maximum entropy modeling of species geographic distributions. *Ecological Modelling* 190, 231–259.

Phillips, S.J., Dudik, M., Elith, J., Graham, C.H., Lehmann, A., Leathwick, J. and Ferrier, S. (2009) Sample selection bias and presence-only distribution models: implications for background and pseudo-absence data. *Ecological Applications* 19, 181–197.

Pyron, R.A., Burbrink, F.T. and Guiher, T.J. (2008) Claims of potential expansion throughout the US by invasive python species are contradicted by ecological niche models. *PLoS One* 3, e2931.

Rodda, G.H., Jarnevich, C.S. and Reed, R.N. (2009) What parts of the US mainland are climatically suitable for invasive alien pythons spreading from Everglades National Park? *Biological Invasions* 11, 241–252.

Rodda, G.H., Jarnevich, C.S. and Reed, R.N. (2011) Challenges in identifying sites climatically matched to the native ranges of animal invaders. *PLoS One* 6, e14670.

Swets, J.A. (1988) Measuring the accuracy of diagnostic systems. *Science* 240, 1285–1293.

Thuiller, W., Lafourcade, B., Engler, R. and Araujo, M.B. (2009) BIOMOD – a platform for ensemble forecasting of species distributions. *Ecography* 32, 369–373.

VanDerWal, J., Shoo, L.P., Graham, C. and William, S.E. (2009) Selecting pseudo-absence data for presence-only distribution modeling: how far should you stray from what you know? *Ecological Modelling* 220, 589–594.

Warren, D.L. and Seifert, S.N. (2011) Environmental niche modeling in MAXENT: the importance of model complexity and the performance of model selection criteria. *Ecological Applications* 21, 335–342.

Warren, D.L., Glor, R.E. and Turelli, M. (2010) ENMTools: a toolbox for comparative studies of environmental niche models. *Ecography* 33, 607–611.

Warton, D. and Aarts, G. (2013) Advancing our thinking in presence-only and used–available analysis. *Journal of Animal Ecology* 82, 1125–1134.

Webber, B.L., Yates, C.J., Le Maitre, D.C., Scott, J.K., Kriticos, D.J., Ota, N., Mcneill, A., Le Roux, J.J. and Midgley, G.F. (2011) Modelling horses for novel climate courses: insights from projecting potential distributions of native and alien Australian acacias with correlative and mechanistic models. *Diversity and Distributions* 17, 978–1000.

Wisz, M.S., Hijmans, R.J., Li, J., Peterson, A.T., Graham, C.H., Guisan, A. and Distribut, N.P.S. (2008) Effects of sample size on the performance of species distribution models. *Diversity and Distributions* 14, 763–773.

Zurell, D., Elith, J. and Schröder, B. (2012) Predicting to new environments: tools for visualizing model behaviour and impacts on mapped distributions. *Diversity and Distributions* 18, 628–634.

6 The NCSU/APHIS Plant Pest Forecasting System (NAPPFAST)

Roger D. Magarey,[1]* Daniel M. Borchert,[2] Glenn A. Fowler[2] and Steven C. Hong[1]

[1]*Center for Integrated Pest Management, North Carolina State University, Raleigh, North Carolina, USA;* [2]*Plant Epidemiology and Risk Analysis Laboratory, Center for Plant Health Science and Technology, Plant Protection and Quarantine, Animal Plant Health Inspection Service, USDA, Raleigh, North Carolina, USA*

Abstract

This chapter describes the North-Carolina-State-University/Animal-and-Plant-Health-Inspection-Service Plant Pest Forecasting System (NAPPFAST). NAPPFAST, developed for pest risk modelling and mapping, was formerly used to support pest detection, emergency response and risk analysis for the US Department of Agriculture. NAPPFAST employs an internet-based graphical user interface to link weather databases with interactive biological model templates. The weather databases include historical daily weather databases for North America and the world. The templates include degree-days, generic empirical models, infection periods and the Generic Pest Forecast System (GPFS). The GPFS, currently in development, is a model that uses hourly inputs and includes modules for development rate, hot and cold mortality, population and potential damage. In this chapter, three examples illustrate the capabilities of NAPPFAST: (i) pathway analysis for *Lymantria dispar asiatica* (Asian gypsy moth); (ii) epidemiological modelling for *Phytophthora ramorum* (the cause of sudden oak death and other plant diseases); and (iii) simple population modelling for *Bactrocera dorsalis* (oriental fruit fly). One advanced feature of NAPPFAST is cyber-infrastructure that supports the sharing of products and data between modellers and end users. The infrastructure includes tools for managing user access, uploading and correcting geographic coordinates for pest observations, and an interactive geographic information system environment for viewing input data and model products. NAPPFAST was used by the US Department of Agriculture, Animal and Plant Health Inspection Service, Plant Protection and Quarantine, although access has been granted to government and university cooperators working on risk analysis of invasive alien species.

Introduction and Scope

In this chapter, the North-Carolina-State-University/Animal-and-Plant-Health-Inspection-Service Plant Pest Forecasting System (NAPPFAST; Magarey *et al.*, 2007b) is described and the roles of modellers and users are discussed. In early 2014, funding for NAPPFAST was discontinued. Although the tool is not currently in use, there is potential for the tool to be reactivated for use by another organization. Since NAPPFAST contains a unique combination of tools and data sets, this chapter provides a general description that might be useful to other

* Corresponding author. E-mail: rdmagare@ncsu.edu

developers who seek to build similar decision support products. NAPPFAST is designed to help plant protection organizations: (i) collect, manage and share data; (ii) create pest risk models and analyses; (iii) interpret data and risk products for national and local needs; and (iv) manage the distribution of information products to end users, including managers, field staff and stakeholders (Magarey et al., 2009b). The extent of access to NAPPFAST data, products and tools can be customized for different users depending on their role, geographic location or organization (Sandhu and Coyne, 1996). Role-based access allows modellers to create and publish risk products which can then be viewed by users in other roles. The cyber-infrastructure to facilitate the sharing of information in NAPPFAST is based on a similar system developed for soybean rust in the USA (Isard et al., 2006). NAPPFAST was used by only a few organizations, i.e. the US Department of Agriculture, Animal and Plant Health Inspection Service, Plant Protection and Quarantine (USDA-APHIS-PPQ) and its approved cooperators. NAPPFAST is a joint venture between USDA-APHIS-PPQ, North Carolina State University and the information technology company ZedX, Inc.

Uses of NAPPFAST

NAPPFAST is used in several ways. NAPPFAST users include programme managers, survey specialists and risk analysts. One of the primary uses of NAPPFAST is to generate risk maps for the most significant invasive alien pest targets in the APHIS Cooperative Agriculture Pest Survey programme (Magarey et al., 2011). The NAPPFAST system also supports pest risk analysis, emergency programmes, and pest-survey and detection activities of USDA-APHIS-PPQ. Some applications of NAPPFAST include forecasting: (i) areas at risk for establishment by invasive alien species; (ii) the frequency of years in which plant epidemics or insect outbreaks might occur or cause crop losses; (iii) geographic variation in the phenology of insect of life stages; (iv) the required duration of mitigation treatments based on elapsed pest generations; and (v) the extent of injury caused by pests or pathogens to host plants.

The NAPPFAST system employs an internet-based graphical user interface to link interactive templates with weather databases. NAPPFAST currently includes three modelling templates: (i) a degree-day template for creating phenology models for arthropod pests and plants; (ii) an infection model template for plant pathogens; and (iii) a generic template for creating simple empirical models, e.g. hot and cold exclusion. Each template follows a simple fill-in-the-blank design. A fourth template, the Generic Pest Forecast System (GPFS), is currently under construction and will allow users to create risk maps that can be exported to a geographic information system (GIS) for further analysis more easily than NAPPFAST currently allows. All templates in NAPPFAST are generic (i.e. applicable to many species), so they can be used to meet the needs of diverse users.

Weather and climatology in NAPPFAST

NAPPFAST uses high-resolution, historical and near-real-time weather databases for North America and the world. Three types of data are available: (i) empirical observations from North American weather stations; (ii) interpolated grid surfaces of weather observations for North America; and (iii) an interpolated grid surface of weather observations for the world. The North America station database compiles observations from approximately 2000 stations supplied by government and commercial sources. One important source of North American data is NOAAPORT, a broadcast from the National Oceanic and Atmospheric Administration (NOAA) of near-real-time environmental data from multiple sources, including weather stations, radar, satellites, rawinsonde and others (Russo, 1999). Station data in NAPPFAST are interpolated to generate a 10 km × 10 km resolution grid. Two-dimensional (2D) interpolation (Barnes, 1964) uses multivariate regression to estimate climate conditions between weather stations without accounting for the potential effects of elevation change. Three-dimensional (3D) interpolation is similar but does account for elevation effects (Splitt and Horell, 1998). Both interpolation options are available in NAPPFAST. Another North American grid database is derived from the Real-Time Mesoscale Analysis (RTMA) database at 5 km × 5 km resolution and is available from the year 2007 onwards (De Pondeca et al., 2011). Although the station data set is typically used for North American data requests, the quality and resolution of RTMA are expected eventually to exceed what can be provided by station data. The global database is from the National Center for Environmental Prediction's (NCEP) Climate Forecast System Reanalysis (CFSR) at a 38 km × 38 km spatial resolution (Saha et al., 2010). Most databases use a time step of 1 day, but the CFSR data have been interpolated from a 6 h time step to a 1 h time step. A second global database, the NCEP Reanalysis 2 (R2) database, is also available in NAPPFAST but is no longer recommended as it has inferior resolution and quality to CFSR.

ZedX, Inc. uses a series of proprietary quality control algorithms for error checking and approximating missing station values. Quality assurance/quality control checks of the data include a plausibility assessment, deviation from climate normals and evidence of twin reports. Missing station values are estimated by using time-series interpolation, nearest-neighbour regression, nearest-neighbour interpolation, station-grid regression and physics (J.M. Russo, Pennsylvania, 2013, personal communication).

The weather variables in each database include daily maximum temperature, minimum temperature, total precipitation, average relative humidity, evaporation, leaf wetness hours, and 2 and 4 inch (i.e. 5 and 10 cm) depth daily average soil temperatures. The derived variables, including leaf wetness, evaporation and soil temperature, are calculated using proprietary algorithms. Algorithms to estimate leaf wetness were validated by Magarey et al. (2007b).

Other resources from NAPPFAST

The www.NAPPFAST.org website also includes other data sets that are useful to generate forecasts for invasive alien species including global plant hardiness zones (Magarey et al., 2008) and an insect development database (Nietschke et al., 2007). Project reports and training materials are also available from this website.

A significant historical limitation of NAPPFAST has been the reliance on daily data and simple models that cannot adequately describe complex biological processes. For example, hourly relative humidity can be helpful for describing sporulation responses of fungal pathogens, whereas daily average humidity is useless for this purpose. This limitation is being addressed with the development of the GPFS template, which has greater complexity and uses hourly weather data.

Tools for Modellers

'Modeller' is an assigned user role in NAPPFAST. Modellers use generic tools in the system to develop models for specific pests and pathogens.

Model set-up tool

Modellers can use three interactive templates in the model set-up tool to create quickly new models for individual pests and pathogens. Speed is an advantage because models are frequently needed by end users with little to no advanced warning. For example, APHIS-PPQ programme managers may need a map when a new invasive alien pest has been detected in the USA or domestic quarantines are contemplated to close pathways by which a species might be moved. Often little to no information exists about an invasive alien pest's epidemiology and, as a result, a system is needed to create 'first guess' pest risk maps from models with few biological inputs (Magarey et al., 2005). NAPPFAST is designed to meet this need.

The interactive templates in NAPPFAST include a degree-day template, primarily for arthropod phenology modelling, a generic template with canned (i.e. pre-programmed) equations to create empirical models and an infection template for plant pathogens. The templates include fill-in-the-blanks to enter species-specific parameters and a colour-ramp selection tool to customize map or graph outputs. The generic template and generic infection template also allow modellers to set up an accumulated variable, where model outputs for each day are accumulated between user-specified start and end dates. For example, models to characterize annual pest dynamics might run from 1 January to 31 December for the northern hemisphere and from 1 July to 30 June for the southern hemisphere. By default, degree-days are an accumulated output variable. Models also can be copied to facilitate easy adaptation for other species or parameters.

Degree-day template

The generic degree-day template was developed to model insect pests, but can also be used to model phenological development of other organisms, e.g. weeds or crops. Degree-days are a measure of thermal development time and are used for poikilotherms, organisms that cannot maintain a constant body temperature. The degree-day calculation is based on the amount of time temperatures remain above a specific base threshold and below an upper developmental threshold. Users enter the base temperature, i.e. the temperature at which development starts, and the upper threshold, i.e. the temperature above which degree-days accumulate at the same rate as the upper threshold temperature. Developmental data including thresholds and degree-day requirements are published for a large number of insects and plant pests (Nietschke et al., 2007; Jarosik et al., 2011). For example, Japanese beetle (*Popillia japonica*) has a base and upper developmental threshold of 10°C and 34°C, respectively, and on average requires 524 degree-days from 1 January before adults will begin to emerge (Ludwig, 1928; Regniere et al., 1981). Japanese beetle would accumulate 20 degree-days after 2 days when daily temperatures average 20°C. After 2 days with average daily temperatures of 34°C, Japanese beetle would accumulate 48 degree-days. If, however, average temperatures increased to 36°C over the same time interval, Japanese beetle would still only accumulate 48 degree-days, a consequence of the upper thermal threshold.

When degree-days are calculated from daily weather summaries, users have the option to base the calculation on Allen's (1976) sine curve method or simple daily averages of the minimum and maximum temperature (i.e. $(T_{min} + T_{max})/2$). The sine curve method provides an approximation of hourly temperatures when only daily highs and lows are given. Estimates of accumulated degree-days are more accurate using the sine curve method than a simple daily average temperature, particularly when temperatures are below lower developmental thresholds or above upper developmental thresholds for part of a day.

The degree-day template is flexible and allows modellers to select different numbers of phenological events or numbers of generations. In addition to the base temperature threshold for the species, modellers must provide the number of accumulated degree-days required to observe each phenological event or generation. The upper temperature threshold is an optional input.

Generic template

The generic template contains a series of equations used to build empirical models. The equations include logical relationships (e.g. 'X and Y', 'X or Y', 'X < C') and polynomial, logistic or exponential functions. Once an equation is selected, modellers can choose a climate variable from a drop-down menu and enter values for the condition(s) that must be met (e.g. minimum temperature < 0°C). The generic template allows for quick and easy creation of a variety of models. Common examples include models that predict the survival of organisms, or their hosts, based on maximum or minimum temperatures. Soybean rust overwintering survival is an example that uses the generic template (Magarey et al., 2007b). One limitation of the template is that the daily output is binary. So, the criterion 'minimum temperature < 0°C' would return a value of 0 if the temperature never fell below the freezing point of water within a specified period, but would give a value of 1 if it did. If the model had been set to accumulate these values, the model outputs would be interpreted as the number of days with freezing temperatures.

Infection template

The generic infection template is based on a temperature–moisture response function (Magarey et al., 2005) that uses daily weather data as its input. Several foliar plant pathogens require leaves to be wet for a certain length of time within a certain range of temperatures to initiate infection. Some pathogens also have rain splash requirements. Splashing from rain drops moves infective propagules from soil, water or infected plants on to healthy plants. For example, ascospores of *Uncinula necator*, the causal agent of grape powdery mildew, must be splashed from the bark of infected plants to susceptible, new growth of uninfected hosts for the pathogen to spread (Gadoury and Pearson, 1990). Parameters for the generic infection template include cardinal temperatures, leaf wetness requirements (hours), rain splash requirement and degree-day initiation. Information to support parameter selection can be obtained from crop compendia, primary literature, laboratory studies or comparisons with related organisms (Magarey et al., 2005). Examples of pathogens for which NAPPFAST infection templates have been completed include *Phytophthora ramorum* (causal agent of sudden oak death); *Phakospsora pachyrhizi* (causal agent of soybean rust); *Phragmidium violaceum* (causal agent of blackberry rust); and *Elsinoe australis* (causal agent of sweet orange scab). Magarey et al. (2007b) provide additional details about infection templates for *P. ramorum* and *P. pachyrhizi*.

Generic Pest Forecast System (GPFS) template

The GPFS template is being developed for NAPPFAST to provide more advanced capabilities than the current templates. The GPFS will have modules to estimate: developmental rates estimated from cardinal temperatures and relative humidity; mortality from cold (Kaliyan et al., 2007), heat (Dentener et al., 1996) or flooding; infection and sporulation events for plant pathogens; relative population densities based on developmental rate and mortality; simultaneous pest and host growth based on degree-days; and potential damage based on pest population, growth stage susceptibility and pest stage.

The GPFS template will use hourly, not daily, weather data inputs. Hourly data provide for a much more refined description of biological processes, such as infection and heat mortality, which often happen on a smaller time step than a day. The new template also allows some variables such as relative humidity to be incorporated more easily, especially when average daily data remove diurnal fluctuations that may be important to biological processes.

Model run tool

The model run tool allows users, primarily designated 'modellers', to create graphs or maps from a completed model template for any of the NAPPFAST weather databases. The graphing function is intended to allow a user to obtain results for a single station or grid cell or a set of stations or grid cells. A simple drawing feature allows users to select multiple stations or grid cells by sketching on a map surface. If multiple stations or a set of grid cells are selected, a composite result will be reported for all stations. Users can select one input and one output variable to be reported in the summary. The output variable can also be exported as a text file for additional analysis or graphing. For example, fruit fly eradication programmes require time for the completion of three generations after the last detection before a successful eradication can be declared. 'Real-time' degree-day calculations are estimated from NAPPFAST recent historical weather records, from the specified start of the model run to the current date. If the projected accumulation of degree-days is of interest, users can choose 10-, 20- or 30-year climatological data as a forecast of future weather conditions.

Users can request to have results plotted on a map. Users specify a model (i.e. a completed template), date range, data source (i.e. North American station, RTMA or CFSR), region of interest (e.g. world, North America) and variable to map, which may be an input or output variable. For maps made with grids created from station data, users can choose either a 2D or 3D interpolation. For output variables, the colours displayed in the map are those selected in the legend using the model set-up tool. We recommend creating a new model when output legends are modified to avoid confusion. Users must also select one of three map types: history, average history or probability.

A history map is best described as a 'snapshot in time' that displays all of the model information for a specific day or period in a specific year. For example, a history map for a pest insect displays all of the generations and stages for that insect on that date. A history map is useful for comparing model output with pest activity (e.g. the appearance of a given pest stage) with observations for a particular period (e.g. capture of adult insects in traps during July 2012 in Iowa).

The average history map is similar to a history map, but the output for the selected dates reflects an average over 10, 20 or 30 years. The map is most meaningful when it is based on degree-days, life stage or an accumulated output variable. One practical way to use the average-history-map request is to compare the average values with historical pest abundance, outbreaks or known levels of pest damage or prevalence. A good example is a study on the causal agent of citrus black spot, *Guignardia citricarpa* (Magarey et al., 2009a). NAPPFAST was used to identify days with climate conditions that would be suitable for the infection of citrus fruits by ascospores. These infection days were accumulated for the period when citrus fruits should be susceptible. An average value map was requested and this output was compared with locations where black spot had been observed to be prevalent or absent in the field. Black spot was prevalent in areas with ≥12 infection-days (i.e. days in which climate was suitable for infection) during the period when fruits were susceptible. This threshold could then be used to create a probability map defining the frequency of years in which black spot would be forecast to be prevalent.

A probability-map request displays the frequency of occurrence of a specific event within a common block of time, such as the presence of first-generation adult insects or a threshold number of days, each year, for the past 10, 20 or 30 years. Users can also choose a particular number of occurrences, for example, 5 days with temperatures below freezing. Probability maps are useful for risk analysis because they provide output that can be readily incorporated into probabilistic models (e.g. an event that occurred in 6 of the past 10 years would be estimated to have a 60% chance of occurring in a future year). These types of probabilistic models are often used for specific situations where qualitative estimates of risk will not meet the needs of the decision maker.

Model view tool

Maps and graphs can be viewed and downloaded from the model view tool. Models are sorted alphabetically and individual model runs by the date of the run. Clicking on the model name displays the map or graph. Users have the choice to enlarge the map, view the legend or export the maps as a GeoTIFF file which can be imported into a GIS. Users can choose to create a session, which allows all the settings to be saved including those for auto-scheduling graphs or maps. Auto-scheduling allows products to be created every day or every week and sent to e-mail addresses on a distribution list.

Tools for Global Analysts

The 'Global Analyst' role is for inductive modelling, i.e. predictive mapping from a species' distribution records. The global analyst tool has four components: GBIF upload tool, the data analyst tool, the Venn tool and the BAMM tool. The Global Biodiversity Information Facility (GBIF; http://www.gbif.org) provides access to species distribution records from around the world, and these records can be downloaded in XLS and CSV formats. Risk analysts then can upload files from GBIF, or user-created files with the same structure, into NAPPFAST with the GBIF upload tool. Uploaded data pass through a number of quality assurance/quality control procedures, including removal of duplicates, checking the validity of scientific names in the Species 2000 database (http://www.sp2000.org) and checking to ensure coordinates match country boundaries. Records that pass these tests are displayed in a phylogenetic tree in the data analyst tool. Modellers can select observations for editing. A hyperlink is available to view the original GBIF record. The 'Data Manager' role allows users to construct 'patches' that will automatically correct identical errors and automatically reject specific error types. The data manager tool provides manual and automatic quality assurance/quality control processes for all data going into the global analyst tool.

The Venn tool allows risk analysts to select data layers, such as custom-drawn areas, elevation, plant hardiness zones (calculated for the most recent 10 years), growing degree-days (with a base temperature 10°C, the threshold used for many insects), Koppen climate classes (Peel et al., 2007), land-use classes and annual precipitation. For classification layers (e.g. land-use classes), users can view one or more specific classes (e.g. grasslands) by clicking on the map or a legend class. For numerical layers, users can also enter lower and upper limits to limit the display to a range of values. After users select two or more layers, union (i.e. intersection) maps can be created. Risk analysts are also able to select existing products, e.g. maps created in the NAPPFAST modelling tool. The power of the Venn tool is that it can enable a risk analyst to create a specialized, summary risk map for a given country, pest or host. A companion tool, the climate profile tool, functions in a similar way but has more climate data layers.

The Bio-environmental Appraisal and Mapping Model (BAMM) tool allows risk analysts to run a simple species distribution model (Schlegel, 2010) based on tolerance limits calculated from Anderson–Darling tests (Anderson and Darling, 1954). Like other species distribution models, BAMM relates species distribution data (i.e. occurrences such as those reported in GBIF) with information on the environmental and/or spatial characteristics of those locations (Elith and Leathwick, 2009). The current simple version of BAMM does not allow users any control over the selection of environmental variables. Results are available as hyperlinks inside the tool and include the observed species distribution map, the potential species distribution map and model performance statistics. It is hoped that in the future the global analyst tool may include a more sophisticated version of BAMM and/or the ability to select from a suite of species distribution models.

Additional NAPPFAST Capabilities

NAPPFAST has three additional roles – 'Observer', 'Programme Manager' and 'Administrator' – that provide functionality for the operational use of risk maps by a phytosanitary agency. These operations include controlling and facilitating the flow of observations and products from data collection and product generation to product distribution and map interpretation (Isard et al., 2006).

The observer tool allows users such as pest scouts or inspectors to upload pest-survey data either in a spreadsheet or using an online form. Currently, NAPPFAST does not contain pest-survey data, although a similar system developed for the Mexican National Plant Protection Organization (NPPO) has been used to organize and store thousands of pest-survey records.

The Programme Manager role is for managers and survey specialists who work with pest detection or pest management programmes. This role includes the Exotic Pest Targeting (EPT) tool to combine risk maps, dynamic pest models and pest data in an online interactive GIS environment. The EPT tool is similar to the global analyst tool but the former is designed for field operation staff to view products and data, whereas the latter is designed for risk analysis. The EPT tool has potential to assist survey specialists in survey planning, including trap deployment. The tool provides quick access to risk maps and weekly phenologies for a number of pests and pathogens. The EPT tool is not currently being used by APHIS-PPQ; however, the functionality of the tool has been incorporated into the Intergrated Pest Information Platform for Extension and Education (iPiPE), an enhanced version of the ipmPIPE (Isard et al., 2006). The iPiPE is designed to include data sharing between industry, extension and government.

The Administrator role controls user access and the dissemination of information. The administrator tool includes the user management tool, the programme management tool and the product management tool. The user management tool allows an administrator to assign particular system roles and associated tools to designated users. The available roles that can be assigned to users are: Modeller, Global Analyst, Observer, Data Manager, Programme Manager and Administrator. Using the programme management tool, an administrator can assign hosts and pests to a particular pest programme. A pest programme is a set of pests (e.g. pests of citrus) that are grouped together for ease of data collection in the observer tool and ease of display in the EPT tool. The product management tool assigns finished map products to specific pest programmes so that products can be viewed alongside data in the EPT tool.

Examples

Three examples illustrate the capabilities of NAPPFAST: (i) a pathway analysis for *L. dispar asiatica* (Asian gypsy moth); (ii) predictive mapping for establishment of *P. ramorum* (the causal agent of sudden oak death); and (iii) simple population modelling for *B. dorsalis* (oriental fruit fly).

Asian gypsy moth

Asian gypsy moth is included as an example of the use of the generic degree-day template and average history requests.

Asian gypsy moth is a major forest pest that occurs in China, Japan, Russia and South Korea. Asian gypsy moth has been introduced into the USA via maritime ships on several occasions. So far, all introductions have been successfully eradicated. In 1993, a cooperative pre-shipment inspection programme was implemented with Russia to help mitigate the movement of Asian gypsy moth on maritime shipments. USDA-APHIS-PPQ was interested in the implementation and maintenance of a similar cooperative programme with Japan. To facilitate this programme and engage in technical discussions with Japan, PPQ analysed the risk posed to the USA from Asian gypsy moths that might be moving on Japanese maritime shipments (Fowler et al., 2008).

A critical component of the analysis was an evaluation of the annual approach rate of infested ships from Japan. The first step was to determine which Japanese ports hosted US-bound ships, using the Lloyds maritime intelligence unit shipping database (Informa, PLC, New York). Longitude and latitude for these ports were obtained and imported into a GIS. Next, the NAPPFAST generic degree-day template was used to identify the period when moth flight would occur. The model used a start date of 1 January, a base temperature of 3°C and a degree-day accumulation of ≥1142 degree-days (Sheehan, 1992). The average-history-map request was used to create weekly maps of flight initiation in Asia, based on averages from 10 years of weather data (1998–2007) (Fig. 6.1; see colour plate section). The NCEP R2 global data set was used because the CFSR data set was not available at the time the maps were produced. Separate maps were created for each week from 15 March to 15 August. The NAPPFAST outputs were then imported into a GIS containing the Japanese ports of interest and were clipped to the outline of Japan. Maps were overlaid in ascending chronological order (Fig. 6.1). PPQ modellers assumed moths would be flying for 2 months after the forecasted flight initiation date (Wallner et al., 1984). PPQ modellers then queried the Lloyds maritime intelligence unit shipping database to determine the number of ships that called at Japanese ports when Asian gypsy moths were flying. Next, interception records from Customs and Border Patrol were used to estimate the probability of ships becoming infested if they called at Japanese ports when Asian gypsy moths were flying. This probability was estimated as the number of ships found to be infested with gypsy moth divided by the total number of ships leaving Japanese ports for the USA within the same period. With that probability and information about the number of ships leaving Japan each year, PPQ modelled the annual number of Asian-gypsy-moth-infested ships coming to the USA from Japan with a binomial distribution (Vose, 2000; Fig. 6.2). The model, fully described in Fowler et al. (2008), has performed well from 2006 to 2012 with the annual number of vessels from Japan with Asian gypsy moth all falling within the predicted distribution (USDA-APHIS-PPQ Port Operations, 2006–2012, unpublished results). Japan and the USA entered into a cooperative pre-shipment inspection programme in 2007, which has since been maintained.

Sudden oak death

P. ramorum is included as an example of the use of the generic infection template, the generic template and probability requests.

Sudden oak death, caused by *P. ramorum*, is an important disease of oak and tanoak; the pathogen also infects some species of rhododendron and camellia (Werres et al., 2001; Rizzo et al., 2002). Regulations currently control the movement of nursery

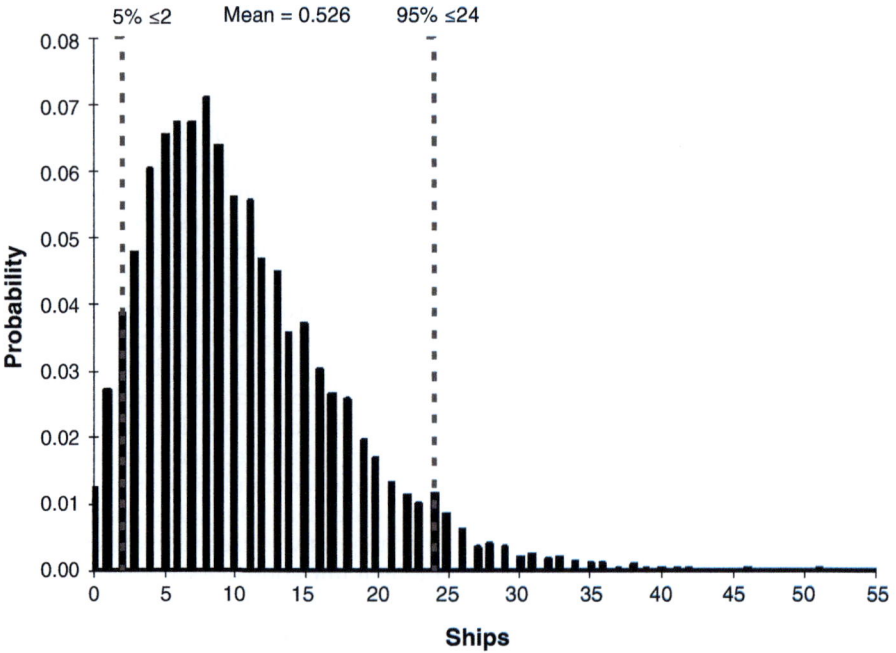

Fig. 6.2. Annual number of infested ships coming to the USA from Japan.

stock and plant material from 12 counties in California and an area in south-western Oregon with the pathogen. In 2003, PPQ programme managers needed a map to help conduct a national survey following the shipment of infected nursery stock from a number of West Coast nurseries nationwide. A forecast map was made based on the results from an infection model and a cold temperature exclusion model in NAPPFAST. Host data were then incorporated to visualize areas where *P. ramorum* could establish.

This chapter describes an updated version of earlier forecast mapping work by Magarey et al. (2007a,b). This update uses recent host data and an alternative approach for combining NAPPFAST outputs. This model assumes that *P. ramorum* would be climatically limited by moisture requirements for infection, temperature requirements for growth and an inability to survive certain extreme minimum temperatures. In the model set-up tool, the generic infection template was used to create an infection model. The temperature thresholds (minimum, optimum, maximum) for *P. ramorum* infection were 3°C, 20°C and 28°C, respectively (Werres et al., 2001; Tooley et al., 2009). The minimum and maximum moisture requirements for infection were 12 h and 24 h of leaf wetness (Tooley et al., 2009). The occurrence of at least 60 favourable days for infection during the year was arbitrarily chosen as an indicator of establishment. The model run tool was used to request a probability map based upon the accumulated number of days that were favourable for infection during the past 10 years. For each 10 km × 10 km grid cell, each day was assigned a value of 0 (i.e. unfavourable for infection) or 1 (i.e. favourable for infection) and these values were accumulated over the year. The resulting map represents the probability of having a sum of at least 60 favourable days per year, based on 10 years of weather data (Fig. 6.3a; see colour plate section).

The generic model template was used to develop a cold exclusion model. Temperatures below −25°C reduce survival of sporangia and chlamydospores of *P. ramorum* in laboratory tests (Turner et al., 2005). So, the logical expression ($X < C$) was used to identify days that might be too cold for the pathogen, where X was the estimated average daily soil temperature at 5 cm depth and C was −25°C. NAPPFAST estimates soil temperature from air temperatures but does not account for the effect of snow cover on soil temperature. (Snow cover depth could be included in NAPPFAST in the future.) From the model run tool, a probability request was made to identify the frequency of years in a 10-year period with at least one day with a minimum temperature <−25°C. Because buffering effects of snow and soil water on soil temperatures are not considered, this model likely overestimates the area where cold might exclude *P. ramorum* from the soil.

The NAPPFAST outputs were exported to a GIS (ArcMap; ESRI, Redlands, California) using the export GeoTIFF function and converted to a projected coordinate system. By using the raster calculator in ArcMap, the cold exclusion raster with integer values from 0 to 10 years, inclusive, was subtracted from 10 years to determine the number of years in each pixel that *P. ramorum* could survive the cold (Fig. 6.3b; see colour plate section). The infection (Fig. 6.3a) and cold survival (Fig. 6.3b) rasters were multiplied by using the raster calculator to create a final climate establishment map in percentage terms (not shown). This raster was then resampled to a 900 m × 900 m resolution. An alternative method that was used in the earlier version of the forecast map to combine rasters is to mask the infection output using areas that had one or more years meeting the cold exclusion requirement (Magarey et al., 2007b).

Maps visualizing *P. ramorum* deciduous and understorey hosts were made at a 900 m × 900 m resolution. The deciduous host presence map was based on the sum of 100% of the deciduous and 50% of the mixed forest density (Fry et al., 2011; M. Colunga-Garcia, Michigan, 2013, personal communication). Understorey host presence was modelled using the distribution of seven Ericaceaeous hosts of *P. ramorum* and brought into ArcMap as a shapefile, which was subsequently converted to raster format (Natureserve, 2013; W. Smith, North Carolina, 2013, personal communication). The understorey host distribution data were converted to a percentage by dividing by seven (to express as proportion of the seven selected hosts present) and multiplying by 100.

A likelihood of establishment map was generated from climate and host data by averaging the NAPPFAST infection/cold survival raster, the deciduous host presence raster and the understorey host raster. Before this calculation was made, the rasters were converted to a USA Contiguous Albers Equal Area Conic projection to tailor the analysis to the area of interest, i.e. the contiguous USA, and to ensure geospatial data alignment. Areas without deciduous hosts, understorey hosts or suitable climate for infection and cold survival were masked to create a final likelihood of establishment map (Fig. 6.3c; see colour plate section). The final forecast map indicates that *P. ramorum* is likely to establish along the west coast and in the eastern United States.

Oriental fruit fly

Oriental fruit fly is included as example of the new GPFS modelling template that uses hourly weather inputs.

The oriental fruit fly belongs to the family Tephritidae (i.e. the 'true fruit flies') and attacks a wide range of fruits and vegetables (Clarke *et al.*, 2005). *Bactrocera* species are significant pests because of their dispersal ability, high reproductive and developmental rate and wide host range (Peck *et al.*, 2005; Leblanc *et al.*, 2011). Oriental fruit fly is distributed throughout most of Asia.

Oriental fruit fly is being used to evaluate the GPFS template at selected sites. For the model to be constructed, literature was used to estimate parameters for developmental temperatures, cold and high temperature mortality, soil moisture mortality, pest and host stages, population growth rate, phenological susceptibility and potential damage. The minimum, optimum and maximum temperature thresholds for oriental fruit fly development were 13.3°C, 24°C and 34°C, respectively (Vargas *et al.*, 1996). The thresholds for cold and high temperature mortality were set as 13.3°C and 39°C (Christenson and Foote, 1960). Pupae were expected to survive for approximately 2 days when the soil was saturated (Xie and Zhang, 2007).

One of the selected sites for model validation was a guava and mango orchard in Bangalore, India (Jayanthi and Verghese, 2011). Hourly weather data inputs (i.e. temperature, relative humidity and precipitation) were obtained from the NAPPFAST CFSR database and used to run the GPFS model. The hourly outputs were summarized to daily values and matched with observed sampling data. Aspects of the simulated and observed population dynamics matched well (Fig. 6.4). In particular, population peaks in the simulations are almost exactly matched with observed peaks in 2001 and 2002. The scatter plot (Fig. 6.5) suggests a linear relationship between the simulated and observed population densities, and the concordance correlation coefficient of 0.22 suggests a modest agreement between the two variables (Tedeschi, 2006). Our preliminary result suggests that the GPFS model for *B. dorsalis* may be a useful tool for forecasting population densities based on weather data. Additional studies with other invasive alien pests are ongoing.

Discussion

One of NAPPFAST's unique features is that it is designed as more than a stand-alone modelling tool. The system also possesses a cyber-infrastructure that could support data collection, integration, risk analysis, modelling, interpretation and dissemination (Magarey *et al.*, 2009b). Although this vision is yet to be fully implemented, it represents an important future direction. Importantly, modelling relies upon pest observations to train and validate models, and the cyber-infrastructure could readily import and analyse these types of data. Other advantages of the platform include internet-based delivery, which enables real-time

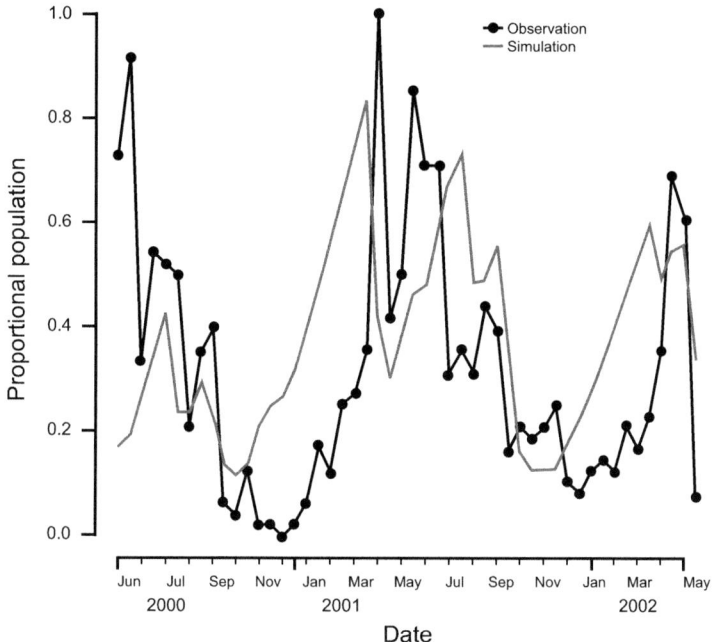

Fig. 6.4. Time series of observed and Generic Pest Forecast System (GPFS)-simulated population of oriental fruit fly (*Bactrocera dorsalis*) in Bangalore, India.

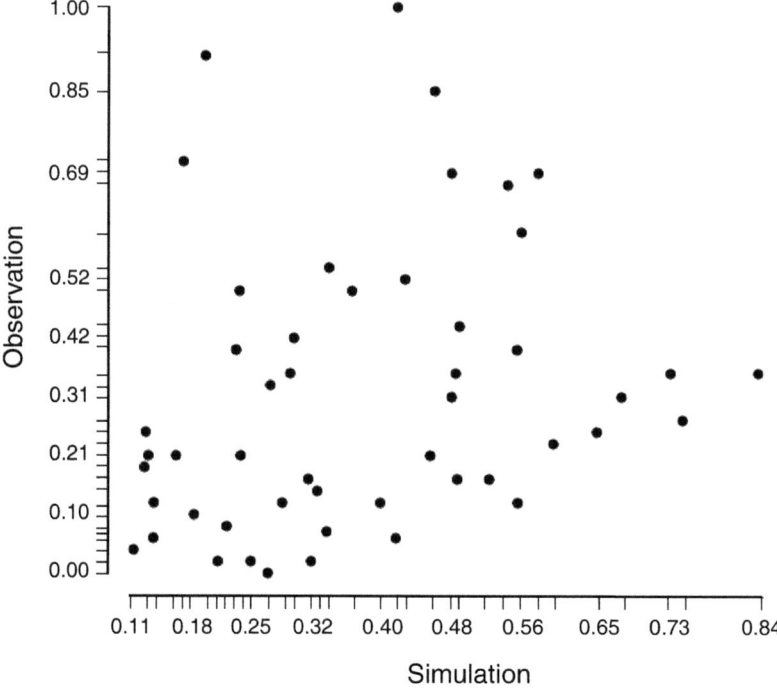

Fig. 6.5. Scatter plot comparison of observed and Generic Pest Forecast System (GPFS)-simulated populations of *Bactrocera dorsalis* in Bangalore, India.

updates and saves users from downloading software. The platform could also potentially be linked to other data collection systems such as the Pest Information Platform (Isard et al., 2006).

Another key feature of NAPPFAST is that its generic platform can be readily adapted to the needs of other NPPOs. For example, NAPPFAST has been adapted for use by the Mexico NPPO. Adding additional NPPOs to the NAPPFAST family helps build economies of scale and lowers developmental costs.

A valuable feature of NAPPFAST is the incorporation of high-resolution, historical and near-real-time weather databases. Station and grid databases provide sources of high-quality near-real-time and historical weather data for North America and the globe. The availability of these data allows more sophisticated modelling than can be achieved with climate averages. Additionally, this saves the user from having to purchase, store, edit, update and process massive quantities of climate data.

NAPPFAST utilizes a simple interface that is based on fill-in-the-blank templates. Although this approach is not unique to NAPPFAST, it provides a simple method for modellers to parameterize models and requires no specialized programming skills. NAPPFAST includes both inductive and deductive modelling processes, allowing for pests to be modelled based on either laboratory or distribution data.

In the past, NAPPFAST modellers had been forced to use simple models with daily data and without important covariates. These limitations are being overcome with the development of the GPFS model and other analytical tools in NAPPFAST. Many biological events that are pivotal to invasions by invasive alien species happen on the scale of minutes and hours, not days. Daily summaries of weather data are too crude to capture periods when temperature and moisture are just right for certain fungi to sporulate or initiate infection.

New refinements for NAPPFAST are continually being explored. Possible future improvements of NAPPFAST include the development of formal tools for uncertainty and sensitivity analysis and a refined interface to simplify model development and testing, particularly for new users.

The long-term future of NAPPFAST, like any software, is uncertain. NAPPFAST has been criticized because it is owned by a private company and access was limited to the license holder, USDA-APHIS-PPQ, and select cooperators. If NAPPFAST models are to be more widely used and understood, more people will need access to the system. Administrators will need to balance the benefits of supporting a broader user base against the potential costs of supporting those users. The use of NAPPFAST may increase in the future as additional organizations purchase licenses, or it may be superseded by more advanced systems than are available now. Other future scenarios include NAPPFAST replacement by freeware or other desktop applications if organizations feel the costs, contractual issues or information technology needed for maintaining a sophisticated platform become too prohibitive. NAPPFAST models may become incorporated into commercially available pest information platforms used by industry. The experience gained from NAPPFAST development can provide guidance and insights for increasing development of cyber-infrastructure to support pest risk modelling and mapping.

References

Allen, J.C. (1976) A modified sine wave method for calculating degree days. *Environmental Entomology* 5, 388–396.

Anderson, T.W., and Darling, D.A. (1954) A test of goodness of fit. *Journal of the American Statistical Association* 49, 765–769.

Barnes, S.L. (1964) A technique for maximizing details in numerical weather map analysis. *Journal of Applied Meteorology* 3, 396–409.

Christenson, L.D. and Foote, R.H. (1960) Biology of fruit flies. *Annual Review of Entomology* 5, 171–192.

Clarke, A.R., Armstrong, K.F., Carmichael, A.E., Milne, J.R., Raghu, S., Roderick, G.K. and Yeates, D.K. (2005) Invasive phytophagous pests arising through a recent tropical evolutionary radiation: the *Bactrocera dorsalis* complex of fruit flies. *Annual Review of Entomology* 50, 293–319.

Dentener, P.R., Alexander, S.M., Lester, P.J., Petry, R.J., Maindonald, J.H. and McDonald, R.M. (1996) Hot air treatment for disinfestation of lightbrown apple moth and longtailed mealy bug on persimmons. *Postharvest Biology and Technology* 8, 143–152.

De Pondeca, M.S., Manikin, G.S., DiMego, G., Benjamin, S.G., Parrish, D.F., Purser, R.J., Wu, W.-S., Horel, J.D., Myrick, D.T. and Lin, Y. (2011) The real-time mesoscale analysis at NOAA's National Centers for Environmental Prediction: current status and development. *Weather and Forecasting* 26, 593–612.

Elith, J. and Leathwick, J.R. (2009) Species distribution models: ecological explanation and prediction across space and time. *Annual Review of Ecology, Evolution, and Systematics* 40, 677–697.

Fowler, G., Takeuchi, Y., Sequeira, R., Lougee, G., Fussell, W., Simon, M., Sato, A. and Yan, X. (2008) *Pathway-Initiated Pest Risk Assessment: Asian Gypsy Moth (Lepidoptera: Lymantriidae: Lymantria dispar (Linnaeus)) from Japan into the United States on Maritime Ships*. USDA-APHIS-PPQ-CPHST-PERAL, Raleigh, North Carolina.

Fry, J., Xian, G., Jin, S., Dewitz, J., Homer, C., Yang, L., Barnes, C., Herold, N. and Wickham, J. (2011) Completion of the 2006 national land cover database for the conterminous United States. *Photogrammetric Engineering and Remote Sensing* 77, 858–864.

Gadoury, D.M. and Pearson, R.C. (1990) Ascocarp dehiscence and ascospore discharge in *Uncinula necator*. *Phytopathology* 80, 393–401.

Isard, S.A., Russo, J.M. and DeWolf, E.D. (2006) The establishment of a National Pest Information Platform for extension and education. *Plant Health Progress* (September), doi: 1094/PHP-2006-0915-01-RV.

Jarosik, V., Honek, A., Magarey, R.D. and Skuhrovec, J. (2011) Developmental database for phenology models: related insect and mite species have similar thermal requirements. *Journal of Economic Entomology* 104, 1870–1876.

Jayanthi, P.D.K. and Verghese, A. (2011) Host–plant phenology and weather based forecasting models for population prediction of the oriental fruit fly, *Bactrocera dorsalis* Hendel (*sic*). *Crop Protection* 30, 1557–1562.

Kaliyan, N., Carrillo, M.A., Morey, R.V., Wilcke, W.F. and Kells, S.A. (2007) Mortality of indianmeal moth (Lepidoptera: Pyralidae) populations under fluctuating low temperatures: model development and validation. *Environmental Entomology* 36, 1318–1327.

Leblanc, L., Vargas, R.I., Mackey, B., Putoa, R. and Pinero, J.C. (2011) Evaluation of cue-lure and methyl eugenol solid lure and insecticide dispensers for fruit fly (Diptera: Tephritidae) monitoring and control in Tahiti. *Florida Entomologist* 94, 510–516.

Ludwig, D. (1928) The effects of temperature on the development of an insect (*Popillia japonica* Newman). *Physiological Zoology* 1, 358–389.

Magarey, R.D., Sutton, T.B. and Thayer, C.L. (2005) A simple generic infection model for foliar fungal plant pathogens. *Phytopathology* 95, 92–100.

Magarey, R., Fowler, G., Colunga, M., Smith, B. and Meentemeyer, R. (2007a) Climate–host mapping of *Phytophthora ramorum*, causal agent of sudden oak death. In: Frankel, S.J., Kliejunas, J.T. and Palmieri, K.M. (eds) *Proceedings of the Sudden Oak Death Third Science Symposium*. USDA Forest Service and California Oak Mortality Task Force, Santa Rosa, California, pp. 269–275.

Magarey, R.D., Fowler, G.A., Borchert, D.M., Sutton, T.B., Colunga-Garcia, M. and Simpson, J.A. (2007b) NAPPFAST: an internet system for the weather-based mapping of plant pathogens. *Plant Disease* 91, 336–345.

Magarey, R.D., Borchert, D.M. and Schlegel, J.W. (2008) Global plant hardiness zones for phytosanitary risk analysis. *Scientia Agricola* 65, 54–59.

Magarey, R.D., Chanelli, S. and Holtz, T. (2009a) *Validation Study and Risk Assessment for Guignardia citricarpa (Citrus Black Spot)*. USDA-APHIS-PPQ-CPHST-PERAL, Raleigh, North Carolina.

Magarey, R.D., Colunga-Garcia, M. and Fieselmann, D.A. (2009b) Plant biosecurity in the United States: roles, responsibilities, and information needs. *Bioscience* 59, 875–884.

Magarey, R.D., Borchert, D.M., Engle, J.S., Colunga-Garcia, M., Koch, F.H. and Yemshanov, D. (2011) Risk maps for targeting exotic plant pest detection programs in the United States. *EPPO Bulletin* 41, 46–56.

NatureServe (2013) *NatureServe Web Service*. NatureServe, Arlington, Virginia. Available at http://services.natureserve.org (accessed 15 December 2013).

Nietschke, B.S., Magarey, R.D., Borchert, D.M., Calvin, D.D. and Jones, E. (2007) A developmental database to support insect phenology models. *Crop Protection* 26, 1444–1448.

Peck, S.L., McQuate, G.T., Vargas, R.I., Seager, D.C., Revis H.C., Jang, E.B. and McInnis, D.O. (2005) Movement of sterile male *Bactrocera cucurbitae* (Diptera: Tephritidae) in a Hawaiian agroecosystem. *Journal of Economic Entomology* 98, 1539–1550.

Peel, M.C., Finlayson, B.L. and McMahon, T.A. (2007) Updated world map of the

Köppen–Geiger climate classification. *Hydrological Earth Systems Science* 11, 1633–1644.

Regniere, J., Rabb, R.L. and Stinner, R. (1981) *Popillia japonica*: simulation of temperature-dependent development of the immatures, and prediction of adult emergence. *Environmental Entomology* 10, 290–296.

Rizzo, D.M., Garbelotto, M., Davidson, J.M., Slaughter, G.W. and Koike, S.T. (2002) *Phytophthora ramorum* as the cause of extensive mortality of *Quercus* spp. and *Lithocarpus densiflorus* in California. *Plant Disease* 86, 205–214.

Russo, J.M. (1999) Weather forecasting for IPM. In: Kennedy, G.G. and Sutton, T.B. (eds) *Emerging Technologies for Integrated Pest Management: Concepts, Research, and Implementation*. APS Press, St Paul, Minnesota, pp. 453–473.

Saha, S., Moorthi, S., Pan, H.L., Wu, X.R., Wang, J.D., Nadiga, S., Tripp, P., Kistler, R., Woollen, J., Behringer, D., Liu, H.X., Stokes, D., Grumbine, R., Gayno, G., Wang, J., Hou, Y.T., Chuang, H.Y., Juang, H.M.H., Sela, J., Iredell, M., Treadon, R., Kleist, D., Van Delst, P., Keyser, D., Derber, J., Ek, M., Meng, J., Wei, H.L., Yang, R.Q., Lord, S., Van den Dool, H., Kumar, A., Wang, W.Q., Long, C., Chelliah, M., Xue, Y., Huang, B.Y., Schemm, J.K., Ebisuzaki, W., Lin, R., Xie, P.P., Chen, M.Y., Zhou, S.T., Higgins, W., Zou, C.Z., Liu, Q.H., Chen, Y., Han, Y., Cucurull, L., Reynolds, R.W., Rutledge, G. and Goldberg, M. (2010) The NCEP climate forecast system reanalysis. *Bulletin of the American Meteorological Society* 91, 1015–1057.

Sandhu, R.S. and Coyne, E.J. (1996) Role-based access control models. *Computer* 2, 38–47.

Schlegel, J.W. (2010) *A Description of BAMM, the Bio-Environmental Appraisal and Mapping Model (Formerly the Climate Matching Tool): A NAPPFAST User Manual*. USDA-APHIS-PPQ-CPHST-PERAL, Raleigh, North Carolina.

Sheehan, K.A. (1992) *User's Guide for GMPHEN: Gypsy Moth Phenology Model*. US Department of Agriculture, Forest Service, Northeastern Forest Experiment Station, Radner, Pennsylvania.

Splitt, M.E. and Horell, J. (1998) Use of multivariate linear regression for meteorological data analysis and quality assessment in complex terrain. *Proceedings of the Tenth Symposium on Meteorological Observations and Instrumentation*. American Meteorological Society, Phoenix, Arizona, pp. 359–362.

Tedeschi, L.D. (2006) Assessment of the adequacy of mathematical models. *Agricultural Systems* 89, 225–247.

Tooley, P.W., Browning, M., Kyde, K.L. and Berner, D. (2009) Effect of temperature and moisture period on infection of rhododendron 'Cunningham's White' by *Phytophthora ramorum*. *Phytopathology* 99, 1045–1052.

Turner, J., Jennings, P. and Humphries, G. (2005) *Phytophthora ramorum* Epidemiology: Sporulation Potential, Dispersal, Infection, Latency and Survival. *DEFRA Project Report PH0194*. Department for Environment, Food and Rural Affairs, London. Available at: http://randd.defra.gov.uk/Document.aspx?Document=PH0194_2004_FRP.pdf (accessed 8 July 2011).

Vargas, R.I., Walsh, W.A., Jang, E.B., Armstrong, J.W. and Kanehisa, D.T. (1996) Survival and development of immature stages of four Hawaiian fruit flies (Diptera: Tephritidae) reared at five constant temperatures. *Annals of the Entomological Society of America* 89, 64–69.

Vose, D. (2000) *Risk Analysis: A Quantitative Guide*, 2nd edn. Wiley, New York.

Wallner, W.E., Cardé, R.T., Xu, C., Weseloh, R.M., Sun, X., Yan, J. and Schaefer, P.W. (1984) Gypsy moth (*Lymantria dispar* L.) attraction to disparlure enantiomers and the olefin precursor in the Peoples Republic of China. *Journal of Chemical Ecology* 10, 753–757.

Werres, S., Marwitz, R., Man in' t Veld, W.A., De Cock, A.W.A.M., Bonants, P.J.M., De Weerdt, M., Themann, K., Ilieva, E. and Baayen, R.P. (2001) *Phytophthora ramorum* sp. nov., a new pathogen on rhododendron and viburnum. *Mycological Research* 105, 1155–1165.

Xie, Q. and Zhang, R. (2007) Responses of oriental fruit fly (Diptera: Tephritidae) third instars to desiccation and immersion. *Journal of Agricultural and Urban Entomology* 24, 1–11.

7 Detecting and Interpreting Patterns within Regional Pest Species Assemblages using Self-organizing Maps and other Clustering Methods

Susan Worner[1]*, Rene Eschen,[2] Marc Kenis,[2] Dean Paini,[3,4] Kari Saikkonen,[5] Karl Suiter,[6] Sunil Singh,[3,4] Irene Vänninen[5] and Mike Watts[7]

[1]Bio-Protection Research Centre, Lincoln University, Lincoln, New Zealand; [2]CABI, Delémont, Switzerland; [3]CSIRO Biosecurity Flagship, Canberra, Australian Capital Territory, Australia; [4]Plant Biosecurity Cooperative Research Centre, Bruce, Australian Capital Territory, Australia; [5]MTT Agrifood Research, Jokioinen, Finland; [6]NSF Centre for Integrated Pest Management, North Carolina State University, Raleigh, North Carolina, USA; [7]Information Technology Programme, AIS St Helens, Auckland, New Zealand

Abstract

This chapter highlights quantitative methods designed to identify and rank exotic species with potential risk to cause economic and/or environmental harm if they establish in a new area. Until now, pest risk assessments have tended to be qualitative and reactive instead of quantitative and proactive. Here, a computational-intelligence technique called a self-organizing map (SOM) is described that can be used to analyse regional profiles or assemblages of pest species to determine their potential for establishment in new regions. In addition to the SOM, two other useful clustering or classification algorithms, k-means and hierarchical analysis, are also demonstrated to provide a quantitative framework to the risk assessment process. The examples described for each method illustrate how a pest risk analyst can identify, from a large list of potential hazards, which species present the most risk to target areas. Furthermore, examples are given of how such analyses may indicate donor and recipient regions for pest invasion and can highlight previously unknown or ignored threats for further investigation. Finally, cautions are provided and limitations of SOMs and other clustering methods applied to the area of pest risk assessment are discussed.

Introduction to the Need for Prioritization

Increasingly, the world faces challenges from the impact of invasive alien species driven largely by tourism, trade and climate change that accelerate opportunities for invasive species to disperse and establish in new

* Corresponding author. E-mail: Susan.Worner@lincoln.ac.nz

regions of the globe. Invasive alien species can do irreparable harm to both indigenous and managed biological ecosystems as well as to human and animal health. While the traditional international imperatives to protect productive ecosystems, as well as preserve biodiversity, are clear, pressure to maintain world food security is growing. Despite decades of study on invasive alien species and enactment of international regulations to prevent invasive species' spread, little progress has been made to develop quantitative methods that help identify and rank the many hundreds of potential pests before they arrive in a new area.

For greater preparedness, risk assessors and policy makers need to identify and prioritize which species out of a long list are likely to establish in a target region if given a chance to do so. The formalized procedures for pest risk assessment are often reactive. For example, species often are selected for pest risk analysis that have high impact in another region, are intercepted frequently at the border, or are associated with new commodities or pathways and are regarded as emerging threats. Furthermore, risk assessments routinely carried out by trading nations on new traded commodities often require the evaluation of several hundreds of species that are known to associate with that commodity. Such assessments often take several years to complete. No country or region wants to be surprised by a new incursion. One way to minimize surprises is to recognize potential pest threats before they arrive by ranking the thousands of possible invasive alien pests with respect to their risk of establishment. The methods described in this chapter can assist the recognition process.

Worner and Gevrey (2006) proposed the first fully quantitative method for ranking large numbers of insect crop pests with respect to their potential establishment. Their method was based on the application of a computational-intelligence analysis method to a database of global distributions of known pest species (Gevrey and Worner, 2006; Worner and Gevrey, 2006). The method is more generally known as a self-organizing map or SOM (Kohonen, 1982, 2001). A SOM is designed for exploratory data analysis using, as its basis, an artificial neural network that mimics the pattern recognition capabilities of the vertebrate brain. Essentially, it is a machine learning algorithm that clusters, classifies and visualizes high-dimensional or complex data without supervision or input from the user.

The basis of the SOM approach used by Worner and Gevrey (2006) is that the specific combination of pest species already established in the target region reflects local climate, host plants and other prevailing conditions which are suitable for those species by definition. Comparisons with regions that share similar pest profiles can identify species not yet established that might prosper in the target environment (Gevrey and Worner, 2006; Worner and Gevrey, 2006). In other words, certain species of plant pests tend to co-occur in particular regions around the world because of specific combinations of environmental factors. Such similarities can be used to identify missing species from a target profile and indicate a potential risk of establishment. SOM analyses of invasive species groupings have been criticized because such groupings have resulted from anthropogenic activity and may be random, but Watts and Worner (2009) have shown that the pest groupings are clearly not random.

Worner and Gevrey (2006) provide an example of the application of SOM analysis to a database of global crop pests. They found that New Zealand has a large number of invasive pest species in common with Italy. This result suggests that the two countries have similar characteristics that have allowed similar pest profiles to develop. Each country may act as a potential species donor or species recipient for the other. Some species in Italy but not in New Zealand may pose a special threat if they reach New Zealand. Yet, good reasons may exist to explain why those species are not present in New Zealand. The most likely explanation is that a suitable pathway does not exist and particular species may never have had the chance to get there. Nevertheless, by using SOM analysis, Worner and Gevrey (2006) were able to sift efficiently through a large list of potential hazards and highlight previously unknown or ignored threats for further assessment.

The use of species assemblages to give information about environmental complexity is not new. Petroleum geologists routinely use assemblages of fossil organisms to indicate fossil fuel deposits (Gregory et al., 2007) and the study of fossil plant assemblages has been used over many decades to reconstruct past climates (e.g. Savin, 1977; Heiri and Lotter, 2005). In ecology, and particularly ecosystem studies, changes in the composition of species assemblages are used as indicators of changes in natural conditions and anthropogenic influences (e.g. Chon et al., 1996; Céréghino and Park, 2009).

At the global scale, identifying regions that have the most similar pest profiles requires some sort of clustering of a large number of geographic regions using a large number of species. In the New Zealand example, the database has 459 regions and, therefore, 459 pest profiles (Figs 7.1 and 7.2; see below and colour plate section). With this many regions, a simple analysis to find the similarity of each pest profile with every other pest profile would require 105,111 comparisons (Worner and Gevrey, 2006). While conventional clustering or similarity analyses could be used, the resulting clusters may not be easy to interpret. However, a SOM converts data with many samples (e.g. 459 global regions) and many elements (e.g. the presence or absence of 844 species in each global region) to cluster similar regional pest profiles on to a two-dimensional map that gives rich information and is easily interpreted (Fig. 7.1).

While the emphasis in this chapter is on SOM analyses, two recent studies have shown that other clustering algorithms also

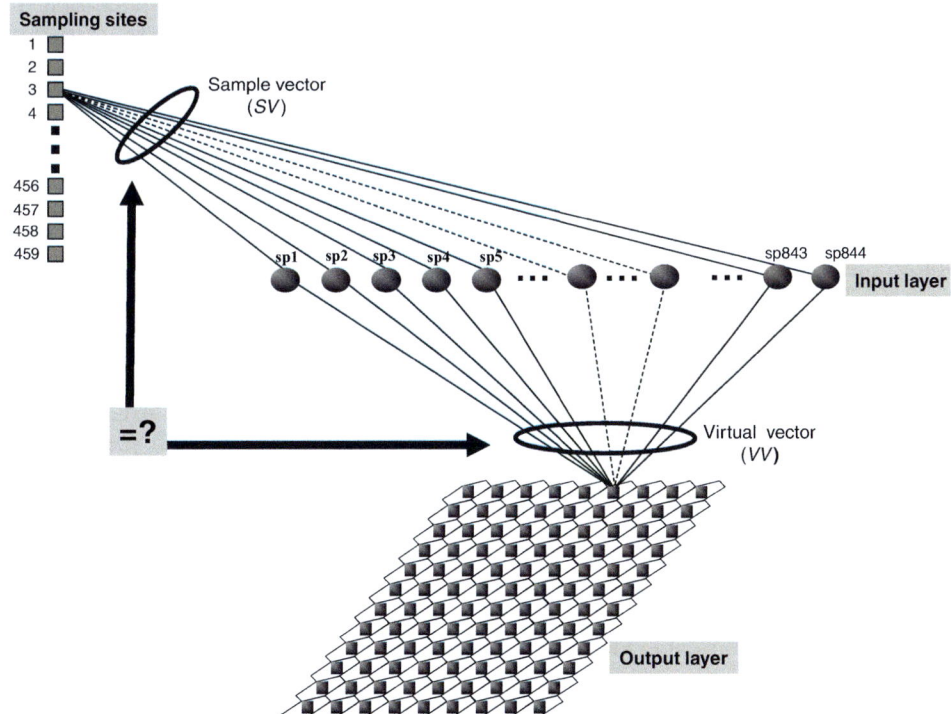

Fig. 7.1. Self-organizing map architecture. The input layer, by connections called weights that define the virtual assemblages of the species, is linked to the cells of the output layer. (Reprinted with permission from Gevrey et al., 2006.)

may be useful to estimate risk of establishment of potential pests. The first study compared SOM analysis with *k*-means clustering (Watts and Worner, 2009). The second study applied hierarchical cluster analysis (Worner et al., 2013). This chapter demonstrates how different algorithms such as SOMS, *k*-means and hierarchical analysis can be used to analyse pest distribution data. All techniques are demonstrated with applications.

Data Requirements and Useful Databases

The data in the studies reviewed here are comprised of presence records for pest species in different countries and regions of the world. A species is presumed to be absent if no published record confirms its presence. Records from CAB International (CABI), a not-for-profit, science-based, development and information organization, were extracted with permission from its interactive multimedia encyclopaedias: Crop Protection Compendium (CABI 2003, 2007; http://www.cabi.org/CPC/), Plantwise Knowledge Bank (http://www.plantwise.org/knowledgebank) and Invasive Species Compendium (http://www.cabi.org/ISC/). Other databases such as the European and Mediterranean Plant Protection Organization (EPPO) Plant Quarantine Data Retrieval System (PQR; http://www.eppo.int/DATABASES/pqr/pqr.htm) and the Global Pest and Disease Database (GPDD; https://www.gpdd.info/) are also available. The GPDD has approximately 120,000 pest distribution records from around the world. This database is maintained by the US Department of Agriculture, Animal and Plant Health Inspection Service, Plant Protection and Quarantine Division (USDA-APHIS-PPQ) and, like the other databases, gathers pest distribution data from several sources such as journals, other refereed publications, pest risk assessments, border interception records and high-quality, publicly available, web-based source material. Other useful databases are North American-oriented but are open access: the National Agricultural and Pest Information System (NAPIS; http://pest.ceris.purdue.edu/index.php) for the USA and the Invasive and Exotic Species Profiles & State, Regional and National Lists (http://www.invasive.org/species.cfm) for North America.

Data must be extracted from these databases and reorganized such that pest presence or absence for each site is represented by binary data in rows or columns, with 0 corresponding to the absence and 1 corresponding to the presence of each species at each site. Incomplete data from the database should be discarded, as well as species with low prevalence, e.g. Worner and Gevrey (2006) excluded species that were present in <5% of the sites.

The SOM Algorithm

Kohonen (1982, 1984) designed the SOM algorithm for use in speech recognition. The SOM quickly found wide application in many disciplines that require statistical analyses for classification and clustering. Oja et al. (2003), cited in Sarlin (2011), noted that the SOM algorithm had many applications in engineering, medicine and other scientific disciplines. A Web of Knowledge™ (Thomson Reuters) title search for 'SOM' or 'self-organizing map' yielded 215 articles and a topic search identified 892 articles published from January 2012 to June 2013. The tool has been used in macroeconomics (Sarlin, 2011), manufacturing (Chattopadhyay et al., 2012), language analysis (Zhao et al., 2011), genome sequencing (Chan et al., 2008), meteorology and oceanography (Lui and Weisberg, 2011), metagenomics and biodiversity (Weber et al., 2011), transcriptomics and metabolomics (Milone et al., 2012) and ecology (Chon, 2011). Variations of SOM analysis have been used, for example, for stock price prediction (Hsu, 2011), growing self-organizing maps for protein sequence classification (Ahmad et al., 2010), batch learning self-organizing maps for microbiome analysis of ticks (Nakao et al., 2013) and hybrid self-organizing maps for river flow forecasting (Ismail et al., 2012).

A SOM is a type of artificial neural network that uses the power of modern computation to mimic the way the vertebrate brain organizes information to learn or discover patterns from sensory inputs (Kohonen, 1982, 1990). The trained SOM provides an ordered mapping of originally complex data, thereby clarifying interpretation (Liebscher et al., 2012). SOMs differ from other approaches of exploratory data analysis such as multidimensional scaling by providing simple, intuitive clustering that is projected on to a two-dimensional map for easy visualization (Fig. 7.2; see colour plate section), while still preserving the relationships between the samples in the original data set (Arsuaga and Díaz, 2005; Sarlin, 2011).

SOM structure

A SOM consists of two layers of artificial neurons, a layer that represents the input data and a layer for the output or map. The map is usually arranged as a two-dimensional structure of cells (Figs 7.1 and 7.2). For the New Zealand example, the input layer is comprised of the 459 sample sites or regions of the world. For each region, a sample vector (i.e. pest profile) indicates the presence and absence of each of the 844 pest species considered in the analysis (Figs 7.1 and 7.2). The SOM organizes the data by applying a machine learning scheme to adapt weight values connecting the vectors of the input layer to the array of neurons (i.e. nodes or cells) of the output layer, where each neuron is represented by a virtual (i.e. reference) vector. Every input neuron (i.e. geographic site or region) is connected to every output neuron (i.e. cell or node), and each connection has a weight attached to it. The SOM algorithm can be summarized as:

1. Initialize the values of the virtual vectors (VV_i, $1 \leq i \leq c$), where i is each virtual vector VV and c is the total number of virtual vectors.
2. Read all the sample vectors (SV) one at a time.
3. Compute the Euclidean distance between SV and VV.
4. Assign each SV to the nearest VV according to the distance results.
5. Modify each VV with the mean of the SV that were assigned to it.
6. Repeat steps 3–5 until convergence or until connection weights change very little.

When the sample (i.e. input) vectors for our New Zealand example are presented to the SOM, the SOM sets up a corresponding virtual (i.e. reference) vector with 844 elements (i.e. one element for each of the 844 species) at each neuron of the map and assigns each element a random weight between 0 and 1, inclusive. For each sample vector (i.e. a site's pest profile composed of 0's and 1's for the absence and presence, respectively, of each of the 844 species), the Euclidean distance between the sample vector and the virtual vector of each output neuron is calculated. Each sample vector is assigned to the closest virtual vector, also known as the best matching unit, according to the Euclidean distance. Each virtual vector is updated during many iterations of the learning process, where weights are calculated according to:

$$w_{i,j}(t+1) = w_{i,j}(t) + h(t)[x_i - w_{i,j}(t)] \quad (7.1)$$

where $w_{i,j}(t)$ is the connection weight from input i to map neuron j at iteration t and x_i is element i of input vector x. The variable h is the neighbourhood function and is defined as:

$$h(t) = \alpha \exp\{-d^2 / [2\sigma^2(t)]\} \quad (7.2)$$

where α is the learning rate, which decays towards zero as time progresses; d is the Euclidean distance between the best matching unit and the current sample vector j; and σ is the neighbourhood width parameter, which also decays towards zero (Watts and Worner, 2009). Eventually, all sample vectors (i.e. pest profiles) that are most similar are associated with the same neuron or nearby neurons on the two-dimensional SOM map.

Generally, SOMs are run through existing software. The latest version of the MATLAB (2013) SOM toolbox V2.1beta, comprising numerous software programs and extensive documentation, can be downloaded from http://research.ics.aalto.fi/software/somtoolbox/. A report by Vesanto et al. (2000) is also available on this site and provides excellent documentation with examples. The toolbox includes programs for other data analysis methods such as *k*-means, principal components analysis and Sammon's projection. Additionally, SOM analyses can be readily completed using R programming (Wherens and Buydens, 2007; Wherens, 2013). Many internet sites including YouTube (http://www.youtube.com) provide background information and dynamic visualization of how SOMs work (e.g. http://davis.wpi.edu/~matt/courses/soms/).

Training SOMs

SOMs identify patterns in the data by analysing training examples (i.e. samples of data), similar to the way vertebrates learn. A SOM can be trained with sequential or batch training. With the sequential training algorithm, only one sample vector (i.e. pest profile) at a time is submitted for processing. With batch training the complete data set (i.e. all pest profiles) is presented to the algorithm before the weight vectors are processed or updated. In our experience, use of the batch training algorithm with linear initialization described below gives reproducible results. Batch training is also more effective than sequential training, because the results of sequential training depend on the order in which samples are presented to the map (Vesanto et al., 2000). Hence batch training is used to train the SOMs in all the examples reported here.

To start SOM training, the initial position of each virtual vector that represents each neuron on the map (e.g. 9 × 12 map with 108 neurons for our New Zealand example) needs to be set (Fig. 7.1). These virtual vectors provide the reference location in multidimensional space to which each of the sample vectors (i.e. pest profiles) is clustered depending on their similarity to that reference vector. Sometimes random locations are assigned to the virtual vectors but this might give different maps on repeated runs. We recommend a linear initialization that distributes the virtual vectors corresponding to the first two eigenvalues of a principal component analysis. This type of initialization distributes virtual vectors in a way that well represents the original data in multidimensional space. Linear initialization also improves the speed of training the network (Kohonen, 2001). Linear initialization is an option in most SOM software. With this option selected, the software automatically calculates eigenvalues for the user.

Map size and quality

A SOM reduces dimensions of the data by mapping sample vectors on to an array of output neurons that are represented as cells on a two-dimensional map. For a SOM to give a good representation of the data, an appropriate number of neurons that comprise the map need to be assigned. The size of the map (i.e. the output layer, Fig. 7.1) depends on the objectives of the study. A large map can reveal details about the data, but too large a map makes differences among cells small and difficult to interpret. A small map may show general patterns in the data but may not illustrate important differences (Chattopadhyay et al., 2012). A common rule for optimum map size is $C = 5\sqrt{n}$, where C is the number of neurons in the map and n is the number of sample vectors in the input layer (Vesanto et al., 2000). In our example for New Zealand, the number of sample vectors was 459, so the recommended number of neurons for the map is $C = 5\sqrt{459} = 107.12$, which, when rounded to the next largest whole number, suggests a map size of 108 cells. Worner and Gevrey (2006) used a 12 × 9 neuron map for their analysis (Figs 7.1 and 7.2). The organization of the map should be such that the length is larger than the width (Vesanto et al., 2000).

The SOM toolbox also provides algorithms to calculate quantization error (QE) and topographic error (TE). Different map sizes are investigated and these errors calculated. The optimal map size is the one in which these errors are the least. In our experience, selecting a map size based on the Vesanto et al. (2000) rule and the minimum QE and TE gives a good result. However, Chattopadhyay et al. (2012) suggest that TE is not a convincing criterion for choosing SOM size and can be ignored. Usually the number of neurons chosen for a map is less than the number of sample vectors; however, that is not an absolute constraint, and it is possible to have more neurons than sample vectors.

Neighbourhood function and learning rate

Default values for the neighbourhood function (i.e. neighbourhood width, σ) and learning rate (α) (Eqn 7.2) are given in the software. The neighbourhood width parameter decreases with the number of iterations and distance (i.e. Euclidean distance) from the virtual vector. The neighbourhood width determines a 'region of influence' of each node or neuron such that it influences nearby vectors or samples more than far away ones. The final virtual vector is also called the best matching unit for a cluster of sample vectors when the SOM has converged. The neighbourhood width declines towards zero with each iteration (i.e. epoch). So, for each iteration, the virtual vector, m, for neuron i on the map is updated this way:

$$m_i(t+1) = m_i(t) + (t)hc_i(t)[x(t) - m_i(t)]$$
(7.3)

with $x(t)$ being a sample vector; $hc_i(t)$ is the neighbourhood kernel or function with respect to Euclidean distance around the best matching unit c, the final virtual vector comprising weights that represent the cluster; $\alpha(t)$ is the learning rate; and t is the iteration (Vesanto and Alhoniemi, 2000). The learning rate decreases with time. At the beginning of training, the learning rate has a high value and the SOM tries to fit the input data approximately with large weight changes; then it fine-tunes the map over smaller changes determined by the learning rate, which continues to decrease until the map converges.

Number of iterations or epochs

By default, the 'optimal' number of learning iterations in the SOM_toolbox is defined by: $trainlen = 50m/n$, where $trainlen$ represents the number of iterations; m is the number of neurons; and n the number of sample vectors. Unfortunately, there is little information about this formula in the SOM_toolbox manual. Using this formula often means that the number of iterations is small. For our New Zealand example, $trainlen = 50(108)/459 = 11.8$, so, when rounded to the next largest whole number, 12 iterations are suggested. A better rule of thumb is that, for good convergence and statistical accuracy, the number of iterations should be at least 500 times larger than the number of neurons (Kohonen, 1990). For our New Zealand example, convergence and good accuracy should be achieved in $500(m) = 500(108) = 54,000$ iterations.

SOM interpretation

While the SOM algorithm is essentially a clustering algorithm, the detail within each cluster is very useful for questions concerning invasive alien species. For example, Worner and Gevrey (2006) found in their analysis of insect pest profiles that New Zealand shares a large number of pest species with Italy (59%), France (58%), Turkey (46%) and Morocco (48%). This similarity exists even though these countries do not appear to have analogous climates and conditions to New Zealand. Clearly, further study of the pests from these countries to determine potential pathways by which serious invasive species could enter New Zealand seems warranted. The most important result of the SOM analysis is that SOM weights can be used to create an index of risk of establishment for species not already found in a region. The SOM weight assigned to each species (i.e. an element in the virtual vector of the best matching unit) can be used to indicate those species most likely to establish in the target area if given a suitable pathway. The weights for each species provide a measure of association with the other species in the profile or, in other words, how frequently the target species co-occurs with the others in the pest profile around the world. In this way, a large number of species can be ranked. Those with the highest weight (i.e. risk index) but are not yet established in the target region comprise a subset of species that can be targeted for more in-depth risk assessment (Fig. 7.3).

Sensitivity analysis of SOMs

All databases contain errors. There is concern that such errors could significantly affect the reliability of any analysis based on the database. For example, Paini *et al.* (2010a) used impact risk assessments generated by the Australian Government's Department of Agriculture, Forestry, and Fisheries (http://www.daff.gov.au/ba/ira/final-plant) as an independent source to estimate the error rate in the species distribution data set used by Worner and Gevrey (2006). The error rate was approximately 8%. Furthermore, Paini *et al.*

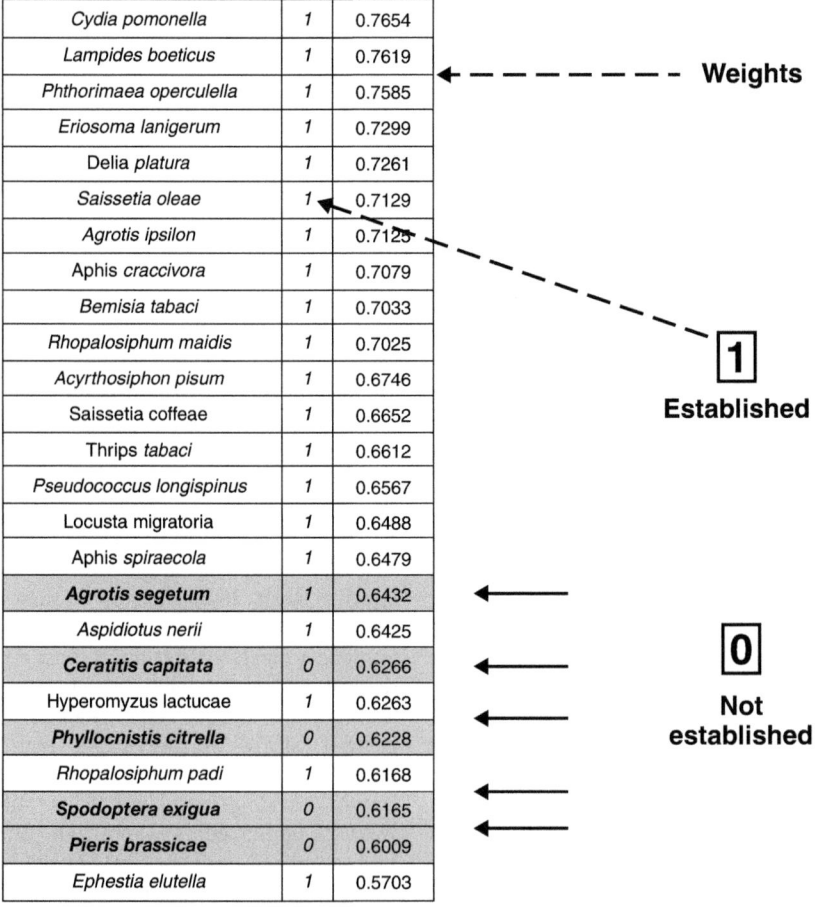

Fig. 7.3. Portion of the pest profile showing connection weights for species that are or are not (bold) established in New Zealand.

(2010a) evaluated the sensitivity of the SOM method by simulating error in the database. They reversed an increasing percentage of the presence/absence data for each site and compared newly generated SOM outputs with the original output. Data from 459 sample vectors (i.e. pest profiles) were altered by 5%, 10%, 20% or 30%. For each test, the desired percentage of species was randomly selected from each pest profile and their presence or absence records reversed. Each region was altered separately so that no two regions were altered in the same way. Paini *et al*. (2010a) found that the species' original rankings remained unaffected by alterations of up to 20% of data.

Clearly, no data set is complete. The impact of potentially incomplete data was tested in another study where species profiles were bootstrapped (i.e. resampled with replacement) 1000 times and the change in each species rank (highest weight to the lowest) was recorded (Watts and Worner, 2009). The New Zealand regional pest profile was used. For the top 50 most highly ranked species that were not established in New Zealand, their ranks changed on average only 14 places out of a possible 800, a result that gave confidence in the method.

SOM validation

One approach to validate a SOM analysis is to address the question: if the SOM analysis had been performed earlier, would it have helped to identify those pest species that actually established in a target region? Worner and Soquet (2010) carried out a SOM analysis on a database extracted from the Crop Protection Compendium (CABI, 2007). New Zealand's pest profile again was used, where the status of each currently established invasive alien pest species was reversed one at a time. In other words, if a species was present in the region (1) its status was changed to absent (0). The objective of the analysis was to determine whether changing species status from present to absent changed its risk index significantly. After the status of a single species was changed from present to absent, a new SOM was created using the modified data and the new risk index for the target species recorded. Following that, the species status was reinstated to its original form prior to repeating the process with the next species in the pest profile.

High correlation between the before (i.e. altered) and after (i.e. unaltered) data (Spearman's rank correlation r = 0.99) showed that the data alterations did not have a significant influence on risk assessment. Ranks changed, on average, 14 places (Fig. 7.2) for the 120 established (i.e. present) species when their status was reversed, and the average change in risk values for the top 100 pest species was 0.07 (Worner and Soquet, 2010). These results illustrate that the SOM analysis would have helped identify those species as high risk before they established in New Zealand. However, while these results may be reassuring, the time of arrival of the species and the particular profile of those already comprising that profile would likely influence results. Additionally, a change of status of four of the 120 species currently present in New Zealand resulted in a change of cluster. For these four species, risk values also changed considerably upward. Those species had low initial risk. Some species have low initial risk simply because of low prevalence. Any interpretation of risk for low-prevalence species requires caution and should be based on additional information. It is clear, however, that this tool is robust enough and stable enough not to be influenced by even quite large variations for a large number of known global crop pests. These results, again, illustrate the usefulness of the method.

In another validation study, data from the GPDD and the CABI Crop Protection Compendium were used in a comparative analysis. Suiter (2011) included in his SOM analyses all pest taxa from bacteria to weeds recorded in the respective databases through 2010. Pre-processing of extracted data eliminated uncertain data and records categorized as 'Unverified', 'Uncertain', 'Eradicated', 'Intercepted' or 'Questionable'.

With only records marked as 'Present', only 10.8% of all species were recorded in both databases. A 15 × 10 SOM map was used for the analysis and the databases were analysed separately. The aim was to determine the number of species classified as high risk by the SOM analysis and not established in the USA in 2007 that subsequently established by 2011.

The analysis of the GPDD database found six species with high risk indices that had not been in the USA in 2007 but were established by 2011, and another six species with high risk indices were found from the Crop Protection Compendium. The species were not the same, so 12 high-risk species that were not present in the USA in 2007 had subsequently established by 2011. It is not known whether any of the species was on any agency risk list, but it appears that the SOM analysis can provide a proactive mechanism for risk assessors to indicate potential threats for detailed analysis that may not come to their attention by other means.

Further validation for the SOM method was obtained using a virtual world to simulate invasions and test the predictions of the SOM (Paini et al., 2011). The authors created a virtual world in which each invasive species was assigned a parameter that defined its environmental requirements, and each region of this virtual world contained a range of environmental characteristics determining which invasive species could invade and establish. This virtual world thus determined exactly which species could invade which region. To test the SOM, each species was distributed over predetermined proportions (i.e. 20% and 50%) of the regions where it could establish at a point in time when not all species had reached or invaded all the regions they are able to. The virtual pest profiles were analysed using a SOM to generate predictions of which species were most likely to establish in each region. These predictions were then compared with the species known to be able to establish in a particular region. They found the SOM was extremely efficient at predicting those virtual species that could establish in a region, especially those that they had not yet invaded, compared with those virtual species that could not.

SOM application: an example from Finland

Vänninen et al. (2011) updated the list of invasive alien invertebrate pests in Finland and identified potential new pests using a SOM analysis. A comprehensive list of indigenous and invasive alien agricultural pests had been compiled nearly 50 years ago (Vappula, 1962) and was considered out of date because crop species, production methods and world trade had undergone substantial changes since that time. Furthermore, climate warming has resulted in longer growing seasons and milder winters in Northern Europe, which is likely to increase yields and create opportunities to produce new crop species and varieties. Future changes in climate and agricultural production practices will increase the need for plant protection against current and novel pests.

Vänninen et al. (2011) focused their analysis on pest species that feed on horticultural or agricultural plants, partly because approximately 15% of invasive alien species that are expanding or shifting polewards in Europe have colonized agricultural environments (Hickling et al., 2006; Roques et al., 2009). The species fell into three categories defined by the International Plant Protection Convention (FAO, 2002, 2004).

The list of invasive alien pest species – both occasionally detected and naturalized species – recorded at MTT Agrifood Research Finland was compared with the list of pest species in Finland compiled by Vappula (1962) and the handbook of alien species in Europe (DAISIE, 2008b) compiled by the European Commission-funded Sixth Framework Programme, Delivering Alien Invasive Species Inventories for Europe (DAISIE; DAISIE, 2008a). The list of potential pests was determined using a SOM that ranked species in terms of their risk indices. The SOM analysis used the CABI Crop Protection

Compendium (CABI, 2003), a global database of pests and pathogens, and clustered geographical areas that have similar pest species assemblages to Finland. Statistical weights were calculated for the species in the pest profiles. These weights were used to determine the risk of invasion for species that were not yet established in Finland.

The inventory of species recorded 77 invasive alien invertebrate pest species established within Finnish agricultural and horticulture systems (Vänninen et al., 2011). Those species included 67 insects, five nematodes, two mites and three slugs. The insect pest species included 39 Hemiptera, ten Thysanoptera, eight Coleoptera, four Lepidoptera, three Diptera and one Hymenopteran species. Of the Hemiptera, the majority were aphids (19 species) and soft or hard scales (11 species). Thirty-five of the recorded 77 pest species were mentioned in Vappula (1962), so nearly half of the species that have invaded Finland have done so during the last 50 years.

Risk indices produced by the SOM analyses were generally high for indigenous species and alien species already established in Finland. This pattern suggests that SOM analysis may be a feasible tool for predicting the likelihood of successful invasions of future invasive alien pests. The SOM analysis predicted high establishment risk indices (>0.5) for eight pests of outdoor crops that, at the time of the study, were not recorded in Finland. One of these species was *Leptinotarsa decemlineata*, a quarantine pest. Another species was *Eriosoma lanigerum*, regulated to protect fruits, berries and woody ornamentals. A third high-risk species was *Ostrinia nubilalis*, a pest primarily of maize, which is not widely cultivated in Finland. *O. nubilalis* is, however, a migrant in Finland and lives there on *Artemisia* spp. and several Poaceae. This insect potentially will become a serious pest if maize is widely cultivated in Finland under climate change. SOM analysis suggests that the greatest risk from alien pests in Finland in the changing climate falls on horticultural crops. In short, Vänninen et al. (2011) recommended SOM analysis as a useful complement to local expert knowledge to assess risks posed by invasive alien species.

SOM application using multi-criteria analysis

To prioritize the biosecurity risks from economically important plant-parasitic nematodes to Australia, Singh et al. (2012) analysed distributions of 250 species from 355 regions worldwide using a SOM. Similar to previous studies, Singh et al. (2012) compared the presence and absence of pest species in Australia with other regions of the world by clustering regions with species assemblages similar to those of Australia and its jurisdictions. The SOM clustering was used to determine other countries and regions which could act as a donor of potential invasive species. The SOM index (interpreted as the likelihood of a species becoming established, if it arrives) was used to select 97 out of the 250 species for more detailed multi-criteria evaluation.

Multiple biotic, abiotic and anthropogenic factors contribute to the chances that an invasive alien species will arrive, establish and cause a negative impact in a new range (Lodge et al., 2006; Hayes and Barry, 2008; Blackburn et al., 2011; Leung et al., 2012). Singh et al. (2012) evaluated risks posed by invasive alien nematodes with a multi-criteria evaluation scheme. The scheme included SOM indices as measures of the likelihood of a species becoming established and nine other criteria: (i) existence of particular pathways; (ii) survival adaptations; (iii) pathogenicity; (iv) host range; (v) whether the species is an emerging pest; (vi) its taxonomy; (vii) the existence of particular pathotypes; (viii) association in disease complexes; and (ix) the level of knowledge that exists about the species. For each of the nine criteria, a probability rubric was developed to indicate the level of risk. Species were evaluated against each of the nine criteria based on information from the literature and expert judgements. The resulting scores for all criteria and the SOM index for each species were combined using weighted averages to determine an overall biosecurity risk.

By integrating SOM analysis into the multi-criteria evaluation, Singh et al. (2012) were able to explicitly incorporate heterogeneous data into their analysis of biosecurity risks from plant-parasitic nematodes to Australia. The SOM clustering of regions with similar species assemblages and potentially similar ecological niches indicated that Australia's major trading partners could also act as donors of species most likely to pose the greatest biosecurity risks. The species rankings based on SOM analysis had a significant positive correlation with the overall risks determined after multi-criteria evaluation. After assessing additional information about the nine criteria and combining with the SOM index, the majority of the 97 species had a greater overall risk than that estimated by the SOM index alone. For instance, by using the SOM index risk scale outlined in Paini et al. (2010a), *Bursaphelenchus xylophilus*, *Heterodera glycines* and *Heterodera carotae* with SOM indices of 0.37, 0.40 and 0.13 would be classified as medium-, medium- and low-risk, respectively. However, after multi-criteria evaluation of additional information, the overall risk value was much higher than that estimated by the SOM index alone (Table 7.1), reflecting additional expert judgement.

For all three species in Table 7.1, the increased overall risks were justified after considering information on pathogenicity, potential economic impacts, records of interceptions indicating the availability of pathways, good survival adaptations and also evidence of their spread into new areas. The study by Singh et al. (2012) re-emphasizes the importance of careful interpretation of results from SOM. Relying only on SOM analysis could lead to under- or overestimation of risks depending on the species (Worner and Gevrey, 2006; Eschen et al. 2012). While SOM analysis is useful for initial prioritization of species for further assessment, Singh et al. (2012) suggest a species' biosecurity risk can be better estimated by integrating SOM results with expert opinion in a multi-criteria evaluation rather than using each method on its own.

Clustering with *k*-Means

An alternative method of clustering species assemblages is *k*-means (Lloyd, 1982), an unsupervised clustering algorithm that determines to which of *k* clusters each sample vector should be assigned. While the *k*-means algorithm is iterative, it is usually much faster than the SOM algorithm, allowing for more replications of the clustering with less computing resources. A number of programming and analysis platforms including R (R Development Core Team, 2013), MATLAB (MATLAB, 2013) and SAS (SAS Institute, Inc., 2011) provide *k*-means algorithms.

The general *k*-means algorithm is as follows (Worner et al., 2013):

Table 7.1. Weighted contributions of criteria to risk scores for plant-parasitic nematodes.

		Score		
Criterion	Criterion weight	*Bursaphelenchus xylophilus*	*Heterodera glycines*	*Heterodera carotae*
Species distribution (SOM index)	0.2	0.37	0.40	0.13
Pathways	0.1	0.80	0.60	0.80
Survival adaptations	0.1	0.70	0.80	0.80
Pathogenicity	0.1	0.90	0.90	0.80
Host range	0.1	0.60	0.50	0.30
Emerging pest	0.1	0.80	0.80	0.50
Taxonomy	0.1	0.50	0.70	0.60
Knowledge	0.1	0.50	0.50	0.60
Pathotypes	0.05	0.50	0.80	0
Disease complex	0.05	0.80	0.60	0
Overall risk Σ(score × weight)		0.62	0.63	0.47

1. Choose k initial centres (these centres represent vectors in multidimensional space and can be generated randomly, or they can be vectors that are randomly selected without replacement from the data set).
2. For each data vector (e.g. regional pest profile), calculate the distance to each of the k cluster centres.
3. Assign each data vector (e.g. pest profile) to its nearest cluster.
4. Calculate new cluster centres that correspond to the mean of all vectors in each cluster.
5. Repeat steps 2–4 until a stopping condition is reached, usually when vectors no longer change the cluster they are assigned, so the clusters are stable.

The k-means algorithm was first applied to the analysis of regional pest and species profiles the same way as the SOMs (Watts and Worner, 2009). Geographical regions represented by their species profiles were clustered and the contents of each cluster were examined. The risk of a species establishing was determined as its frequency in the cluster. High frequency indicated a high level of co-occurrence with the other species in the pest profiles in the same cluster.

Several analyses using k-means have been reported (Watts and Worner, 2009, 2011, 2012). Watts and Worner (2009) compared the results of clustering insect pest profiles using SOM with the results of clustering profiles using k-means. Cluster quality was measured using the QE, which is the mean distance between each vector and the centre of its cluster (Hansen and Jaumard, 1997). The comparison showed the k-means algorithm produced clusters of the same quality and highly similar risk weightings as the SOM. The study did not examine the effects of noise or small random changes to the species' presence.

The k-means and SOM algorithms were again compared with the species profiles being bacteria that cause crop diseases (Watts and Worner, 2012). While some differences in performance of the clustering algorithms existed, in most instances the differences were not significant. The exception was computational efficiency, where k-means was orders of magnitude faster. More importantly, the different algorithms assigned high to medium risk indices to essentially the same species: only twelve species of the top eighty were not in both the SOM- and k-means-generated risk lists.

Hierarchical Clustering

In hierarchical clustering, each item in the data set is initially considered a cluster and in each subsequent step, two clusters are merged based on similarities or dissimilarities until all clusters are merged into a single cluster (Borgatti, 1994; Hastie et al., 2009). Clusters can be merged by different methods including the most similar pair of observations in two clusters (i.e. single linkage), the most dissimilar pair of observations (i.e. complete linkage) or the dissimilarity between the average of the observations in each cluster (i.e. group average; Hastie et al., 2009). The method determines cluster sizes and relationships among clusters. The optimal number of clusters can be determined using a statistic such as the Davies–Bouldin Index that describes the ratio of variation within and among clusters (Davies and Bouldin, 1979). A dendrogram provides a graphical representation of the relationship among the clusters. Like other clustering techniques, the results of hierarchical clustering do not imply causal relationships and should be interpreted with caution.

Eschen et al. (2012) used hierarchical clustering to detect regional patterns in pest species distributions to investigate trade in plants for planting in Europe as a major pathway for the introduction of invasive alien forest pests and diseases. The study aimed to develop an objective assessment of risk posed by individual pests and diseases and to identify potential sources of invasive species not yet present in Europe, based on known species distributions. For their assessment, Eschen et al. (2012) analysed distribution data from CABI's Plantwise Knowledge Bank (http://www.plantwise.org/knowledgebank) for 1009 invertebrate pests and pathogens of woody hosts in 351 global regions within 183 countries. Seven large countries were subdivided into regions. The 1009 taxa were divided into twelve groups: four microorganism and eight invertebrate. Countries and regions with similar pest species assemblages were identified for each organism group using hierarchical cluster analysis with the Ward (1963) minimum variance method to determine which clusters were merged. The likelihood of establishment was calculated as the proportion or frequency of countries within each cluster containing European Union (EU) and European Free Trade Association (EFTA) countries where each species was present.

Regions with pest species assemblages most similar to EU countries were North America, the Mediterranean region, the northern part of Eurasia and Australia/New Zealand, which have similar climates to the EU and a long history of live plant importation. On average, the clusters for microorganisms were twice as large as clusters for invertebrate groups (111 versus 61 regions per cluster), which may reflect better distribution data for invertebrates which are easier to detect and identify than are microorganisms. This resulted in better-defined clusters for invertebrates than microorganisms. For the Oomycetes, European countries were spread over all clusters, but otherwise no interpretable microbial clusters were formed. Most pest species in the database including non-European species were recorded in one or more EU countries, suggesting that risk posed by some exotic species in Europe primarily comes from the spread within the EU. Similarly, Paini et al. (2010b) found that most 'new' species that might affect US agriculture were recorded already in one or more states.

Cautions and Limitations of SOMs and Other Clustering Methods

Cautions and caveats for SOMs and other clustering methods need to be acknowledged. Species of low prevalence generally are not represented sufficiently in their global distribution to show any significant association or co-occurrence with other species and most often have low risk values. The analyst should pre-process the data to exclude endemic or invasive alien species that have very low global prevalence. Paini et al. (2011) noted that regions with a pest profile comprised of fewer than eight species can become significantly more unstable or difficult to predict, and they recommended that any analysis should not generate lists for regions with so few species. A SOM analysis might be used to identify outliers on complex data sets (Liebscher et al., 2012) and the SOM analysis might cluster such vectors (i.e. pest profiles) together. However, Suiter (2011) cautioned that the outcome of SOM analysis might be affected by sampling artefacts. Countries such as the USA, China and Australia that have good records or have been heavily sampled for invasive pests consistently cluster together in analyses. Suiter (2011) suggested that this clustering was the result of the high probability of overlap in pest assemblages for countries with a large number of pests, the potential climatic similarity in parts of large countries with a wide range of climates, heavy historical trade volumes among countries with high pest numbers may create similar pest assemblages, and the extensive capacity to survey and document invasive alien species in these countries. Analysts should consider the potential effects of unequal pest sampling and reporting on clustering outcomes. However, similarities in climate, hosts and trading history should be responsible for many similarities in pest profiles (Worner and Gevrey, 2006; Paini et al., 2010a) and the goal is to reveal these similarities through clustering techniques.

Direct comparison of risk values for species in different countries or profiles from different databases is not possible. These risk values depend on the details of which species have been recorded as invasive or endemic in the target regions (Paini et al., 2010b; Eschen et al., 2012). Suiter (2011) found that many nations that were found to be similar based on their pest species assemblages, as described in the GPDD, continued to be similar when assemblages were based on information from the Crop Protection Compendium. However, the high-risk species profiles for the USA for both data sets were very different. Thus the analysis of each database should be interpreted separately.

Specific pathways that bring potential pests into new regions are not considered here. Eschen *et al.* (2012) suggested that combining pest distribution data with recent trade data could provide a better indication of the likely origin of invasive alien species. Such pathways would certainly be considered when high-risk species are subject to a full risk analysis. However, changes in global commerce may not immediately equate to change in pest assemblages. Significant lags occur between pest establishment, subsequent detection and publication of the findings. Thus, the choice of which trade statistics to combine with pest distribution data remains problematic.

Users of *k*-means and hierarchical analyses should proceed with caution. *k*-Means has some weaknesses: it tends to be dependent on the initial number of clusters chosen by the user; it is sensitive to outliers; the analysis does not easily capture high-level (i.e. inter-cluster) relationships; and it does not preserve data topology (i.e. the relationships among neighbours) in the clustering. Hierarchical analysis does not require the number of clusters to be predefined and its hierarchical structure allows explanation of inter-cluster relationships. However, hierarchical analysis is not tolerant to noise (i.e. random errors in the data) and it does not preserve data topology in the clustering. Clearly more research is needed to evaluate the suitability of these methods to provide fine-scaled analysis of pest risk data.

Conclusions

Clustering pest profiles has potential to rank and prioritize the hundreds, even thousands, of potentially invasive alien pest species that could threaten a country. Indeed, Paini *et al.* (2011) claimed that up to 10,000 species could be analysed on a desktop computer, which would bring considerable functionality to the assessment of so many species. For research purposes, such large-number assessment could be useful; but in practice, elucidation of this many species could add considerable noise to outputs (reports, maps) that must be considered by decision and policy makers. Since most potentially invasive pests would be considered low priority, there is value in addressing the subset that present the greatest danger. In this chapter, the process for developing SOMs has been carefully elucidated but the merits of other clustering methods are discussed. Furthermore, the study of high-risk species does not stop with a cluster analysis and examination of a pest profile. Close analysis of similarity using a standard measure, such as percentage similarity or a simple matching coefficient that accounts for shared and non-shared species, is required. Other data must be taken into account for a full assessment of a species' potential to establish. For example, Saikkonen *et al.* (2012) suggested that the risk assessment of potential invasive species across latitudes would be more reliable if, for example, photoperiodism and species' responses to seasonal changes in day length and light quality could be included in the models forecasting species' range shifts.

Pest establishment is only one component of overall risk of any invasive alien species to any country or region. The probability of entry and the potential to spread and cause impact are also part of a full pest risk assessment. Lack of data often constrains pest risk assessments, and much research remains to be completed. The analysis of pest profiles by first clustering, followed by a similarity analysis, in combination with the characterization of the climate or climate profile may offer some early insights. In the absence of other data, analysis of pest assemblages, hosts and climate might provide sufficient information to support characterizations of risks posed by invasive alien species about which we know very little. The incorporation of SOMs into a multi-criteria analysis illustrates the value of SOMs, and possibly other clustering techniques, to complement a full risk assessment process and to elicit expert interpretation.

References

Ahmad, N., Alahakoon, D. and Chau, R. (2010) Cluster identification and separation in the growing self-organising map: application in protein sequence classification. *Neural Computing and Applications* 19, 531–542.

Arsuaga, U.E. and Díaz, M.F. (2005) Topology preservation in SOM. *International Journal of Applied Mathematics and Computer Science* 1, 19–22.

Blackburn, T.M., Pysek, P., Bacher, S., Carlton, J.T., Duncan, R.P., Jarosik, V., Wilson, J.R.U. and Richardson, D.M. (2011) A proposed unified framework for biological invasions. *Trends in Ecology & Evolution* 26, 333–339.

Borgatti, S.P. (1994) How to explain hierarchical clustering. *Connections* 17, 78–80.

CABI (2003) *Crop Protection Compendium, Global Module*, 2nd edn. CAB International, Wallingford, UK.

CABI (2007) *Crop Protection Compendium, Global Module*, 5th edn. CAB International, Wallingford, UK.

Céréghino, R. and Park, Y.-S. (2009) Review of the self-organising map approach in water resources: commentary. *Environmental Modelling and Software* 24, 945–947.

Chan, C.-K.K., Hsu, A.L. Tang, S.-L. and Hagamuge, S.K. (2008) Using growing self-organising maps to improve the binning process in environmental whole-genome shotgun sequencing. *Journal of Biomedicine and Biotechnology* 2008, 513701.

Chattopadhyay, M., Dan, P.K. and Mazumdar, S. (2012) Application of visual clustering properties of self organizing map in machine-part cell formation. *Applied Soft Computing* 12, 600–610.

Chon, T.-S. (2011) Self-organising maps applied to ecological sciences. *Ecological Informatics* 6, 50–61.

Chon, T.-S., Park, Y.-S., Monn, K.H. and Cha, E.Y. (1996) Patternizing communities by using an artificial neural network. *Ecological Modelling* 90, 69–78.

DAISIE (2008a) *Delivering Alien Invasive Species Inventories for Europe Database*. Available at: http://www.europe-aliens.org/default.do (accessed 31 December 2013).

DAISIE (2008b) *Handbook of Alien Species in Europe*. Springer, Dordrecht, The Netherlands.

Davies, D.L .and Bouldin, D.W. (1979). A cluster separation measure. *IEEE Transactions on Pattern Analysis and Machine Intelligence* 1, 224–227.

Eschen, R., Holmes, T., Smith, D., Roques, A. and Kenis, M. (2012) The risk of invasion of Europe by forest pests and diseases, based on their worldwide occurrence. *Proceedings of the 3rd Meeting of IUFRO Working Unit*. International Union of Forest Research Organizations, Tokyo, Japan. Available at: http://hyoka.nenv.k.u-tokyo.ac.jp/alien_reports.html (accessed 31 December 2013).

FAO (2002) *Regulated Non-Quarantine Pests: Concept and Application. International Standard for Phytosanitary Measures No. 16*. Food and Agriculture Organization of the United Nations, Rome. Available at: https://www.ippc.int/sites/default/files/documents/20131009/ispm_16_2002_en_2013-08-26_2013100911%3A09--247.09%20KB.pdf (accessed 31 December 2013).

FAO (2004) *Pest Risk Analysis for Quarantine Pests, Including Analysis of Environmental Risks and Living Modified Organisms. International Standard for Phytosanitary Measures No. 11*. Food and Agriculture Organization of the United Nations, Rome. Available at: http://www.fao.org/docrep/008/y5874e/y5874e00.HTM (accessed 31 December 2013).

Gevrey, M. and Worner, S.P. (2006) Prediction of global distribution of insect pest species in relation to climate by using an ecological informatics method. *Journal of Economic Entomology* 99, 979–986.

Gevrey, M., Worner, S.P., Kasabov, N. and Giraudel, J.-L. (2006) Estimating risk of events using SOM models: a case study on invasive species establishment. *Ecological Modelling* 197, 361–372.

Gregory, F.J., Copestake, P. and Pearce, J.M. (2007) *Key Issues in Petroleum Geology*. The Geological Society Publishing House, Bath, UK.

Hansen, P. and Jaumard, B. (1997) Cluster analysis and mathematical programming. *Mathematical Programming* 79, 191–215.

Hastie, T., Tibshirani, R. and Friedman, J. (2009) *The Elements of Statistical Learning*, 2nd edn. Springer, Berlin.

Hayes, K.R. and Barry, S.C. (2008) Are there any consistent predictors of invasion success? *Biological Invasions* 10, 483–506.

Heiri, O. and Lotter, A.F. (2005) Holocene and late glacial summer temperature reconstruction in the Swiss Alps based on fossil assemblages of aquatic organisms: a review. *Boreas* 34, 506–516.

Hickling, R., Roy, D.B., Hill, J.K., Fox, R. and Thomas, C.D. (2006) The distributions of a wide range of taxonomic groups are expanding polewards. *Global Change Biology* 12, 450–455.

Hsu, C.-M. (2011) A hybrid procedure for stock price prediction by integrating self-organising map. *Expert Systems with Applications* 38, 14026–14036.

Ismail, S., Shabri, A. and Samsudin, R. (2012) A hybrid model of self organizing maps and least square support vector machine for river flow forecasting. *Hydrology and Earth System Sciences* 16, 4417–4433.

Kohonen, T. (1982) Self-organized formation of topologically correct feature maps. *Biological Cybernetics* 43, 59–69.

Kohonen, T. (1984) *Self-Organization and Associative Memory.* Springer, Berlin.

Kohonen, T. (1990) The self-organizing feature map. *Proceedings of the IEEE* 78, 1464–1480.

Kohonen, T. (2001) *Self-Organizing Maps.* Springer, Berlin.

Leung, B., Roura-Pascual, N., Bacher, S., Heikkilä, J., Brotons, L., Burgman, M.A., Dehnen-Schmutz, K., Essl, F., Hulme, P.E., Richardson, D.M., Sol, D. and Vilà, M. (2012) Teasing apart alien species risk assessments: a framework for best practices. *Ecology Letters* 15, 1475–1493.

Liebscher, S., Kirschstein, T. and Becker, C. (2012) The flood algorithm – a multivariate, self-organising-map-based, robust location and covariance estimator. *Statistical Computing* 22, 325–336.

Lloyd, S.P. (1982) Least squares quantization in PCM. *IEEE Transactions on Information Theory* 28, 128–137.

Lodge, D., Williams, S., MacIsaac, H., Hayes, K., Leung, B., Loope, L., Reichard, S., Mack, R., Moyle, P., Smith, M., Andow, D., Carlton, J. and McMichael, A. (2006) Biological invasions: recommendations for policy and management. *Ecological Applications* 16, 2035–2054.

Lui, Y. and Weisberg, R.H. (2011) A review of self-organising map applications in meteorology and oceanography. In: Mwasiagi, J.I. (ed.) *Self Organising Maps – Applications and Novel Algorithm Design.* InTech, Rijeka, Croatia, pp. 253–272.

MATLAB (2013) *MATLAB 8.2 R2013.* The MathWorks, Inc., Natick, Massachusetts.

Milone, D.H., Stegmayer, G., Kamenetzky, L., López, M. and Carrari, F. (2012) Clustering biological data with SOMs: on topology preservation in non-linear dimensional reduction. *Expert Systems with Applications* 40, 3841–3845.

Nakao, R., Abe, T., Nijhof, A.M., Yamamoto, S., Jongejan, F., Ikemura, T. and Sugimoto, C. (2013) A novel approach, based on BLSOMs (batch learning self-organizing maps), to the microbiome analysis of ticks. *The ISME Journal* 7, 1003–1015.

Oja, E., Kask, S. and Kohonen, T. (2003) Bibliography of self organizing map (SOM) papers 1998–2001. *Neural Computing Surveys* 3, 1–156.

Paini, D.R., Worner, S.P., Thomas, M.B. and Cook, D.C. (2010a) Using a self-organising map to predict invasive species: sensitivity to data errors and a comparison with expert opinion. *Journal of Applied Ecology* 47, 290–298.

Paini, D.R., Worner, S., Cook, D., De Barro, P. and Thomas, M. (2010b) Threat of invasive pests from within national borders. *Nature Communications* 1, 115.

Paini, D.R., Bianchi, F.J.J.A., Northfield, T.D. and De Barro, P.J. (2011) Predicting invasive fungal pathogens using invasive pest assemblages: testing model predictions in a virtual world. *PLoS ONE* 6, e25695.

R Development Core Team (2013) *R: A Language and Environment for Statistical Computing.* R Foundation for Statistical Computing, Vienna.

Roques, A., Rabitsch, W., Rasplus, J.-Y., Lopez-Vaamonde, C., Nentwig, W. and Kenis, M. (2009) Alien terrestrial invertebrates of Europe. *Handbook of Alien Species in Europe.* Springer, Dordrecht, The Netherlands, pp. 63–79.

Saikkonen, K., Taulavuori, K., Hyvonen, T., Gundel, P.E., Hamilton, C.E., Vanninen, I., Nissinen, A., and Helander, M. (2012) Climate change-driven species' range shifts filtered by photoperiodism. *Nature Climate Change* 2, 239–242.

Sarlin, P. (2011) Evaluating a self-organising map for clustering and visualising optimum currency area criteria. *Economics Bulletin* 31, 1483–1495.

SAS Institute, Inc. (2011) *SAS/STAT® 9.3 User's Guide.* SAS Institute, Inc., Cary, North Carolina.

Savin, S.M. (1977) The history of the Earth's surface temperature during the past 100 million years. *Annual Review of Earth and Planetary Sciences* 5, 319–355.

Singh, S., Paini, D.R., Ash, G.J. and Hodda, M. (2012) Prioritising plant parasitic nematode species biosecurity risks using self organising maps. *Biological Invasions* 16, 1515–1530.

Suiter, K. (2011) Progress on SOM analysis: a comparison of pre-emergent invasive pests lists using distribution data obtained from the GPDD and CABI databases. *Proceedings of the International Pest Risk Mapping and Modelling Workshop V.* IPRMW, Raleigh, North Carolina. Available at: http://www.pestrisk.org/PDF/Suiter_IPRMW_5.pdf (accessed 28 December 2013).

Vänninen, I., Worner, S., Huusela-Veistola, E., Tuovinen, T., Nissinen, A. and Saikkonen, K. (2011) Recorded and potential alien invertebrate pests in Finnish agriculture and horticulture. *Agricultural and Food Science* 20, 96–114.

Vappula, N.A. (1962) Pests of cultivated plants in Finland. *Annales Agriculturae Fenniae* 1 (Suppl. 1), 1–239 (English edition).

Vesanto, J. and Alhoniemi, E. (2000) Clustering of the self-organizing map. *IEEE Transactions on Neural Networks* 11, 586–600.

Vesanto, J., Himberg, J., Alhoniemi, E. and Parhankangas, J. (2000) *SOM Toolbox for MATLAB 5*. Helsinki University of Technology, Espoo, Finland. Available at: http://www.cis.hut.fi/somtoolbox/package/papers/techrep.pdf (accessed 29 December 2013).

Ward, J.H. Jr (1963) Hierarchical grouping to optimize an objective function. *Journal of the American Statistical Association* 48, 236–244.

Watts, M.J. and Worner, S.P. (2009) Estimating the risk of insect species invasion: Kohonen self-organising maps versus *k*-means clustering. *Ecological Modelling* 220, 821–829.

Watts, M.J. and Worner, S.P. (2011) Improving cluster-based methods for investigating potential for insect pest species establishment: region-specific risk factors. *Computational Ecology and Software* 1, 138–145.

Watts, M.J. and Worner, S.P. (2012) Using artificial neural networks to predict the distribution of bacterial crop diseases from biotic and abiotic factors. *Computational Ecology and Software* 2, 70–79.

Weber, M., Teeling, H., Huang, S., Waldmann, J., Kassabgy, M., Fuchs, B.M., Klindworth, A., Klockow, C., Wichels, A., Gerdts, G., Amann, R. and Glöckner, F.O. (2011) Practical application of self-organizing maps to interrelate biodiversity and functional data in NGS-based metagenomics. *The ISME Journal* 5, 918–928.

Wherens, R. (2013) *Package 'kohonen' ver. 2.0.14. Supervised and Unsupervised Self-Organising Maps*. Available at: http://cran.r-project.org/web/packages/kohonen/index.html (accessed 31 December 2013).

Wherens, R., and Buydens, L.M.C. (2007) Self- and super-organising maps in R: the Kohonen package. *Journal of Statistical Software* 21, 1–19.

Worner, S.P. and Gevrey, M. (2006) Modelling global insect species assemblages to determine risk of invasion. *Journal of Applied Ecology* 43, 858–867.

Worner, S.P. and Soquet, A. (2010) A retrospective analysis of the use of ecological theory and pest species assemblages to prioritise pests (abstract). *Proceedings of the 4th International Pest Risk Modelling and Mapping Workshop*. IPRMW, Raleigh, North Carolina. Available at: http://www.pestrisk.org/Workshop2010Report.pdf (accessed 28 December 2013).

Worner, S.P., Gevrey, M., Eschen, R., Kenis, M., Paini, D., Singh, S., Suiter, K. and Watts, M.J. (2013) Prioritizing the risk of plant pests by clustering methods; self-organising maps, *k*-means and hierarchical clustering. *NeoBiota* 18, 83–102.

Zhao, X., Ping, L. and Kohonen, T. (2011) Contextual self-organizing map: software for constructing semantic representations. *Behavior Research Methods* 43, 77–88.

8 Modelling the Spread of Invasive Species to Support Pest Risk Assessment: Principles and Application of a Suite of Generic Models

Christelle Robinet,[1]* Hella Kehlenbeck[2] and Wopke van der Werf[3]

[1]*INRA, UR633 Zoologie Forestière, Orléans, France;* [2]*Julius Kühn-Institute, Institute for Strategies and Technology Assessment, Kleinmachnow, Germany;* [3]*Centre for Crop Systems Analysis, Wageningen University, Wageningen, The Netherlands*

Abstract

The estimation of rates and patterns of spread is one of the key steps in a pest risk assessment. Pest risk analysts across the world wish to make quantitative, scientifically defensible assessments of likely spread by invasive alien species. However, data and time to develop detailed models for pest invasions are usually lacking and the resources to test those models in practice are not available. Therefore, generic and simple models are needed. A generic spread module composed of four models has been developed to assess the spread of plant pests. Four different models were developed to represent differences in objectives, available data and assumptions underlying the assessment of spread. The most complex of the models simulates spread in time and space and has four biological parameters, representing population growth and dispersal. The simplest of the models has only one parameter and considers only geographic range expansion. A third model assumes logistic growth of invaded area and a fourth model assumes logistic growth of population density in invaded cells. All models consider climatic suitability and presence of hosts. Consideration of economic value is optional. This chapter describes concepts and application of these models. They are illustrated by case studies for western corn rootworm, *Diabrotica virgifera virgifera*, in Europe.

Scope

The objective of the generic spread module is to quantify and map the potential spread or temporal increase of invasive pests in a way that does not require too much work in data collection, model development and model fitting. It was developed to help risk assessors rapidly assess the likely spread or temporal increase of invasive alien plant pests.

The module consists of four models that differ in scope and sophistication (Table 8.1). Models A and B estimate presence/absence whereas models C and D estimate

* Corresponding author. E-mail: christelle.robinet@orleans.inra.fr

© CAB International 2015. *Pest Risk Modelling and Mapping for Invasive Alien Species* (ed. R.C. Venette)

Table 8.1. Description of the four models within the generic spread module. Risk assessors must provide estimates for: r, the relative rate of spatial increase (/year); c, the rate of range expansion (km/year); λ_{max}, the maximum yearly multiplication factor of population density (no units); P_{max}, the carrying capacity (number of individuals/cell); u, the scale parameter of the 2Dt distribution (km); and v (no units), the shape parameter of this distribution (comparable to the number of degrees of freedom of the t-distribution).

Model	Output	Growth model	Dispersal	Parameter(s)	Model scale	Initial condition
A	Presence/ absence	Logistic	Spatially randomized or structured by economic value of the host	R	Growth model describes number of invaded cells, scattered on the map according to economic information	Number of initially invaded cells
B	Presence/ absence	None	Radial range expansion	c	The model describes increase of radial range around initial location(s)	Spatial locations of initially invaded cells
C	Population density	Logistic	None	λ_{max}; P_{max}	Growth model describes population dynamics in individual cells	Initial density assumed homogeneous over the area of potential establishment
D	Population density	Logistic	Dispersal kernel (2Dt)	λ_{max}; P_{max}; u; v	Growth model describes population dynamics in individual cells; dispersal model describes spatial location of dispersing individuals	Spatial locations and initial densities in the initially invaded cells

population density. Models A and C are spatially implicit whereas models B and D are spatially explicit. The appropriateness of these models depends upon the objective of the risk assessor and the data available. One or more models may be used and the results compared.

With this modelling module, it is, for instance, possible to assess the spatial extent at different times after entry (e.g. within 5 years, 10 years or 15 years) from one or more entry points (i.e. see models B and D). If the species is already present in the pest-risk-assessment area, these points might be known and could be effectively used in the models. If the species is absent in the pest-risk-assessment area or already present but the risk assessor would like to test other potential entry points, some arbitrary points can be chosen. Entry near main airports, ports or railway stations can be particularly interesting to test.

In addition, it is possible to simulate the population density expressed as a percentage of the carrying capacity (i.e. models C and D). To determine the area where the density of the invasive pest could reach outbreak levels, it is possible to assume that the population is initially present everywhere but at a very low density and then simulate the population growth at different time horizons according to the environmental conditions (i.e. model C). In model C, population growth is independent of the distance to other populations and dispersal from one location to another is ignored.

Another approach that ignores the spatial distance is calculation of population growth in terms of the number of presence points and then the allocation of the occupied area according to the economic value of the host (i.e. model A). This model goes one step further than the assessment of potential spread as it provides a first estimate of the economic impact.

Two of the models (i.e. B and D) are true spread models in the sense that they describe explicitly the potential range expansion of a species over a territory and their outputs are represented on a map. The other two models, also called 'spread models' in this chapter, refer to models describing the temporal increase of population density, but this increase depends on the spatial location within the territory and thus their outputs can also be represented on a map. These two types of models basically describe different processes and cannot be used to answer the same question.

The strength of this generic spread module is that it is applicable to any plant pest species because only a few estimated parameters, four at most, need to be provided by the assessor. Estimates for these parameters can either be found in published materials or given by experts. All these models take into account the climate suitability of the plant pest species, as well as the habitat distribution and elements of growth and dispersal of the pest population. Climate suitability is commonly assessed in pest risk assessments, so climate data should be available for assessing potential spread. Concerning habitat distribution, geographic information system (GIS) files for some plant or tree species can be downloaded from various websites. The output of these models can be combined with other components of risk maps (Baker et al., Chapter 2 in this volume) and used to estimate the endangered area and potential economic impact.

This module was designed to fulfil pest-risk-assessment needs and assumes that sufficient information about a target pest and the invaded habitat is available. A separate decision support scheme has been created to guide risk assessors in their decision of whether to apply this module (see the online supplement S1 to Chapter 8; Kehlenbeck et al., 2012). Only the main processes involved in growth and dispersal are included in the models. The models proposed here for pest risk assessment have intentional simplifications that may limit their suitability for addressing alternative aims. More detailed studies and more sophisticated models will also be needed, even if they cannot be developed within the time and resource limitations of the pest risk assessment.

The Models

Each of the models is applied to an area of interest as a regular grid either in decimal degrees or metric units. The outputs of the models are given for each cell (i.e. the elementary unit of the grid) and each time step between the start and end of a given time horizon. The climate constraints of the invasive alien species should first be considered; in the current version of the module, the outputs of CLIMEX (Sutherst and Maywald, 1985; Sutherst et al., 2007) can be used directly. In CLIMEX, the Ecoclimatic Index (EI) ranges from 0 to 100 and gives an indication of long-term persistence. The Growth Index (GI) ranges from 1 to 100 and gives an indication of the potential growth during the favourable season (Sutherst and Maywald, 1985; Sutherst et al., 2007). Suitable cells are the cells where EI > 0 within the habitat distribution, and they represent the area of potential establishment. GI is used in models C and D to adjust the growth rate of the pest. The yearly multiplication factor (λ) in the logistic growth functions of models C and D is calculated for each cell as a function of the ratio GI/GI_{max}, where GI_{max} is the greatest value of GI that is realized in the pest-risk-assessment area (Eqn 3 in Robinet et al., 2012). Other climate suitability surfaces can potentially be used as long as they can provide indices similar to EI and GI; if used, the code to implement the models should be modified accordingly.

The four models in the module are described briefly below. The choice of the model depends on the purpose of the analysis and the available data. None of the models is necessarily better than the others, although advantages and drawbacks are identified for each one. Further details about each model are provided in Robinet *et al.* (2012).

Model A: logistic increase of invaded area combined with the economic value of assets

In the first model, the number of cells occupied by the pest species at each time t is calculated with a logistic growth function that depends on the estimated parameter r, the relative rate of increase of the invaded area (Table 8.1). In this model, growth represents the increase of the occupied area with time and not the increase of the population density. Therefore, estimated r should not represent the intrinsic rate of population increase. This growth function is:

$$n_t = \frac{n_0 \exp(rt)}{1 + n_0[\exp(rt) - 1]/100} \quad (8.1)$$

where n_t is the percentage of cells within the area of potential establishment (Baker *et al.*, Chapter 2 in this volume) that are invaded at time t and n_0 is this percentage at time $t = 0$. The location of these occupied cells is then chosen among the suitable cells following three scenarios. In the best case scenario, these occupied cells are chosen among the least valuable cells (i.e. where the value of the host plant is the lowest). In the worst case, these occupied cells are chosen among the most valuable cells (i.e. where the value of the host plant is the highest). In the random scenario, these occupied cells are selected randomly, irrespective of the cell's economic value. For each of these scenarios, it is possible to map the potential spread of the plant pest and to quantify the total host value at risk for different time horizons. Spread stops when all suitable cells on the map are invaded.

Objective

To estimate how rapidly economic impact accrues under contrasting scenarios for invasion.

Required parameters

Only one assessor-estimated parameter is needed: the relative rate of increase (proportion per year) in the number of cells invaded. The risk assessor should estimate this parameter based upon expert judgement and comparison to past invasions.

Advantages

The model gives a direct estimate of the economic impact through time; it draws scenarios (best/worst cases). As it is easy to apply, it can give rapid assessments of the range of potential impacts.

Drawbacks

The model ignores the geographical distance between source and target cells in the calculation of spread, which is likely to be biologically unrealistic. Furthermore, this model disregards the effects of geographic differences in climatic suitability for the growth of the pest population. Thus, the results represent a scenario with weak biological underpinning.

Model B: radial range expansion

This second model simulates a radial range expansion from one or more entry points. The only assessor-provided parameter is the spread rate c (km/year; Table 8.1). All the suitable cells located within the radius $c \times t$ from the entry point(s) are occupied by the plant pest species at time t. This model arises from the Fisher–Skellam theory based on reaction-diffusion models (Fisher, 1937; Skellam, 1951) where dispersal of the population is derived from random walk theory and the population growth can take various forms; for instance, it can be exponential. The combined effects of dispersal and growth make the population range advance at a constant velocity. Model B presented here is a simplified version with no explicit consideration of population growth.

Objective

This model describes the likely expansion of the range of an invasive alien species over time.

Required parameters

Only one assessor-estimated parameter is needed: the rate of radial range expansion.

Advantages

This model is easy to use; it generates a simple map of potential dispersal over time and space. This map can be readily combined with spatially explicit data on economic value of assets. The distribution maps over time are readily assessed as to plausibility by risk assessors.

Drawbacks

This model ignores spatial variability in the density of the host and the climatic suitability for the growth of the pest population, and therefore does not account for relationships between population growth and spatial spread. The model assumes that the species can jump over unsuitable areas.

Model C: population growth

In the third model, the plant pest species is assumed to be initially present over all suitable cells albeit at a very low density. A logistic growth model depending on the carrying capacity P_{max} (i.e. potential maximum population density within a cell of the study area) and the maximum yearly multiplication factor λ_{max} gives the population density in terms of percentage of the carrying capacity for any time t. The population density increases heterogeneously over the study area because growth depends on the growth rate derived from GI (Robinet et al., 2012). The maximum yearly multiplication factor λ_{max} is supposed to be recorded where GI is the highest in the study area. The population density p_t, which is expressed as a percentage of the carrying capacity, P_{max}, at time t, is given by the following function:

$$p_t = \frac{p_0 \exp[(GI/GI_{max})\ln(\lambda_{max})t]}{1 + p_0 \{\exp[(GI/GI_{max})\ln(\lambda_{max})t] - 1\}/100}$$

(8.2)

Objective

This model describes how long it will take for an invasive alien species to reach high population densities (e.g. densities sufficient to cause material damage), assuming that very low starting population densities are present at $t = 0$ in each suitable cell.

Required parameters

The risk assessor must estimate the potential yearly increase of the population under optimal conditions found in the pest-risk-assessment area (i.e. λ_{max}). To calculate the initial density, which is expressed on a relative scale from 0 to 1, the risk assessor should estimate both the initial founder population at $t = 0$ and the carrying capacity at the level of an entire grid cell. The estimate of carrying capacity should include estimates of the area of the host per grid cell and the maximum density of the pest per unit area of its host (see example for western corn rootworm, *Diabrotica virgifera virgifera* LeConte (Coleoptera: Chrysomelidae), in section 'Analyses for the Case Study of Western Corn Rootworm' below).

Advantages

This model shows variability in lag time (i.e. time until damaging population densities are reached). The model is based on a CLIMEX output specifying spatial variation in climatic suitability for the pest species in combination with an assessment of the potential growth rate of the population under optimal conditions.

Drawbacks

It assumes that the species is initially present over the whole suitable area and it does not account for the spatial process of dispersal.

Model D: population growth and dispersal kernel

Model D simulates the growth of a pest population with the logistic growth function described for model C and it simulates dispersal with a rotationally symmetric two-dimensional t-distribution (2Dt) dispersal kernel (Clark et al., 1999; Robinet et al., 2012). The 2Dt kernel is given by the following formula:

$$K(x,y) = \frac{1}{u^2 \pi v} \frac{\Gamma[(v+1)/2]}{\Gamma[(v-1)/2]} \left(1 + \frac{1}{v}\frac{x^2+y^2}{u^2}\right)^{-(v+1)/2} \quad (8.3)$$

$K(x,y)$ gives the probability of dispersion into location (x,y) given that it comes from $(0,0)$ and Γ is the gamma function. This kernel depends on two parameters: u, the scale parameter, which defines the width of the dispersal distribution; and v, the shape parameter, which defines the fatness of the tail of the distribution. The shape of the distribution varies according to the value of v from the fat-tailed Cauchy distribution for $v \to 1$ to the thin-tailed normal distribution for $v \to +\infty$. The shape parameter v defines the fatness of the tail of the two-dimensional t-distribution and is identical in meaning and effect to the 'degrees of freedom' parameter in Student's one dimensional t-distribution, familiar to most readers from the t-test of statistics. It is well known that t-distributions with a low number of degrees of freedom have fatter tails, i.e. higher frequency of extreme values, than t-distributions with high number of degrees of freedom, which tend to the normal distribution. The model simulates the spread of the species from the entry point(s) with given population densities.

Objective

Starting from one or more initial points of establishment, and using the local density at those points, this model can show the likely spatial spread of the organism over time and space, as affected by spatial variability in climatic suitability and the biological parameters provided by the user.

Required parameters

This model requires four assessor-estimated parameters, two for population growth and two for dispersal (Table 8.1). Population growth parameters are the same as for model C. Parameters for dispersal are those of the 2Dt distribution. These are a length scale (u; km) and a shape factor (v).

Advantages

It is biologically a more realistic model than the other models as both population growth and dispersal are simulated.

Drawbacks

It is conceptually more difficult to estimate all the parameters and obtain the data for this model than for model A, B or C. Simulations with this model take substantially more time to run compared with the other models.

Resources to Use the Modules

R software and code

The generic spread module has been coded in R, but theoretically it can be coded in another language. R is a language and environment for statistical computing (free software, available at http://www.r-project.org/; R Development Core Team, 2010). If R is not already on the computer used for risk analysis, it should be installed. We recommend using the R version that is included in the spread module package which is available from a permanent repository at the Royal Dutch Academy of Arts and Sciences (https://easy.dans.knaw.

nl/ui/datasets/id/easy-dataset:51346/ tab/2) in the folder called 'R installer'. Newer versions of the software available on the R website may not recognize all the functions used in the code provided in the package.

The R code necessary to apply the spread module should be retrieved from https:// easy.dans.knaw.nl/ui/datasets/id/easy-dataset:51346/tab/2 in the folder called 'Spread module code'. There are three versions of the module: (i) DD version-Europe; (ii) DD version-world; and (iii) metric version-Europe. Version 1 'DD version-Europe' is the original version of the spread module. It was developed for Europe and the spatial reference of GIS files (see subsection 'Input data' below) should be in decimal degrees. Version 2 'DD version-world' is an extended version for any part of the world and should also be in decimal degrees. For versions 1 and 2, the study area and grid resolution in the spread module are automatically defined by the grid extent and grid resolution used to generate the CLIMEX output files (i.e. EI and GI). Version 3 'metric version-Europe' is the decimal-degree version in a European metric system Lambert-Azimuthal-Equal-Area projection with a spatial resolution of 10 km × 10 km. This version was developed to combine the spread module maps with other risk maps from the European Union (EU)-funded PRATIQUE project (e.g. to determine the endangered areas at risk in Europe) since their risk maps were in this projection system and at this spatial resolution (Baker et al., 2012, Chapter 2 in this volume).

Before the assessor applies the spread module, she or he must create a new folder for each species containing all the files provided in one of the three versions of the module, including the R code. Some files given as examples (e.g. 'ClimexOutput', 'habitat', 'econ' or 'presence') should be replaced by the appropriate ones, depending on each species.

Input data

To apply the generic spread module, the assessor must have enough information about the suitable area where the invasive alien species can potentially persist. This area is defined by climatic constraints and habitat distribution. In the R code, the climatic constraints are the two indices, EI and GI, produced by a CLIMEX model of the invasive alien species applicable to the pest-risk-assessment area. The spatial resolution chosen for the CLIMEX grid will be the spatial resolution at which the spread module will be applied (i.e. it depends on the user's choice). The CLIMEX output is one of the main inputs of the generic spread module; the other inputs being a map of host or habitat distribution and the values for model parameters provided by the user. In essence, the CLIMEX output provides the module with information about the area of potential establishment (i.e. where EI > 0) and about the relative suitability for population growth across this area (i.e. according to the value of GI). All models need EI, while models C and D also need GI. The habitat distribution could be the host plant distribution (i.e. absence and presence of the host plant coded by 0 and 1, respectively) or the host plant density (i.e. presence will be considered where the density is above 0 in this case). Other factors, such as the area where soil is suitable for the plant pest species, can also be used for the habitat distribution (i.e. either as presence/absence of the favourable factor or as the density of favourable factor). Although the habitat distribution is not needed to apply the models, it contributes to a better estimate of the potential spread of the species. Moreover, it is possible to restrict the suitable area by a maximum elevation. For model A, in addition to the CLIMEX output and the habitat distribution, a GIS file in raster format is needed to account for the economic value of the affected plants over the pest-risk-assessment area. The R code can read a variety of raster formats (e.g. .tif, .asc and ESRI grid).

Generating CLIMEX output as input for the models (needed for all models)

In CLIMEX (version 3), use the function 'Compare Locations (1 species)' and the simulation file 'Grid Data' when opening the CLIMEX model for your study species. The grid used in the generic spread module for versions 1 and 2 will be the grid used in CLIMEX. In MetManager, select the study area where the potential spread should be mapped. Do not select 'World' because the grid will be too large for the spread module application. Then click 'Run'. Once the model run is complete, click 'Save to file', select 'New', select the model variables: 'Latitude, Longitude, EI, and GI' (comma-delimited, this order is very important), and 'Save'. Then find the exported file, often in the 'models' folder of CLIMEX, copy this .csv file and paste it in the spread module folder of your study species. This file should be renamed 'ClimexOutput.csv' (writing capital or lowercase letters is very important for R). It is also possible to have this file in text format (.txt) but in this case the header lines should be removed.

Habitat distribution (not compulsory)

If the user chooses to account for habitat distribution, a GIS file in raster format with presence/absence (1/0) of the habitat or the habitat density should be provided. In the latter case, the habitat distribution is defined by the area where the host plant density is above 0. This file can be retrieved from various sources. For instance, the file used for *D. virgifera virgifera* (i.e. distribution of grain and forage maize) was retrieved from McGill University (Monfreda *et al.*, 2008; see Supplementary Material 1 in Robinet *et al.*, 2012). The spatial reference system of this GIS file should be consistent with the version of the module. If the decimal-degree version of the module is used, the habitat distribution should be given in geographic coordinates according to the WGS system of 1984. Then it will be automatically projected on the CLIMEX grid by the spread module code. If the metric version of the model is used (i.e. spatial extent of *x*-axis = 2500–7500 km and *y*-axis = 1000–5500 km), the habitat distribution should also be projected on a metric grid of 10 km × 10 km spatial resolution.

Economic data (for model A only)

To account for the economic value of the host, the assessor should provide a GIS file in raster format. This file can be retrieved from various sources. For instance, the file used for *D. virgifera virgifera* (i.e. value of grain and forage maize) was retrieved from McGill University (Monfreda *et al.*, 2008). The spatial reference of the GIS file with economic data used by model A should be the same as described above for the habitat distribution.

Analyses for the Case Study of Western Corn Rootworm

The western corn rootworm is an economically important pest of maize in the USA and Canada (Krysan and Miller, 1986; EPPO, 2004). This species, supposedly native to Mexico, was first detected in Europe in 1992 in Belgrade, Serbia (Baca, 1994; EPPO, 1994) and it has since spread rapidly throughout central and south-eastern Europe (Edwards, 2011), causing considerable economic impact. Since the growth and dispersal of this insect are well documented, this species is used to illustrate the application of the generic spread module.

This section describes: (i) how to apply the four models of the generic spread module to a particular pest species, i.e. the western corn rootworm; (ii) how the parameters are estimated; (iii) how the R commands are typed; and (iv) how to interpret results. The example presented here was created with the 'DD version-Europe'. No elevation limit was considered for this case study. The CLIMEX model developed by Kriticos *et al.* (2012) was used to define the climate constraints in combination with 1961–1990 mean monthly climate data interpolated over Europe at 0.5° × 0.5° spatial resolution by the Climate Research Unit of the University of East Anglia (Norwich, UK; Mitchell and Jones, 2005). The suitable area was defined by the area where EI > 0 and maize was present. Both grain maize and forage maize were considered for the habitat distribution of *D. virgifera virgifera* and for the economic data on host value, required by model A. Initial time (i.e. $t = 0$) was set to the year the plant pest was first detected in Europe (i.e. 1992). The entry point needed in models B and D was Belgrade (N 44°82'; E 20°30'; EPPO, 1994). An example of the R commands that are used to set up and apply the models is provided in Step 1 of the online supplement S2 for Chapter 8.

Model A: logistic increase of invaded area combined with the economic value of assets

Parameter estimation

The relative rate of spatial increase, r, is estimated based on the increase of the number of cells colonized over time. If the number of cells initially infested is set to $N_0 = 1$ and the number of infested cells (N_t) is known t years later, then the following equation derived from the logistic growth function can be solved to estimate r:

$$r = \ln\left(\frac{N_t}{N_{max} - N_t}\right) - \ln\left(\frac{N_0}{N_{max} - N_0}\right)$$

(8.4)

with N_{max} the number of suitable cells. In the case of *D. virgifera virgifera*, according to the distribution map (Edwards, 2011), the species was present over 358 cells in 2010 (i.e. at $t = 18$ years since $t = 0$ in 1992) and $N_{max} = 3104$ cells. The value for N_{max} is directly calculated and included in the output as 'number of cells in the area of potential establishment' provided by the command 'printinfo()'. Therefore, for this case study, $r = 0.33$/year.

To apply the model, the assessor must provide an initial value for the variable n_t which provides the percentage of cells within the area of potential establishment that have been invaded at time t. In our example, $n_0 = 100 \times N_0/N_{max}$, then $n_0 = 0.03\%$. An example of the command lines to run model A and obtain output can be found in the online supplement S2 to Chapter 8.

Description of the output

Several results are calculated by the function call in R, including: number of cells considered in the model, number of cells within the suitable area, list of economic values used in the model, number of invaded cells at time t, percentage of invaded cells within the suitable area, sum of the economic value over all the invaded cells following the worst case scenario, sum of the economic value over all the invaded cells following the best case scenario and a summary (quantiles) of the sum of economic value over all the invaded cells across the replicate simulations of the random case scenario (see Table S8-2.2 in the online supplement S2 to Chapter 8). In addition, a six-panel window shows spread maps (Fig. 8.1). Three maps represent the spread of the invasive pest within the area of potential establishment under different scenarios (i.e. best case, worst case and random case, depending on the economic value of the host plant). Three other panels show the number of invaded cells according to their economic values. For the random case, this number of invaded cells is accumulated over the number of replicate simulations.

Model B: radial range expansion

Parameter estimation

For this model, only the spread rate (km/year), c, needs to be estimated. Recorded spread rates for *D. virgifera virgifera* vary from 60 to 100 km/year in Europe (Baufeld and Enzian, 2005). Like MacLeod *et al.* (2005), we consider in this example a typical spread rate of 80 km/year.

Other values that should be entered

To apply the model, the assessor must give the coordinates of the first entry point in decimal degrees. We, therefore, enter the coordinates of Belgrade (N 44°82', E 20°30'). An example of the R commands to run the model is provided in Step 3 in the online supplement S2 to Chapter 8.

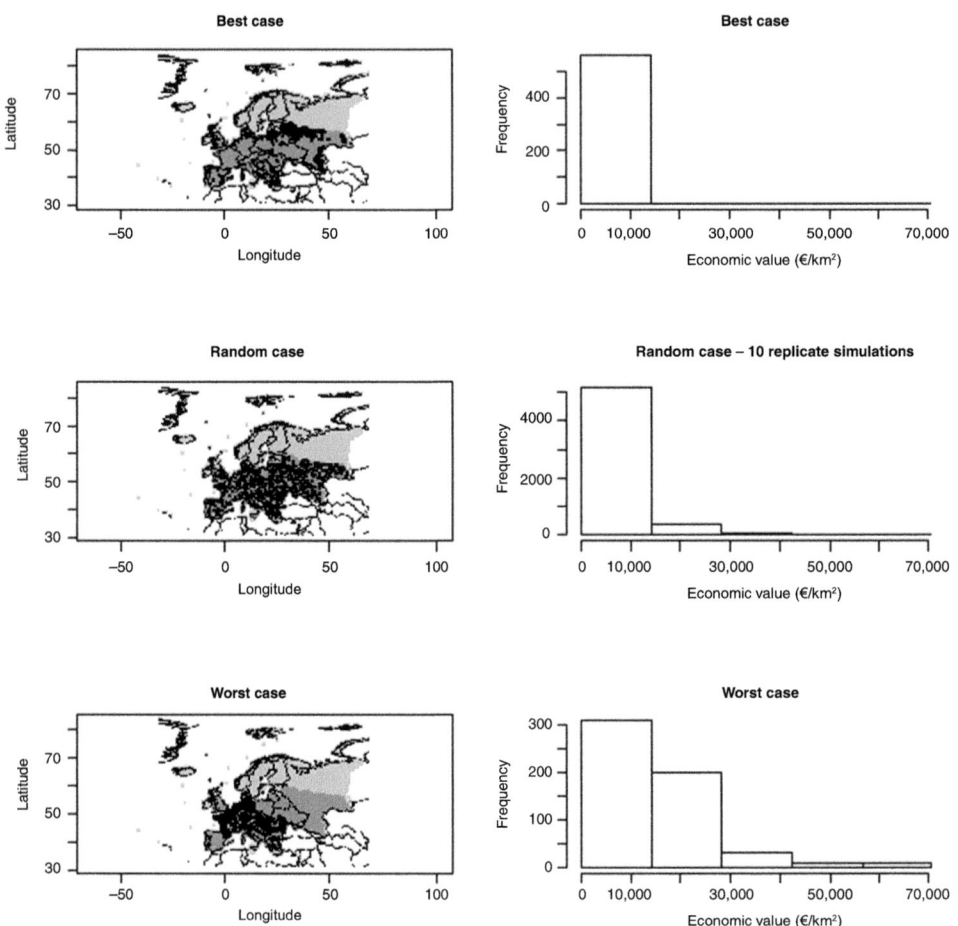

Fig. 8.1. Output of model A. Potential spread of the western corn rootworm at time $t = 20$ (year 2012). On the left side, the potential spread is represented on a map. Black indicates the colonized area; dark grey indicates the suitable area but not colonized; light grey indicates not suitable area; and white means that data are missing (outside the study area). On the right side, a histogram summarizes the economic values of invaded cells. Results are shown for the best case scenario, the random case scenario and the worst case scenario.

Description of the output

Several results are calculated when the function 'radial' is executed, including: number of cells considered in the model, number of cells within the suitable area, number of invaded cells at time t, percentage of invaded cells within the suitable area and coordinates in decimal degrees of the centre of the invaded cells (see Table S8-2.4 in the online supplement S2 to Chapter 8). In addition, the simulated spread map is shown (Fig. 8.2).

Model C: population growth

Parameter estimation

Both the maximum yearly multiplication factor (λ_{max}) and the carrying capacity (P_{max}) should be estimated by the assessor. The maximum yearly multiplication factor is given as the value of the multiplication factor where GI is maximal over the study area; in other words, where the species is thought to grow most rapidly. This maximum value is around λ_{max} = 40 over Europe

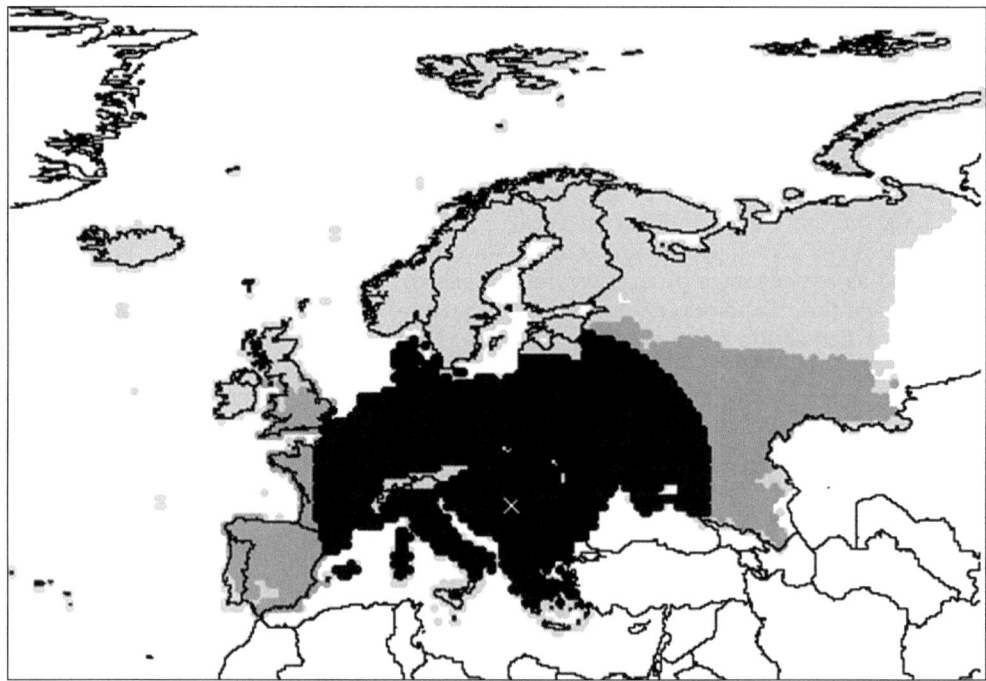

Fig. 8.2. Output of model B. Potential spread of the western corn rootworm at time t = 20 (year 2012). Black indicates the colonized area; dark grey indicates the suitable area but not colonized; light grey indicates not suitable area; and white means that data are missing (outside the study area). The white cross indicates the entry point (Belgrade).

(Hemerik et al., 2004). The maximum population density (i.e. the carrying capacity) of *D. virgifera virgifera* is about 200 individuals/m^2 of maize (P. Baufeld, Julius Kuhn Institute, Germany and Z. Dancsházy, Plant Protection and Soil Conservation, Hungary, 2010, personal communication). To estimate the carrying capacity over one cell, the mean area covered by maize should be approximated. We assume, for instance, that 80% of the landscape is covered with arable land and a four-year rotation for maize, and thus cells where maize is present contain about 20% of maize. If the proportion of habitat is known in each cell, it is possible to have different values for the carrying capacities. Although this estimate of maize area is uncertain, it is supposed to represent the landscape and it should provide a reference value on which to assess the population density in terms of percentage of the carrying capacity. Then, we can calculate the carrying capacity by multiplying the maximum population density per unit area of maize (i.e. 200 × 10^6 individuals/km^2) by the proportion of the cell containing maize (i.e. 0.2) and by the cell area (i.e. approximately 1578 km^2). (Note: cell area varies with latitude as the CLIMEX grid is given in decimal degrees.) The average cell area over the map is calculated by the code 'printinfo()' providing P_{max} = 6.3 × 10^{10} individuals per cell. The population density is then calculated as a percentage of this carrying capacity P_{max}.

To apply the model, the assessor must provide an initial value for the population density, P_0. In this model, we assume that the species is present over the entire suitable area but at very low density. This density is typically found after an accidental introduction. We assume that 100 *D. virgifera virgifera* individuals are present initially in each suitable cell (i.e. P_0 = 100). Since $p_0 = 100 \times P_0 / P_{max}$, we set p_0 = 1.6 × 10^{-7}%. An example of the R commands to run the model is provided in Step 4 in the online supplement S2 to Chapter 8.

Description of the output

Several values are calculated by the output function, including: number of cells considered in the model, number of cells within the suitable area, population density (expressed as a percentage of the carrying capacity) over the grid points and number of cells with a population density (pd) = 0%, >0–25%, >25–50%, >50–75% or >75–100% of carrying capacity (see Table S8-2.6 in the online supplement S2 to Chapter 8). In addition, the spread map is shown (Fig. 8.3; see colour plate section) and eventually the series of spread maps from $t = 1$ to the given time horizon. The output map shows the increase of the population density, assuming that the invasive pest was initially present over all the suitable cells and that it could grow at various rates depending on local climate conditions. This output should not be interpreted as the potential spatial spread over the pest-risk-assessment area but instead as a spatially explicit indicator of the area where the invasive pest could reach outbreak levels at different times in the future, given initial presence.

Model D: dispersal kernel

Parameter estimation

Four parameters should be estimated by the assessor for this model: two parameters associated with the population growth (λ_{max} and P_{max} are already estimated for model C; see subsection 'Model C: population growth' above) and two parameters associated with the population dispersal (u and v). The scale parameter u (km) determines the average spread distance whereas the shape parameter v determines the frequency of long-distance dispersal (see Robinet et al., 2012 for details). Consider that the spread rate c gives an estimate of the scale parameter, and for the example of *D. virgifera virgifera*, u = 80 km. However, no data can be used to estimate v directly. As a result, we recommend testing several values and selecting the one that provides the spread pattern that is most consistent with current knowledge on the organism of interest. Here we consider a relatively low value, $v = 5$, because the frequency of long-distance dispersers among western corn rootworm populations is presumably high.

To apply the model, the assessor must provide an initial value for the population density, p_0, at each entry point. This value is the same as the value estimated in model C but here it should be given in a separate file called 'presence.txt' file. The first column is the longitude and second column the latitude in decimal degrees, and the third column is the population density, p_0. An example of the R commands to run the model is provided in Step 5 in the online supplement S2 to Chapter 8.

Description of the output

Several values are returned by the output function, including: number of cells considered in the model, number of cells within the suitable area, number of invaded cells at time t, percentage of invaded cells within the suitable area, population density (expressed as a percentage of the carrying capacity) over the grid points and a summary of points from 'presence.txt' or a random selection (see Table S8-2.8 in the online supplement S2 to Chapter 8). In addition, the simulated spread map is shown (Fig. 8.4; see colour plate section). Either the population density (Fig. 8.4a) or the population above a given density threshold (Fig. 8.4b) can be requested. The potential spread of the invasive pest resulting from this model depends on local climate conditions, the population growth and the dispersal capabilities over the suitable cells. The overall result is, therefore, a radial range expansion whose rate is modulated by climate suitability and host presence in the area of potential establishment.

Discussion

The generic spread module package provides a new tool to explore relatively easily the potential spread of plant pest species based on a limited number of biological parameters and integration of simple population dynamics models with a tool for assessing potential pest distributions, CLIMEX, that is widely used in the pest-risk-assessment community. Using the spread module requires limited mathematical expertise. The outputs of the four models are not considered to be precise predictions but scenarios that can help analysts assess how geographic distribution and impact of an invasive pest population can change over time and space. Uncertainty in parameter estimation can be approximated by inserting test values for the parameters and drawing multiple maps going from the best case to the worst case. The outputs of the models can provide useful quantitative evidence for pest risk analysis in support of expert assessments on the endangered area (Baker et al., 2012, Chapter 2 in this volume). Pest risk analysts who tested this module were generally positive about its use, but indicated that further development is desirable to improve ease of use and to clarify how biological parameters should be estimated from data, especially those for the dispersal kernel in model D (Robinet et al., 2012). Use of model A would be hampered by lack of spatially explicit economic data.

This version of the spread module is designed to meet basic needs of pest risk assessors. It can be refined and adjusted to particular cases. For instance, this version requires outputs of CLIMEX as input, but many different niche models are available (Guisan and Zimmermann, 2000), and models other than CLIMEX could be used to generate input. The only requirement is that the bioclimatic model provides two indices: (i) an analogue of EI (i.e. an index equal to 0 where the species cannot establish and above 0 where it can survive); and (ii) an analogue of GI (i.e. an index ranging from 0 to any positive value that indicates the potential growth of the species during the suitable season). The results should be structured in a way that is analogous to the 'ClimexOutput' file resulting from CLIMEX. Alternatively, R code could be written to read a different type of input file.

Climate change will affect the long-term spread of pest species. In spread models, it is possible to consider different climate change scenarios and different climatic layers on which to apply spread. However, it is more reasonable to concentrate on short-term climate change scenarios (e.g. 20–30 years) because outcomes of spread models become more uncertain as the prediction horizon is longer (Pitt et al., 2011).

A number of models can be used to simulate and map potential spread of species. Answering some questions may help identify the most suitable modelling approach:

1. Who is going to apply the model(s)? A pest risk assessor will probably choose a generic model that does not require too much mathematical background whereas a modeller will probably use a more sophisticated model that can be adjusted to the particular case study and to a specific question.
2. In which context is the model applied? For a pest risk assessment, generic models that do not require too many data and information about the species such as those presented in this chapter will probably be chosen whereas more sophisticated and pest-specific models should be used for better understanding the potential spread of a well-documented species and to explore the effects of specific factors.
3. Is an estimation of the economic impact needed? Some models focus on potential spread only; others describe both spread and economic impact, and others describe more extensively the economic impact.

The generic spread module presented in this chapter was developed to be used by assessors who prepare pest risk assessments and concepts were, therefore, simplified and data demands reduced to a minimum. Another generic spread model was developed by Waage et al. (2005). Concepts in their modelling framework resemble those of model D in the module presented here, but to our knowledge this model has to date not been incorporated into pest-risk-assessment practice. Many models have been developed outside the context of pest risk assessment and can be used by modellers, such as Modular Dispersal in GIS (MDiG; Pitt, 2008; Pitt et al., 2009, 2011) or some bioeconomic models (Carrasco, 2009; Carrasco et al., 2010). These models usually contain more detailed descriptions of life cycle and biology and require, therefore, greater effort and more time to parameterize than generic models. Some pest spread models have been developed specifically for one species at the country or continental scale to more precisely account for the characteristics of that species' growth, dispersal or economic impact, such as for: the horse chestnut leafminer, *Cameraria ohridella* (Gilbert et al., 2005); the pine wood nematode, *Bursaphelenchus xylophilus*, and pine wilt disease (Togashi and Shigesada, 2006; Robinet et al., 2009, 2011; Soliman et al., 2012); the emerald ash borer, *Agrilus planipennis* (Muirhead et al., 2006); the western corn rootworm, *D. virgifera virgifera* (Carrasco et al., 2012); and the pine processionary moth, *Thaumetopoea pityocampa* (Kriticos et al., 2013; Robinet et al., 2013). The development of these models is very useful for the specific instances; however, they usually require substantial modelling expertise and more time to parameterize as it is sometimes necessary to design and implement a set of experiments to get requisite data. When little information is available about the growth and dispersal of a species, time is limited or standardized tools are preferred (e.g. in the context of pest risk assessment), sophisticated pest-specific models generally cannot be readily developed and generic spread models such as those presented in this chapter appear to be suitable.

Acknowledgments

We gratefully acknowledge support for this work from the EU project PRATIQUE KBBE-2007-212459 (7th Framework Project, 'Enhancements of pest risk analysis techniques') and we also thank the book editor, Rob Venette, for helpful suggestions to clarify the text of the chapter. W.v.d.W. gratefully acknowledges support from the OECD Co-operative Research Programme on Biological Resource Management for Sustainable Agricultural Systems to present this work at the conference 'Advancing risk assessment models for invasive alien species in the food chain: contending with climate change, economics, and uncertainty' from 23 to 26 July 2012 in Tromsø, Norway.

This chapter is enhanced with supplementary resources.

To access the material please visit:

www.cabi.org/openresources/43946/

References

Baca, F. (1994) New member of the harmful entomofauna of Yugoslavia *Diabrotica virgifera virgifera* LeConte (Coleoptera, Chrysomelidae). *Zaštita Bilja* 45, 125–131.

Baker, R., Benninga, J., Bremmer, J., Brunel, S., Dupin, M., Eyre, D., Ilieva, Z., Jarošík, V., Kehlenbeck, H., Kriticos, D.J., Makowski, D., Pergl, J., Reynaud, P., Robinet, C., Soliman, T., van der Werf, W. and Worner, S. (2012) A decision support scheme for mapping endangered areas in pest risk analysis. *EPPO Bulletin* 42, 65–73.

Baufeld, P. and Enzian, S. (2005) Western corn rootworm (*Diabrotica virgifera virgifera*), its potential spread and economic and ecological consequences in Germany. In: Alford, D.V. and Backhaus, G.F. (eds) *Plant Protection and Plant Health in Europe: Introduction and Spread of Invasive Species*. British Crop Protection Council, Alton, UK, pp. 149–154.

Carrasco, L.R. (2009) Modelling for pest risk analysis: spread and economic impact. PhD thesis, Imperial College London, UK.

Carrasco, L.R., Mumford, J.D., MacLeod, A., Knight, J.D. and Baker, R.H.A. (2010) Comprehensive bioeconomic modelling of multiple harmful non-indigenous species. *Ecological Economics* 69, 1303–1312.

Carrasco, L.R., Cook, D., Baker, R., MacLeod, A., Knight, J.D. and Mumford, J.D. (2012) Towards the integration of spread and economic impacts of biological invasions in a landscape of learning and imitating agents. *Ecological Economics* 76, 95–103.

Clark, J.S., Silman, M., Kern, R., Macklin, E. and Hille Ris Lambers, J. (1999) Seed dispersal near and far: patterns across temperate and tropical forests. *Ecology* 80, 1475–1494.

Edwards, C.R. (2011) *Western Corn Rootworm.* Purdue University, West Lafayette, Indiana. Available at: http://extension.entm.purdue.edu/wcr/ (accessed 17 September 2012).

EPPO (1994) *NEW PEST/DIABVI: First record of Diabrotica virgifera in Europe. EPPO Reporting Service 94/001.* Available at: http://archives.eppo.int/EPPOReporting/1994/Rse-9401.pdf (accessed 26 December 2013).

EPPO (2004) Diagnostic protocols for regulated pests: *Diabrotica virgifera. EPPO Bulletin* 34, 289–293.

Fisher, R.A. (1937) The wave of advance of advantageous genes. *Annals of Eugenics* 7, 255–369.

Gilbert, M., Guichard, S., Freise, J., Grégoire, J.-C., Heitland, W., Straw, N., Tilbury, C. and Augustin, S. (2005) Forecasting *Cameraria ohridella* invasion dynamics in recently invaded countries: from validation to prediction. *Journal of Applied Ecology* 42, 805–813.

Guisan, A. and Zimmermann, N.E. (2000) Predictive habitat distribution models in ecology. *Ecological Modelling* 135, 147–186.

Hemerik, L., Busstra, C. and Mols, P. (2004) Predicting the temperature-dependent natural population expansion of the western corn rootworm, *Diabrotica virgifera. Entomologia Experimentalis et Applicata* 111, 59–69.

Kehlenbeck, H., Robinet, C., van der Werf, W., Kriticos, D., Reynaud, P. and Baker, R. (2012) Modelling and mapping spread in pest risk analysis: a generic approach. *EPPO Bulletin* 42, 74–80.

Kriticos, D.J., Reynaud, P., Baker, R.H.A. and Eyre, D. (2012) The potential global distribution of the western corn rootworm (*Diabrotica virgifera virgifera*). *EPPO Bulletin* 42, 56–64.

Kriticos, D.J., Leriche, A., Palmer, D.J., Cook, D.C., Brockerhoff, E.G., Stephens, A.E.A. and Watt, M.S. (2013) Linking climate suitability, spread rates and host-impact when estimating the potential costs of invasive pests. *PLoS One* 8, e54861.

Krysan, J.L. and Miller, T.A. (1986) *Methods for Study of Pest Diabrotica.* Springer, New York.

MacLeod, A., Baker, R.H.A. and Cannon, R.J.C. (2005) Costs and benefits of European Community (EC) measures against an invasive alien species: current and future impacts of *Diabrotica virgifera virgifera* in England & Wales. In: Alford, D. and Backhaus, G.F. (eds) *Plant Protection and Plant Health in Europe: Introduction and Spread of Invasive Species.* British Crop Protection Council, Alton, UK, pp. 167–172.

Mitchell, T.D. and Jones, P.D. (2005) An improved method of constructing a database of monthly climate observations and associated high-resolution grids. *International Journal of Climatology* 25, 693–712.

Monfreda, C., Ramankutty, N. and Foley, J.A. (2008) Farming the planet: 2. Geographic distribution of crop areas, yields, physiological types, and net primary production in the year 2000. *Global Biogeochemical Cycles* 22, GB1022. Available at: http://www.geog.mcgill.ca/landuse/pub/Data/175crops2000/ArcASCII-Zip/ (accessed 30 January 2013).

Muirhead, J.R., Leung, B., van Overdijk, C., Kelly, D.W., Nandakumar, K., Marchant, K.R. and MacIsaac, H.J. (2006) Modelling local and long-distance dispersal of invasive emerald ash borer *Agrilus planipennis* (Coleoptera) in North America. *Diversity and Distributions* 12, 71–79.

Pitt, J.P.W. (2008) Modelling the spread of invasive species across heterogeneous landscapes. PhD thesis, Lincoln University, Lincoln, New Zealand.

Pitt, J.P.W., Worner, S.P. and Suarez, A. (2009) Predicting Argentine ant spread over the heterogeneous landscape using a spatially-explicit stochastic model. *Ecological Applications* 19, 1176–1186.

Pitt, J.P.W., Kriticos, D.J. and Dodd, M.B. (2011) Limits to simulating the future spread of invasive species: *Buddleja davidii* in New Zealand. *Ecological Modelling* 222, 1880–1887.

R Development Core Team (2010) *R: A Language and Environment for Statistical Computing.* R Foundation for Statistical Computing, Vienna.

Robinet, C., Roques, A., Pan, H., Fang, G., Ye, J., Zhang, Y. and Sun, J. (2009) Role of human-mediated dispersal in the spread of the pinewood nematode in China. *PLoS One* 4, e4646.

Robinet, C., Van Opstal, N., Baker, R. and Roques, A. (2011) Applying a spread model to identify the entry points from which the pine wood nematode, the vector of pine wilt disease, would spread most rapidly across Europe. *Biological Invasions* 13, 2981–2995.

Robinet, C., Kehlenbeck, H., Kriticos, D.J., Baker, R.H.A., Battisti, A., Brunel, S., Dupin, M., Eyre, D., Faccoli, M., Ilieva, Z., Kenis, M., Knight, J., Reynaud, P., Yart, A. and van der Werf, W. (2012) A suite of models to support quantitative assessment of spread in pest risk analysis. *PLoS One* 7, e43366.

Robinet, C., Rousselet, J. and Roques, A. (2013) Potential spread of pine processionary moth in France: preliminary results from a simulation model and future challenges. *Annals of Forest Sciences* 71, 149–160.

Skellam, J.G. (1951) Random dispersal in theoretical populations. *Biometrika* 38, 196–218.

Soliman, T., Mourits, M.C.M., van der Werf, W., Hengeveld, G.M., Robinet, C. and Oude Lansink, A.G.J.M. (2012) Framework for modelling economic impacts of invasive species, applied to pine wood nematode in Europe. *PLoS One* 7, e45505.

Sutherst, R.W. and Maywald, G.F. (1985) A computerised system for matching climates in ecology. *Agriculture, Ecosystems & Environment* 13, 281–99.

Sutherst, R.W., Maywald, G.F. and Kriticos, D.J. (2007) *CLIMEX version 3 User's Guide*. Hearne Scientific Software Pty Ltd, Melbourne, Australia. Available at http://www.hearne.com.au/getattachment/0343c9d5-999f-4880-b9b2-1c3eea908f08/Climex-User-Guide.aspx (accessed 23 June 2013).

Togashi, K. and Shigesada, N. (2006) Spread of the pinewood nematode vectored by the Japanese pine sawyer: modeling and analytical approaches. *Population Ecology* 48, 271–283.

Waage, J.K., Fraser, R.W., Mumford, J.D., Cook, D.C. and Wilby, A. (2005) *A New Agenda for Biosecurity*. Department for Environment, Food, and Rural Affairs, London.

9 Estimating Spread Rates of Non-native Species: The Gypsy Moth as a Case Study

Patrick C. Tobin,[1]* Andrew M. Liebhold,[2] E. Anderson Roberts[3] and Laura M. Blackburn[2]

[1]*University of Washington, Seattle, Washington, USA;* [2]*USDA Forest Service, Northern Research Station, Morgantown, West Virginia, USA;* [3]*Department of Entomology, Virginia Polytechnic Institute and State University, Blacksburg, Virginia, USA*

Abstract

Estimating rates of spread and generating projections of future range expansion for invasive alien species is a key process in the development of management guidelines and policy. Critical needs to estimate spread rates include the availability of surveys to characterize the spatial distribution of an invading species and the application of analytical methods to interpret survey data. In this chapter, we demonstrate the use of three methods, (i) square-root area regression, (ii) distance regression and (iii) boundary displacement, to estimate the rate of spread in the gypsy moth, *Lymantria dispar*, in the USA. The gypsy moth is a non-native species currently invading North America. An extensive amount of spatial and temporal distributional data exists for this invader. Consequently, it provides an ideal case study to demonstrate the use of methods to estimate spread rates. We rely on two sources of data: (i) polygonal data obtained from county quarantine records describing the geographical extent of gypsy moth establishment; and (ii) point data consisting of counts of male gypsy moths captured in pheromone-baited traps used to detect and monitor newly established gypsy moth populations. Both data sources were compiled during the gypsy moth's invasion of the Lower Peninsula of Michigan, USA. We show that even with spatially crude county records of infestation, spread rates can still be estimated using relatively simple mathematical approaches. We also demonstrate how the boundary displacement method can be used to characterize the spatial and temporal dynamics of spread.

The Importance of Spread Rates and Patterns

Spread of a non-native species is the process by which an organism expands its range from geographical areas it currently occupies into ones it does not. The rate of spread is most often expressed as the rate of change in the distributional range per unit of time, and can vary considerably among species (Elton, 1958; Shigesada and Kawasaki, 1997; Liebhold and Tobin, 2008) and across spatial and temporal scales within a species (Tobin *et al.*, 2007c). In nearly all biological invasions, spread results from the coupling of local dispersal with population growth (Fisher, 1937; Skellam, 1951). However, in most cases, the spread of an invading alien species includes long-distance 'jumps' in

* Corresponding author. E-mail: pctobin@uw.edu

which new colonies arise far from the established range. The combined process of short- and long-range dispersal is referred to as stratified dispersal (Shigesada et al., 1995). Under stratified dispersal, colonies that successfully establish ahead of the expanding range can grow and eventually coalesce with the established area, greatly increasing the rate of spread over what would be expected in the absence of long-distance jumps (Hengeveld, 1989; Shigesada and Kawasaki, 1997). The ramifications of stratified dispersal have been documented for several non-native species, including the Africanized honeybee (*Apis mellifera scutellata*; Winston, 1992), Argentine ant (*Linepithema humile*; Suarez et al., 2001), emerald ash borer (*Agrilus planipennis*; Muirhead et al., 2006), horse-chestnut leaf miner (*Cameraria ohridella*; Gilbert et al., 2004) and the gypsy moth (*Lymantria dispar*; Liebhold et al., 1992).

Several methods exist to estimate the rate of spread of an invading species (Andow et al., 1990; Sharov et al., 1997; Tobin et al., 2007b; Gilbert and Liebhold, 2010) and the ability to estimate spread rates can be a crucial step in the development of pest risk maps and management strategies. For example, before an invader spreads into a new area, several information needs must be addressed. These include determining susceptible habitats that are most vulnerable to invasion, estimating the time before a new invader spreads to these susceptible areas and predicting the eventual economic and ecological impacts. Spread rate estimates can be used to project future range boundaries and, in some cases, allow management tactics to mitigate expected impacts prior to arrival (Waring and O'Hara, 2005). Thus, it is not surprising that much past work has focused on estimating rates of spread of invading species, including very early studies that were published before the widespread recognition of the importance of biological invasions (Cooke, 1928; Elton, 1958). In this chapter, we describe three analytical methods that can be used to estimate the rate of spread of invading species using data on the spread of the gypsy moth.

Context for a Case Study: Gypsy Moth

One of the more widely studied and documented biological invasions is that of the gypsy moth in the USA. Despite the fact that many non-native pest species are currently invading the USA (Pimentel et al., 2000; Aukema et al., 2010), the gypsy moth is somewhat unique among invaders in that we know when it was introduced (1869), approximately from where it originated (France or Germany), where it was introduced (27 Myrtle Street, Medford, Massachusetts, USA) and by whom (Etienne Léopold Trouvelot; Riley and Vasey, 1870; Forbush and Fernald, 1896; Liebhold et al., 1989). The gypsy moth is univoltine and its larvae are polyphagous folivores that can feed on over 300 host plants including the preferred genera of *Betula* (birch), *Crataegus* (hawthorn), *Larix* (larch), *Populus* (aspen), *Quercus* (oak), *Salix* (willow) and *Tilia* (basswood) (Elkinton and Liebhold, 1990; Liebhold et al., 1995).

Larvae hatch from overwintering egg masses in spring and undergo five (male) or six (female) instars over approximately 8 weeks. The pupal period is approximately 2 weeks, followed by adult emergence. Females of the European strain, which is the strain established in North America, are not capable of sustained flight (Keena et al., 2008). Males locate calling females through a sex pheromone and mate; females oviposit a single egg mass containing 250–500 eggs. Although gypsy moth populations are innocuous and barely noticed in most years, populations can periodically erupt in spatially widespread outbreaks that occur over 2–3 years (Haynes et al., 2009). Ramifications of gypsy moth outbreaks include host tree mortality, loss of ecosystem services, detrimental effects to native species and public nuisance (Gansner and Herrick, 1984; Leuschner et al., 1996; Redman and Scriber, 2000). Since 1924, over 360,000 km^2 of forests in the USA have been defoliated by the gypsy moth (USDA Forest Service, 2013).

Since its introduction in 1869 in Medford, Massachusetts, the gypsy moth has slowly expanded its range in North America such that it now occupies a range from Nova Scotia to Wisconsin, and Ontario to Virginia (Tobin et al., 2007b). Spread can be facilitated by larval ballooning and adult male flight, both of which are considered to occur over short distances (Mason and McManus, 1981; Elkinton and Liebhold, 1990). Longer-distance dispersal is believed to occur primarily through the anthropogenic movement of life stages (Lippitt et al., 2008; Hajek and Tobin, 2009; Bigsby et al., 2011). Despite the fact the gypsy moth has been established in North America for over 140 years and currently occupies >900,000 km^2, almost three-quarters of forested areas considered to be susceptible to gypsy moth outbreaks remain uninfested (Morin et al., 2005). Thus, efforts to estimate the rate of gypsy moth spread remain of critical importance. Moreover, this species provides an ideal example for demonstrating methods of estimating invasion spread due to the extensive amount of spatial and temporal data collected on it. In this chapter, we estimate gypsy moth spread using both point data and polygonal data.

Resources to Estimate Spread Rates

There are two broad data types that can be used to estimate the rate of spread in an invading species: (i) point data; and (ii) polygonal data. Point data can include the number of individuals collected from sampling devices, such as traps baited with semiochemical attractants and placed at a specific point in space. Point data can also include a record of the observed presence of the species. In addition to such records historically collected by regulatory officials or the scientific community, citizen scientists have contributed, more recently, to observation records (Ingwell and Preisser, 2011). In fact, in New Zealand, approximately half of new plant pest detections are first reported by the general public (Froud et al., 2008). It is also believed that every known Asian longhorned beetle (*Anoplophora glabripennis*) infestation in the USA was first discovered by a citizen. In some cases, the species need not be observed directly but rather its presence implied by a specific indication of damage. For example, in the case of wood and subcortical phloem feeders, the architecture of larval tunnelling and associated symbionts are often species-specific (Paine et al., 1997), which can reveal the presence of a specific invasive alien pest even if the pest is absent; in some cases, feeding injury can reveal when the invasive alien species was first present when analysed through dendroecological techniques (Siegert et al., 2010).

In contrast to point data where a species is considered to be present at a specific point in space, polygonal data encompass an area considered to be infested by a non-native species. Among the more common examples of polygonal data are those that are defined by geopolitical boundaries, such as county, state or territory boundaries. In many countries, species that are regulated under domestic quarantines have their range boundaries defined by polygonal data. In most cases, point data are essentially used in the construction of polygonal data with the assumption that a species detected through, for example, a trapping device at a specific point in space is, in actuality, distributed over a larger area. Geopolitical boundaries are often used to define this larger area because they facilitate regulatory responses, such as the restriction of potentially infested material from being transported from an infested county or state without proper phytosanitary measures.

Many data resources and repositories exist that contain point and polygonal data on the presence of a non-native species. Many governments maintain quarantines against established non-native species to limit their movement to uninfested areas and publish these records in government documents. In the USA, for example, quarantine regulations are codified by the USA Code of Federal Regulations, Title 7, Chapter III, Part 301, which is divided into subparts by species, and includes non-native insects, plants, nematodes and pathogens (Table 9.1). Within each subpart is a section on 'generally infested areas' that lists states (in whole), or specific counties or townships,

Table 9.1. Non-native species in the USA currently included in subparts of the USA Code of Federal Regulations, Title 7, Chapter III, Part 301.

Name	Species	Subpart
Fruit flies	Several	301.32
Black stem rust	Puccinia graminis	301.38
Gypsy moth	Lymantria dispar	301.45
Japanese beetle	Popillia japonica	301.48
Pine shoot beetle	Tomicus piniperda	301.50
Asian longhorned beetle	Anoplophora glabripennis	301.51
Pink bollworm	Pectinophora gossypiella	301.52
Emerald ash borer	Agrilus planipennis	301.53
South America cactus moth	Cactoblastis cactorum	301.55
Plum pox	Potyvirus spp.	301.71
Citrus canker	Xanthomonas axonopodis pv. citri	301.75
Asian citrus psyllid	Diaphorina citri	301.76
Witchweed	Striga spp.	301.80
Imported fire ant	Solenopsis invicta and Solenopsis richteri	301.81
Golden nematode	Globodera rostochiensis	301.85
Pale cyst nematode	Globodera pallida	301.86
Sugarcane diseases	Xanthomonas albilineans	301.87
Karnal bunt	Tilletia indica	301.89
European larch canker	Lachnellula willkommi	301.91
Sudden oak death	Phytophthora ramorum	301.92

that are considered infested at the time of publication. By going through the various years in which USA Code of Federal Regulations has been published, beginning with its first publication in 1938, it is possible to generate a space–time series of the presence of a regulated non-native species (Fig. 9.1).

When using either point or polygonal data to estimate spread rates, the minimum required details are the spatial and temporal distribution of the data. In many cases, point or polygonal data are considered binary (i.e. presence only), but in some cases, estimates of density are available; regardless, binary and continuous measurements of the non-native species can be used to estimate rates of spread. Based upon where and when a species is detected, or when an area is considered to be infested, there are a number of quantitative methods available. Given these time–space data, all of the estimation methods can be accomplished through statistical packages such as SAS (SAS Institute, Inc., 1999) or R (R Core Team, 2013), and some methods can be accomplished using more basic software packages such as Microsoft® Excel.

In this chapter, we use both polygonal and point gypsy moth data collected from the Lower Peninsula of Michigan to demonstrate three methods in estimating spread rates. This region provides an ideal case study of gypsy moth spread for several reasons. First, the introduction of gypsy moth life stages in Michigan was spatially disjunct from the established area at the time. Although Michigan has a long history of management efforts against the gypsy moth (Hanna, 1982; Dreistadt, 1983), the first counties were declared to be infested and included in the USA Code of Federal Regulations in 1981; at this time, the closest infested areas were in western New York and Pennsylvania. Thus, the invasion dynamics of the gypsy moth in Michigan would be comparable to those expected in a new invasion. From 1981 to 1994, counties from the Lower Peninsula of Michigan were added to the regulated area, which allows for the construction of a time series based upon these polygonal data (Fig. 9.2a). Second, when Michigan initially became infested, standardized pheromone-baited traps, which are sensitive monitoring tools that are effective even at low population densities

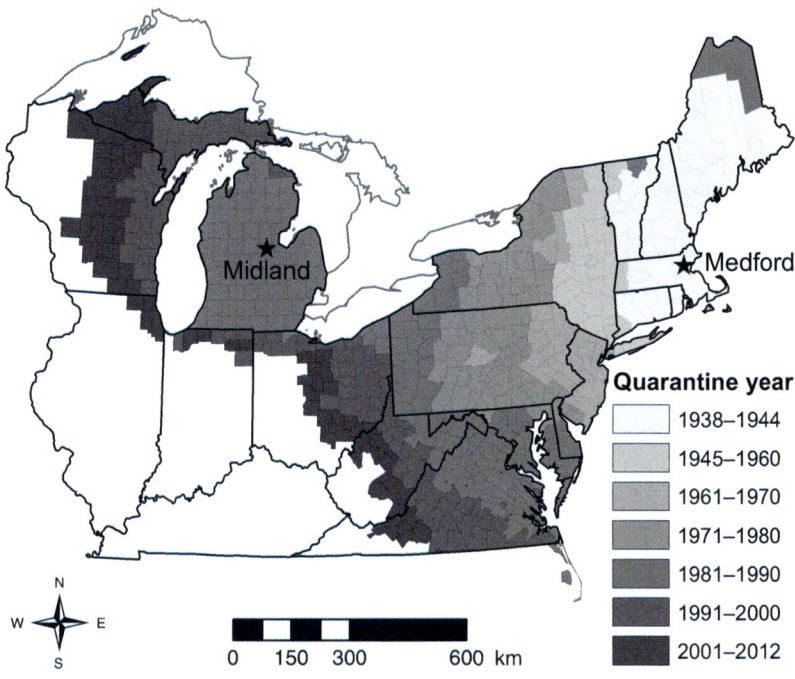

Fig. 9.1. Distribution of the gypsy moth in the USA, 1938–2012, based upon county quarantine records. Medford, Massachusetts and Midland, Michigan represent the initial and a subsequent site of introduction, respectively.

(Mastro et al., 1977; Elkinton and Childs, 1983; Thorpe et al., 1993), were available and allowed for the collection of point data (Tobin et al., 2012). From 1985 to 1996, the entire Lower Peninsula of Michigan was trapped each year, with traps set ~5 km apart (Fig. 9.2b; Gage et al., 1990; Yang et al., 1998). Last, the initial infestation was considered to be in Midland County, Michigan, which is centrally located in the Lower Peninsula of Michigan. Thus, the gypsy moth had the opportunity to spread radially from this centralized point.

Data Analysis

There are three analytical methods that can be used with both point and polygonal data to estimate spread rates: (i) square-root area regression; (ii) distance regression; and (iii) boundary displacement.

Square-root area regression

This method is based on the analysis of distance-to-time and uses successive measurements of the invaded area. For each year, the square root of the total infested area is considered according to:

$$\sqrt{\frac{\text{total infested area}}{\pi}} \qquad (9.1)$$

The values for each year are then regressed as a function of time to estimate the radial rate of spread, which is ascertained by the estimate of the regression slope (Shigesada and Kawasaki, 1997; Gilbert and Liebhold, 2010).

When applying the square-root area regression method to polygonal data from Michigan (Fig. 9.3a; see colour plate section), 1981 is considered as year 1 at which time six counties, encompassing an area of

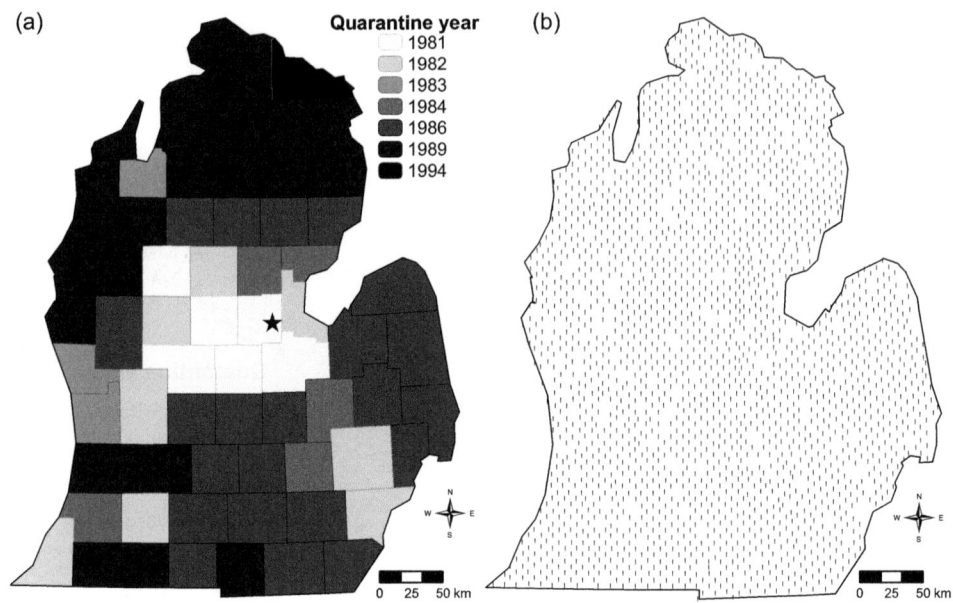

Fig. 9.2. (a) Distribution of the gypsy moth in the Lower Peninsula of Michigan, USA, based upon county quarantine records, 1981–1994; the star indicates Midland, Michigan, which is considered to be the site of the initial introduction into Michigan. (b) Spatial representation of the trapping grid used to record gypsy moth densities, 1985–1996.

9786 km^2, were regarded as infested with the gypsy moth. In 1982, eight additional counties were added to the infested area for a cumulative total infested area of 23,226 km^2. The last county was considered to be infested in 1994, bringing the total infested area to 106,887 km^2. However, by 1989, the infested area was 105,151 km^2 and no additional counties were added to the quarantine until 1994. Thus, when restricting the regression analysis from the initial year of introduction (1981) to the year at which the entire Lower Peninsula was nearly completely infested (1989), the annual rate of spread (as ascertained from the slope estimate) is 17.1 km/year (Fig. 9.4a). The standard error associated with the slope estimate from the linear regression provides an estimate of the variability associated with the spread rate, which in this case is 1.9.

The square-root area regression method can also be applied to the point trapping data from Michigan (Fig. 9.3b; see colour plate section). In this case, population thresholds can also be considered because trapping data provide a continuous measurement of density, as opposed to the presence/absence data that are generally available from polygonal data. We considered three population thresholds in this analysis: (i) an estimate of the area over which trapping records indicate gypsy moth presence (i.e. threshold = 1 moth); (ii) an estimate of the area where at least ten male moths are trapped; and (iii) an estimate of the area where at least 100 male moths are trapped. We chose these population thresholds arbitrarily to demonstrate the method. To estimate the area where populations exceeded these thresholds, we spatially interpolated the raw trapping data (latitude, longitude and male moths trapped at each trapping location) for each year to generate a continuous surface over a network of 1 km × 1 km cells using median indicator kriging (Isaaks and Srivastava, 1989; Deutsch and Journel, 1992). As with the respective slope estimates for each population threshold (Fig. 9.5a). When applied to point data, and again restricting the regression analysis to 1981–1989 (i.e. estimates during the invasion of Michigan), this method estimates spread rates (± SE) of 13.5 (2.3), 21.2 (1.2) and 23.8 (2.2) km/year for the 1-, 10- and 100-moth thresholds, respectively. It is also possible to estimate a composite rate of spread by averaging over the estimates from all population thresholds, which yields a spread rate estimate (± SE) of 19.5 (3.1) km/year.

Distance regression

This method is based on regressing the distance of an infested location, either from polygonal or point data, from a reference point on the year it first became infested (Liebhold et al., 1992; Tobin et al., 2007b; Gilbert and Liebhold, 2010). The reference point can be an arbitrary location, but ideally it should reflect the initial site of introduction or simply the location at which a species was first detected. For the Michigan data, we used the city of Midland, the county seat of Midland County, as a proxy for the initial gypsy moth infestation in Michigan. Using the polygonal data, we first estimated the minimum distance between each infested county and Midland (Fig. 9.3c; see colour plate section). The distance for each county was then regressed on the year it was first infested, and the estimate of the slope of the regression line provided the estimated radial rate (± SE) of spread, which is 9.6 (2.0) km/year (Fig. 9.4b).

When applying the distance regression method to the point trapping data, we again used multiple population thresholds, such as the 1-, 10- and 100-moth thresholds, for each of the years in which trapping data exist (1985–1996). In this case, the distance between Midland, Michigan, and each trapping location that captured at least one, ten and 100 male moths is estimated (Fig. 9.3d; see colour plate section) and then regressed for each year. This method, when applied to point data, estimates spread rates (± SE) of 15.7 (0.3), 19.3 (0.5) and

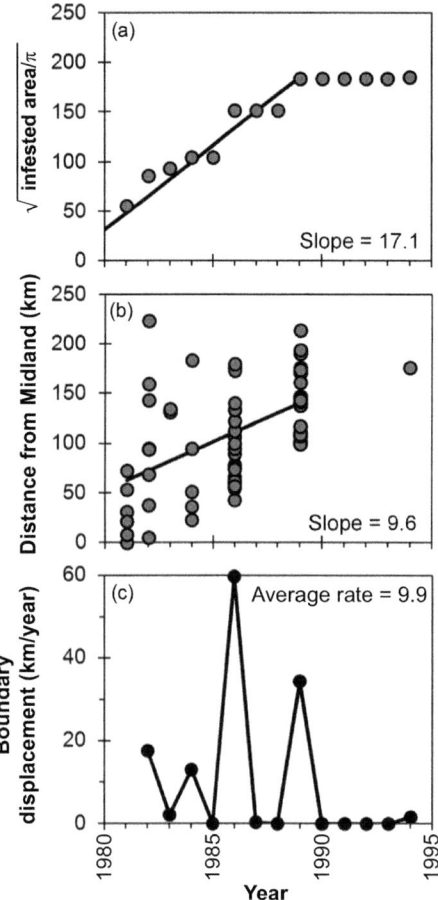

Fig. 9.4. Use of polygonal data to estimate gypsy moth rate of spread in the Lower Peninsula of Michigan using (a) the square-root area method; (b) the distance regression method; and (c) the boundary displacement method. Each circle in (a) represents the square root of the infested area for each year, while each circle in (b) represents the distance between each infested county (by year of infestation) and Midland, Michigan. The slope estimate from least-squares regression provides an estimate of the annual spread rate. The year-to-year boundary displacements in (c) can be averaged to estimate an annual rate of spread.

polygonal data, we then estimated the square root of each population threshold's area ÷ π for each year (Eqn 9.1) and next regressed this against the year. In this case, we also estimated an annual rate of spread for each population threshold based upon

22.8 (0.6) km/year for the 1-, 10-, 100-moth thresholds, respectively, with an overall average (± SE), across the estimates from all three population thresholds, of 19.3 (2.0) km/year (Fig. 9.5b).

Boundary displacement

This method considers the displacement distances between pairs of consecutive invasion boundaries to estimate rates of spread. Typically, displacement is measured along axes radiating from a reference point, which could be the origin of the invasion or a point that falls along a line that is perpendicular to the main invasion front (Sharov et al., 1995; Tobin et al., 2007b; Gilbert and Liebhold, 2010). The first step in this method is to delimit spatially invasion boundaries. One simple approach for constructing boundaries is to use one of a variety of software packages to generate contour lines. Contour lines can be constructed for each year of data, from which the year-to-year displacements in the spatial location of contour lines can be quantified and used as an estimate of spread.

In this chapter, we used several steps to estimate spread from boundary displacements. First, we used indicator kriging to generate a spatially continuous surface using both polygonal and point data (Isaaks and Srivastava, 1989; Deutsch and Journel, 1992). When using polygonal data from Michigan, we overlaid a grid consisting of a network of 2 km × 2 km cells across the state. For each year of polygonal data, we scored each cell by using the centre point of the cell as its spatial coordinates, as 1 or 0, where the former designation indicates that the centre of the cell was in an infested county while the latter indicates an absence of infestation. This resulted in a time series of spatially referenced binary point data based upon the polygonal data. We then used indicator kriging to generate a continuous surface from the spatially referenced binary point data (Fig. 9.3e; see colour plate section).

When using point data, we likewise used kriging (Isaaks and Srivastava, 1989; Deutsch and Journel, 1992) to interpolate a

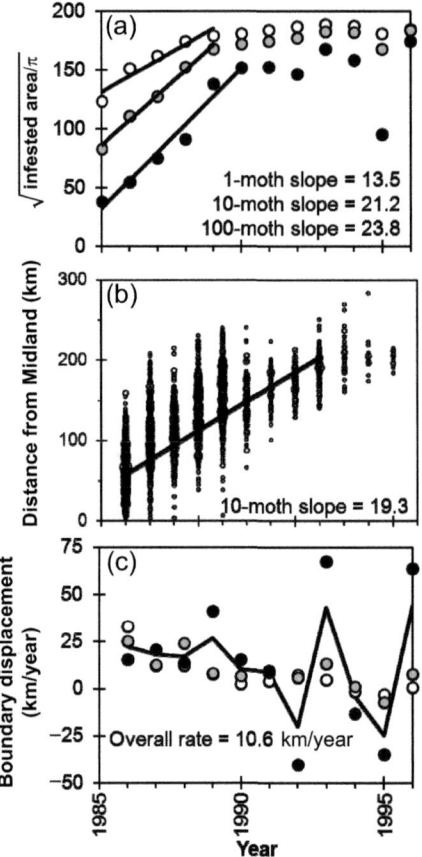

Fig. 9.5. Use of point data to estimate gypsy moth rate of spread in the Lower Peninsula of Michigan using (a) the square-root area method; (b) the distance regression method; and (c) the boundary displacement method. In (a), least-squares regression is fit against the linear portion of the relationship between the infested area and time when the infested area is based upon the 1-moth (open circles), 10-moth (grey circles) and 100-moth (solid circles) population thresholds. In (b), least-squares regression is fit against the linear portion of the relationship between the distance of the trap and Midland for each year when using the 10-moth threshold as an example; circles represent the distance between each trapping record by year and Midland, Michigan, and are proportional in size to the number of records. In (c), the year-to-year displacement for the 1-moth (open circles), 10-moth (grey circles) and 100-moth (solid circles) population thresholds is plotted over time. The solid line represents an average of the displacement across all three population thresholds at each year, which can be averaged across years and population threshold to estimate an overall average spread rate.

spatially continuous surface for each year of trapping data (Fig. 9.3f; see colour plate section). Because point trapping data are a continuous measurement of density, it is also possible, as we have done in the previous methods using point data, to estimate a boundary that reflects the 1-, 10- and 100-moth threshold. Thus, for example, the 10-moth threshold boundary would delineate an area in which traps recorded ≥10 moths within the boundary and <10 moths outside the boundary, much like a presence boundary delineates an area in which a species is present or absent.

We then applied an optimization approach to delimit the location of boundaries from the spatially interpolated surfaces generated from each data source (Sharov et al., 1995). Population boundaries derived from spatially interpolated maps are often irregular, with 'islands', 'lakes' and 'folds' common within and outside the invading species' established area. Because irregular boundaries can be difficult to analyse, we used this optimization approach to construct boundaries that are more regular (Sharov et al., 1995). This method connects populations of similar densities, such as presence or absence, to minimize the inclusion of populations within a boundary that do not satisfy a specific density, while also minimizing the exclusion of populations within a boundary that do.

The final step is to estimate the displacement of boundaries from year to year. To accomplish this step, we measured the distance from a fixed focal point in space to boundaries in consecutive years; in this case, we used transects radiating from the focal point at 0.5° intervals. The year-to-year displacement (i.e. from 1981 to 1982) at each transect can then be measured and averaged to obtain a spread rate for each pair of successive years, which then in turn can be averaged to estimate an overall annual rate of spread across all years. When we used polygonal data from Michigan, annual rates of spread ranged from 0 to 59.7 km/year, while the overall average (± SE) rate of spread (1981–1994) was 9.9 (5.0) km/year (Fig. 9.4c; Table 9.2). When we used point data, estimates of the annual spread rates ranged from –3.1 to 33.2, –7.0 to 25.2 and –40.0 to 67.7 km/year for the 1-, 10- and 100-moth thresholds, respectively, while the overall (1985–1996) average rate of spread (± SE) was 10.6 (3.7) km/year (Fig. 9.5c; Table 9.2).

Discussion

A comparison of the spread rate estimates for all three methods and when using both polygonal and point data is presented in Table 9.2. All three methods provide similar estimates of spread when using polygonal data, which is not surprising given both the coarse nature of polygonal boundaries and the fact that a decision to regard a county as infested is generally never retracted. Point data, in contrast, generally provide greater spatial resolution in the determination of species presence. Moreover, point data can often consist of trapping devices from which a continuous estimate of abundance can be obtained as opposed to merely presence or absence.

Table 9.2. Estimates of gypsy moth spread from three analytical methods when based upon polygonal and point data.

		Spread rate estimate (± SE), km/year		
Data source	Moth threshold	Square-root area regression	Distance regression	Boundary displacement
Polygonal data	NA	17.1 (1.9)	9.6 (2.0)	9.9 (5.0)
Point data	1	13.5 (2.3)	15.7 (0.3)	7.3 (3.0)
	10	21.2 (1.2)	19.3 (0.5)	9.7 (2.8)
	100	23.8 (2.2)	22.8 (0.6)	14.7 (10.5)
	Overall	19.5 (3.1)	19.3 (2.0)	10.6 (3.7)

NA, not applicable.

The use of population thresholds (Fig. 9.5) based upon continuous measurements of abundance can also provide an estimate of spread rates for different population levels, such as in comparisons between the spread rate of initial populations (i.e. the 1-moth threshold) and high-density populations (i.e. the 100-moth threshold). This can be especially critical in management efforts because lower-density populations tend to be more amenable to control tactics than the higher-density populations at which the ecological and economic impacts are also the greatest. Spread rate estimates obtained from different population thresholds could also reflect the roles that other forces, such as stochasticity, abiotic factors and biotic factors, play in the biological invasion process (Hufbauer et al., 2013; Miller and Inouye, 2013; Potapov and Rajakaruna, 2013). For example, low-density populations can be particularly prone to extinction after which reinvasion could occur and be successful; Sharov et al. (1997) revealed high variability in gypsy moth spread when measured by thresholds <10 moths, while intermediate population densities, as measured by the 10- and 30-moth thresholds, tended to be the most stable in space and time. Very-high-density populations can be affected by outbreak dynamics; in the gypsy moth system, outbreaks can be cyclical, and synchronously erupt and collapse across a large landscape due to biotic interactions (Elkinton et al., 1996; Bjørnstad et al., 2010).

Depending on the data source and method used to estimate spread, there can be considerable differences in spread rate estimates (Table 9.2). Also, the estimation of spread rates of many species can be constrained by the lack of adequate survey data. Point data typically tend to be more robust than polygonal data because they are usually replicated at smaller spatial scales and can be used to estimate pest abundance at a specific point as opposed to simple presence or absence of a pest within a political boundary. However, point data are also greatly influenced by the sensitivity of the method used for making measurements.

Semiochemical-based traps that contain species-specific attractants (e.g. sex or aggregation pheromones) often provide a highly sensitive means to detect a species (Elkinton and Cardé, 1981; Suckling and Karg, 2000). However, for many invasive alien species, especially for those that are not economically important in their native range, research on pheromone identification has been insufficient and so sensitive monitoring tools may not be immediately available. For other species, attractants may be difficult to identify or produce synthetically for use in survey programmes (Crook et al., 2008). Moreover, some species, such as the emerald ash borer (A. planipennis), may lack chemically mediated attraction behaviours that can be exploited with trapping systems to detect newly established, low-density populations (Crook and Mastro, 2010). In such cases, point data from poor trapping systems could, in fact, be misleading and provide either an underestimate of population density or the time of initial establishment. Thus, the mere availability of point data may not necessarily equate to a more accurate estimate of spread even when polygonal data are measured over a non-biological scale, such as a county or state boundary.

Regardless of the type of survey data available to estimate spread, 'true' rates of spread can still be challenging to ascertain (Gilbert and Liebhold, 2010). Part of this is due to stochastic processes that can affect spread, including the role of anthropogenic, atmospheric and hydrological transport mechanisms that facilitate long-distance dispersal (Venette and Ragsdale, 2004; Davidson et al., 2005; Tobin and Blackburn, 2008; Bigsby et al., 2011). The spread of invading organisms can also be affected by biological constraints, such as Allee effects, or positive-density dependence (Andow et al., 1990; Lewis and Kareiva, 1993; Taylor et al., 2004; Tobin et al., 2007c). Gilbert and Liebhold (2010) generated synthetic data from simulations based upon a reaction-diffusion model with a known rate of spread and compared different methods for quantifying spread. They found that the distance regression method provided the most reliable estimate of spread, particularly when sample size was limited (Gilbert and Liebhold, 2010). In cases where the invaded area is irregularly constrained, such as by lakes or other geographic barriers, previous work has shown that the square-root area regression method provided biased estimates of radial spread rates and thus would be an undesirable approach under such conditions (Shigesada and Kawasaki, 1997; Gilbert and Liebhold, 2010). As noted above and previously (Gilbert and Liebhold, 2010), an advantage of the boundary displacement method is that it can characterize temporal and spatial variation in spread rates.

We demonstrated three principal methods that are used for quantifying spread rates; however, there are additional methods that could be used to quantify invasion speed. For example, one additional method that has been used to estimate spread rates is the use of the 'Wombling' approach, which is a statistical technique for estimating vector gradients from spatially referenced data (Womble, 1951). This method can be used to estimate local rates of change from a map surface, such as the waiting time associated with the time of first establishment for an invading species. Consequently, these localized slope estimates provide a measure of local spread rate, and can be furthermore used to characterize the spatial and temporal variation in the rate of invasion spread (Fortin et al., 2005; Fitzpatrick et al., 2010). Regardless of the challenges associated with estimating rates of spread, and the limitations associated with various methods, even coarse estimates of spread can still provide guidance to managers, such as in efforts aimed at managing spread (Taylor and Hastings, 2004; Tobin et al., 2007a; McCullough and Mercader, 2012). Estimates of spread can also be useful in identifying and quantifying the role of long-distance dispersal on the overall spread of invaders (Liebhold et al., 1992), thereby providing a basis to formulate management guidelines (Sharov and Liebhold, 1998).

References

Andow, D.A., Kareiva, P.M., Levin, S.A. and Okubo, A. (1990) Spread of invading organisms. *Landscape Ecology* 4, 177–188.

Aukema, J.E., McCullough, D.G., Von Holle, B., Liebhold, A.M., Britton, K. and Frankel, S.J. (2010) Historical accumulation of nonindigenous forest pests in the continental US. *Bioscience* 60, 886–897.

Bigsby, K.M., Tobin, P.C. and Sills, E.O. (2011) Anthropogenic drivers of gypsy moth spread. *Biological Invasions* 13, 2077–2090.

Bjørnstad, O.N., Robinet, C. and Liebhold, A.M. (2010) Geographic variation in North American gypsy moth cycles: subharmonics, generalist predators, and spatial coupling. *Ecology* 91, 106–118.

Cooke, M.T. (1928) *The Spread of the European Starling in North America (to 1928). Circular No. 40*. US Department of Agriculture, Washington, DC.

Crook, D.J. and Mastro, V.C. (2010) Chemical ecology of the emerald ash borer, *Agrilus planipennis*. *Journal of Chemical Ecology* 36, 101–112.

Crook, D.J., Khrimian, A., Francese, J.A., Fraser, I., Poland, T.M., Sawyer, A.J. and Mastro, V.C. (2008) Development of a host-based semiochemical lure for trapping emerald ash borer *Agrilus planipennis* (Coleoptera: Buprestidae). *Environmental Entomology* 37, 356–365.

Davidson, J., Wickland, A.C., Patterson, H.A., Falk, K.R. and Rizzo, D.M. (2005) Transmission of *Phytophthora ramorum* in mixed-evergreen forest in California. *Phytopathology* 95, 587–596.

Deutsch, C.V. and Journel, A.G. (1992) *GSLIB: Geostatistical Software Library and User's Guide*. Oxford University Press, New York.

Dreistadt, S.H. (1983) An assessment of gypsy moth eradication attempts in Michigan (Lepidoptera: Lymantriidae). *The Great Lakes Entomologist* 16, 143–148.

Elkinton, J.S. and Cardé, R.T. (1981) The use of pheromone traps to monitor the distribution and population trends of the gypsy moth. In: Mitchell, E.R. (ed.) *Management of Insect Pests with Semiochemicals*. Plenum, New York, pp. 41–55.

Elkinton, J.S. and Childs, R.D. (1983) Efficiency of two gypsy moth (Lepidoptera: Lymantriidae) pheromone-baited traps. *Environmental Entomology* 12, 1519–1525.

Elkinton, J.S. and Liebhold, A.M. (1990) Population dynamics of gypsy moth in North America. *Annual Review of Entomology* 35, 571–596.

Elkinton, J.S., Healy, W.M., Buonaccorsi, J.P., Boettner, G.H., Hazzard, A.M., Smith, H.R. and Liebhold, A.M. (1996) Interactions among gypsy moths, white-footed mice, and acorns. *Ecology* 77, 2332–2342.

Elton, C.S. (1958) *The Ecology of Invasions by Animals and Plants.* Methuen and Co., London.

Fisher, R.A. (1937) The wave of advance of advantageous genes. *Annals of Eugenics* 7, 355–369.

Fitzpatrick, M.C., Preisser, E.L., Porter, A., Elkinton, J., Waller, L.A., Carlin, B.P. and Ellison, A.M. (2010) Ecological boundary detection using Bayesian areal wombling. *Ecology* 91, 3448–3455.

Forbush, E.H. and Fernald, C.H. (1896) *The Gypsy Moth.* Wright and Potter, Boston, Massachusetts.

Fortin, M.-J., Keitt, T.H., Maurer, B.A., Taper, M.L., Kaufman, D.M. and Blackburn, T.M. (2005) Species' geographic ranges and distributional limits: pattern analysis and statistical issues. *Oikos* 108, 7–17.

Froud, K.J., Oliver, T.M., Bingham, P.C., Flynn, A.R. and Rowswell, N.J. (2008) Passive surveillance of new exotic pests and diseases in New Zealand. In: Froud, K.J., Popay, I.A. and Zydenbos, S.M. (eds) *Surveillance for Biosecurity: Pre-border to Pest Management.* New Zealand Plant Protection Society, Christchurch, New Zealand, pp. 97–110.

Gage, S.H., Wirth, T.M. and Simmons, G.A. (1990) Predicting regional gypsy moth (Lymantriidae) population trends in an expanding population using pheromone trap catch and spatial analysis. *Environmental Entomology* 19, 370–377.

Gansner, D.A. and Herrick, O.W. (1984) Guides for estimating forest stand losses to gypsy moth. *Northern Journal of Applied Forestry* 1, 21–23.

Gilbert, M. and Liebhold, A.M. (2010) Comparing methods for measuring the rate of spread of invading populations. *Ecography* 33, 809–817.

Gilbert, M., Grégoire, J.-C., Freise, J.F. and Heitland, W. (2004) Long-distance dispersal and human population density allow the prediction of invasive patterns in the horse chestnut leafminer *Cameraria ohridella. Journal of Animal Ecology* 73, 459–468.

Hajek, A.E. and Tobin, P.C. (2009) North American eradications of Asian and European gypsy moth. In: Hajek, A.E., Glare, T.R. and O'Callaghan, M. (eds) *Use of Microbes for Control and Eradication of Invasive Arthropods.* Springer, New York, pp. 71–89.

Hanna, M. (1982) Gypsy moth (Lepidoptera: Lymantriidae) history of eradication efforts in Michigan, 1954–1981. *The Great Lakes Entomologist* 15, 191–198.

Haynes, K.J., Liebhold, A.M. and Johnson, D.M. (2009) Spatial analysis of harmonic oscillation of gypsy moth outbreak intensity. *Oecologia* 159, 249–256.

Hengeveld, R. (1989) *Dynamics of Biological Invasions.* Chapman and Hall, London.

Hufbauer, R.A., Rutschmann, A., Serrate, B., Vermeil de Conchard, H. and Facon, B. (2013) Role of propagule pressure in colonization success: disentangling the relative importance of demographic, genetic and habitat effects. *Journal of Evolutionary Biology* 26, 1691–1699.

Ingwell, L.L. and Preisser, E.L. (2011) Using citizen science programs to identify host resistance in pest-invaded forests. *Conservation Biology* 25, 182–188.

Isaaks, E.H. and Srivastava, R.M. (1989) *An Introduction to Applied Geostatistics.* Oxford University Press, Oxford.

Keena, M.A., Côté, M.-J., Grinberg, P.S. and Wallner, W.E. (2008). World distribution of female flight and genetic variation in *Lymantria dispar* (Lepidoptera: Lymantriidae). *Environmental Entomology* 37, 636–649.

Leuschner, W.A., Young, J.A., Waldon, S.A. and Ravlin, F.W. (1996) Potential benefits of slowing the gypsy moth's spread. *Southern Journal of Applied Forestry* 20, 65–73.

Lewis, M.A. and Kareiva, P. (1993) Allee dynamics and the spread of invading organisms. *Theoretical Population Biology* 43, 141–158.

Liebhold, A.M. and Tobin, P.C. (2008) Population ecology of insect invasions and their management. *Annual Review of Entomology* 53, 387–408.

Liebhold, A., Mastro, V. and Schaefer, P.W. (1989) Learning from the legacy of Leopold Trouvelot. *Bulletin of the Entomological Society of America* 35, 20–22.

Liebhold, A.M., Halverson, J.A. and Elmes, G.A. (1992) Gypsy moth invasion in North America: a quantitative analysis. *Journal of Biogeography* 19, 513–520.

Liebhold, A.M., Gottschalk, K.W., Muzika, R.M., Montgomery, M.E., Young, R., O'Day, K. and Kelley, B. (1995) *Suitability of North American Tree Species to the Gypsy Moth: A Summary of Field and Laboratory Tests. General Technical Report No. NE-211.* US Department of Agriculture, Forest Service, Radnor, Pennsylvania.

Lippitt, C.D., Rogan, J., Toledano, J., Sangermano, F., Eastman, J.R., Mastro, V. and Sawyer, A.J. (2008) Incorporating anthropogenic variables into a species distribution model to map gypsy moth risk. *Ecological Modelling* 210, 339–350.

Mason, C.J. and McManus, M.L. (1981) Larval dispersal of the gypsy moth. In: Doane, C.C. and McManus, M.L. (eds) *The Gypsy Moth: Research Toward Integrated Pest Management. Technical Bulletin No. 1584*. US Department of Agriculture, Washington, DC, pp. 161–202.

Mastro, V.C., Richerson, J.V. and Cameron, E.A. (1977) An evaluation of gypsy moth pheromone-baited traps using behavioral observations as a measure of trap efficiency. *Environmental Entomology* 6, 128–132.

McCullough, D.G. and Mercader, R.J. (2012) Evaluation of potential strategies to SLow Ash Mortality (SLAM) caused by emerald ash borer (*Agrilus planipennis*): SLAM in an urban forest. *International Journal of Pest Management* 58, 9–23.

Miller, T.E.X. and Inouye, B.D. (2013) Sex and stochasticity affect range expansion of experimental invasions. *Ecology Letters* 16, 354–361.

Morin, R.S., Liebhold, A.M., Luzader, E.R., Lister, A.J., Gottschalk, K.W. and Twardus, D.B. (2005) *Mapping Host-Species Abundance of Three Major Exotic Forest Pests. Research Paper No. NE-726*. US Department of Agriculture, Forest Service, Newtown Square, Pennsylvania.

Muirhead, J.R., Leung, B., van Overdijk, C., Kelly, D.W., Nandakumar, K., Marchant, K.R. and MacIsaac, H.J. (2006) Modelling local and long-distance dispersal of invasive emerald ash borer *Agrilus planipennis* (Coleoptera) in North America. *Diversity and Distributions* 12, 71–79.

Paine, T.D., Raffa, K.F. and Harrington, T.C. (1997) Interactions among scolytid bark beetles, their associated fungi, and live host conifers. *Annual Review of Entomology* 42, 179–206.

Pimentel, D., Lach, L., Zuniga, R. and Morrison, D. (2000) Environmental and economic costs of nonindigenous species in the United States. *Bioscience* 50, 53–65.

Potapov, A. and Rajakaruna, H. (2013) Allee threshold and stochasticity in biological invasions: colonization time at low propagule pressure. *Journal of Theoretical Biology* 337, 1–14.

R Core Team (2013) *R: A Language and Environment for Statistical Computing*. R Foundation for Statistical Computing, Vienna.

Redman, A.M. and Scriber, J.M. (2000) Competition between the gypsy moth, *Lymantria dispar*, and the northern tiger swallowtail, *Papilio canadensis*: interactions mediated by host plant chemistry, pathogens, and parasitoids. *Oecologia* 125, 218–228.

Riley, C.V. and Vasey, G. (1870) Imported insects and native American insects. *The American Entomologist* 2, 110–112.

SAS Institute, Inc. (1999) *SAS/STAT® User's Guide, Version 8*. SAS Institute, Inc., Cary, North Carolina.

Sharov, A.A. and Liebhold, A.M. (1998) Bioeconomics of managing the spread of exotic pest species with barrier zones. *Ecological Applications* 8, 833–845.

Sharov, A.A., Roberts, E.A., Liebhold, A.M. and Ravlin, F.W. (1995) Gypsy moth (Lepidoptera: Lymantriidae) spread in the central Appalachians: three methods for species boundary estimation. *Environmental Entomology* 24, 1529–1538.

Sharov, A.A., Liebhold, A.M. and Roberts, E.A. (1997) Methods for monitoring the spread of gypsy moth (Lepidoptera: Lymantriidae) populations in the Appalachian Mountains. *Journal of Economic Entomology* 90, 1259–1266.

Shigesada, N. and Kawasaki, K. (1997) *Biological Invasions: Theory and Practice*. Oxford University Press, New York.

Shigesada, N., Kawasaki, K. and Takeda, Y. (1995) Modelling stratified diffusion in biological invasions. *American Naturalist* 146, 229–251.

Siegert, N.W., McCullough, D.G., Williams, D.W., Fraser, I., Poland, T.M. and Pierce, S.J. (2010) Dispersal of *Agrilus planipennis* (Coleoptera: Buprestidae) from discrete epicenters in two outlier sites. *Environmental Entomology* 39, 253–265.

Skellam, J.G. (1951) Random dispersal in theoretical populations. *Biometrika* 38, 196–218.

Suarez, A.V., Holway, D.A. and Case, T.J. (2001) Patterns of spread in biological invasions dominated by long-distance jump dispersal: insights from Argentine ants. *Proceedings of the National Academy of Sciences USA* 98, 1095–1100.

Suckling, D.M. and Karg, G. (2000) Pheromones and other semiochemicals. In: Rechcigl, J.E. and Rechcigl, N.A. (eds) *Biological and Biotechnological Control of Insect Pests*. CRC Press, Boca Raton, Florida, pp. 63–100.

Taylor, C.M. and Hastings, A. (2004) Finding optimal control strategies for invasive species: a density-structured model for *Spartina alterniflora*. *Journal of Applied Ecology* 41, 1049–1057.

Taylor, C.M., Davis, H.G., Civille, J.C., Grevstad, F.S. and Hastings, A. (2004) Consequences of an Allee effect on the invasion of a Pacific estuary by *Spartina alterniflora*. *Ecology* 85, 3254–3266.

Thorpe, K.W., Ridgway, R.L. and Leonhardt, B.A. (1993) Relationship between gypsy moth (Lepidoptera: Lymantriidae) pheromone trap catch and population density: comparison of traps baited with 1 and 500 μg (+)-disparlure lures. *Journal of Economic Entomology* 86, 86–92.

Tobin, P.C. and Blackburn, L.M. (2008) Long-distance dispersal of the gypsy moth (Lepidoptera: Lymantriidae) facilitated its initial invasion of Wisconsin. *Environmental Entomology* 37, 87–93.

Tobin, P.C., Sharov, A.A., and Liebhold, A.M. (2007a) The decision algorithm: project evaluation. In: Tobin, P.C. and Blackburn, L.M. (eds) *Slow the Spread: A National Program to Manage the Gypsy Moth. General Technical Report No. NRS-6*. US Department of Agriculture, Forest Service, Newtown Square, Pennsylvania, pp. 61–76.

Tobin, P.C., Liebhold, A.M. and Roberts, E.A. (2007b) Comparison of methods for estimating the spread of a non-indigenous species. *Journal of Biogeography* 34, 305–312.

Tobin, P.C., Whitmire, S.L., Johnson, D.M., Bjørnstad, O.N. and Liebhold, A.M. (2007c) Invasion speed is affected by geographic variation in the strength of Allee effects. *Ecology Letters* 10, 36–43.

Tobin, P.C., Bai, B.B., Eggen, D.A. and Leonard, D.S. (2012) The ecology, geopolitics, and economics of managing *Lymantria dispar* in the United States. *International Journal of Pest Management* 53, 195–210.

USDA Forest Service (2013) *Gypsy Moth Digest – Defoliation*. US Department of Agriculture, Forest Service, Morgantown, West Virginia. Available at: http://na.fs.fed.us/fhp/gm/defoliation/index.shtml (accessed 20 December 2013).

Venette, R.C. and Ragsdale, D.W. (2004) Assessing the invasion by soybean aphid (Homoptera: Aphididae): where will it end? *Annals of the Entomological Society of America* 97, 219–226.

Waring, K.M. and O'Hara, K.L. (2005) Silvicultural strategies in forest ecosystems affected by introduced pests. *Forest Ecology and Management* 209, 27–41.

Winston, M.L. (1992) The biology and management of Africanized honey bees. *Annual Review of Entomology* 37, 173–193.

Womble, W.H. (1951) Differential systematics. *Science* 114, 315–355.

Yang, D., Pijanowski, B.C. and Gage, S.H. (1998) Analysis of gypsy moth (Lepidoptera: Lymantriidae) population dynamics in Michigan using Geographic Information Systems. *Environmental Entomology* 27, 842–852.

10 Predicting the Economic Impacts of Invasive Species: The Eradication of the Giant Sensitive Plant from Western Australia

David C. Cook,[1]* Andy Sheppard,[2] Shuang Liu[2] and W. Mark Lonsdale[2]

[1]*Department of Agriculture and Food Western Australia, Bunbury, Western Australia, Australia & The University of Western Australia, Crawley, Western Australia, Australia;* [2]*CSIRO Biosecurity Flagship, Canberra, Australian Capital Territory, Australia & Plant Biosecurity Cooperative Research Centre, Bruce, Australian Capital Territory, Australia*

Abstract

Economic analyses can facilitate biosecurity investment decisions in relation to pest risk management. The term biosecurity here refers to activities intended to prevent or control the introduction and spread of alien species that have a net negative effect on social welfare as determined by economic, environmental and social capital. The aim of the chapter is to demonstrate a model that indicates an appropriate level of funding that decision makers should allocate to specific pest management activities based on anticipated returns. We demonstrate the model with a recently introduced weed species to Western Australia, *Mimosa pigra*. This weed has been targeted for eradication and the analysis described here estimates the amount biosecurity managers should spend on eradication before costs begin to outweigh the likely returns.

An Introduction to Economic Perspectives on Eradication

Historically, the decision to attempt eradication of a recently discovered weed infestation has proved to be challenging. While eradication is widely recognized as a more preferable option than weed suppression or a 'do-nothing' approach (Myers *et al.*, 2000; Simberloff, 2003), eradication is hard to achieve when infestations already occupy a substantial area (Rejmánek and Pitcairn, 2002; Cacho *et al.*, 2006). The decision to eradicate must be made quickly but cannot be made solely on the basis of the ecological feasibility and likelihood of success. It must also consider economic costs and benefits of the control investment. The ecology of the plant (e.g. habitat, mode of reproduction, seed dispersal distances, etc.) plays a key role in the likelihood of eradication success. But, even if these factors mean the likelihood of successful eradication is low, the long-term benefits of success may be extremely high. So, the decision to enact and continue eradication actions should include a long-term perspective.

To calculate the returns on investment, a partial budget model is developed. It is 'partial' in the sense that it analyses relationships within a particular industry affected by an invasive species and assumes that changes in this industry have negligible implications for the rest of the economy.

*Corresponding author. E-mail: david.cook@agric.wa.gov.au

Typically, the costs to manage an invasive alien species (e.g. increased intensity of on-farm hygiene practices, pesticide costs, etc.) increase the total investment growers must make to supply a given quantity of product. The effects of these expenses on the budgets of land managers as an invasive alien weed spreads are summed in a partial budget model to give an indication of the aggregate financial impacts of the weed. Spread can take place via natural means described by a diffusion-like process or by large jumps, often mediated by humans. When combined in our model, these means of dispersal produce a stratified diffusion process (Hengeveld, 1989; Shigesada et al., 1995; Pitt et al., 2009) that allow us to forecast the impacts of an invasive alien weed over time.

Mimosa as a Case Study

On 16 November 2009, a biosecurity officer from the Department of Agriculture and Food Western Australia (DAFWA) found several patches of mimosa (*M. pigra*) near Lake Argyle in the Kimberley region of Western Australia. The plants were discovered over approximately 1 ha in a seasonal billabong (i.e. an isolated pond in a dry riverbed) some 12 km from the town of Kununurra in the far north of the state. Based on the size of the plants and the fact that they had set seed, they were estimated to be at least 3 years and possibly up to 5 years old (Lloyd and Vinnicombe, 2010). The true extent of the area occupied by the plants may, therefore, have been considerably larger than was known at the time.

The origin of the outbreak is not known. The plants were discovered in an isolated area not known as a camping or recreation site. A homestead in the area was abandoned some 80 to 90 years earlier and human visitation thereafter is believed to have been infrequent, at best. Since the billabong is not connected to any creeks, seeds could not have been transported there from other waterways (Lloyd and Vinnicombe, 2010). Fortunately, this also means seeds are unlikely to have been carried large distances by water. Nevertheless, an emergency response was initiated by DAFWA with the first ground and aerial surveillance taking place the day after detection. Extension materials were distributed to local businesses, cattle stations and commuters (via the DAFWA quarantine checkpoint at Kununurra). Statements were released via local radio, newspaper and community newsletters. Despite all of the alerts, no further infestations have been reported (Lloyd and Vinnicombe, 2010).

Although not previously found in Western Australia, mimosa has been present in the eastern states of Australia since the late 1800s where it was introduced from tropical America as an ornamental plant and cover crop (Parsons and Cuthbertson, 1992). Often known as the giant sensitive plant, its appeal as an ornamental was particularly strong due to the curious property of the leaves folding upon being touched. As a shrub, mimosa now forms impenetrable thickets over more than 800 km^2 of floodplain in the Northern Territory bordering Western Australia. It represents a threat to Kakadu National Park and other wetland areas in tropical Australia as it profoundly modifies native wetland habitat and restricts recreational use of waterways (Braithwaite et al., 1989; Paynter, 2005). The plant is also of significance to agriculture as it outcompetes pasture species and restricts livestock movements and access to water on such floodplains.

The Kimberley region of Western Australia is located in the far north of the state where mimosa is likely to affect tropical cattle production. This industry, comprised of about 90 properties covering approximately 250,000 km^2, making the average business size over 2700 km^2, is heavily geared towards the live export market, providing feeder cattle into the Indonesian market. Annual rainfall and different soil types influence the number of stock carried, the number of cattle turned off (i.e. marketed) each year, the costs of running each business and the amount of debt held. The region is dependent on monsoonal rainfall during the wet season (i.e. November to April), which dictates the abundance and distribution of vegetation for grazing and the degree to which preferred pasture species will be displaced by mimosa should it become established. Given the presence of the weed in the Northern Territory, its removal and continued eradication from Western Australia may be difficult. Mimosa is dispersed via movement of seedpod segments, which are particularly suited to transport by water, animals, man and machinery (Parsons and Cuthbertson, 1992).

This mix of different impacts of the weed, with resultant uncertainty as to who benefits from, and who pays for, its control, helped initiate the call for a National Weed Strategy in Australia in the early 1990s. The National Strategy sought to improve decision making between agricultural and environmental sectors of society. Now, chemical sprays are the main method of controlling mimosa in areas of low environmental sensitivity. Herbicides with active ingredients glyphosate (e.g. Roundup™ Biactive) or dicamba (e.g. Titan Dicamba®) are effective. Since the mid-1980s, 12 insect species and two fungal pathogens have been released in Australia as biological control agents and are now having a noticeable impact in some areas (Julien et al., 2004). Seven of these insects are well established and abundant, including a flower-feeding weevil (*Coelocephalapion pigrae*), a twig-mining moth (*Neurostrota gunniella*), a stem-mining moth (*Carmenta mimosa*), a seed-feeding beetle (*Acanthoscelides puniceus*), two leaf-feeding beetles (*Chlamisus mimosa* and *Malacorhinus irregularis*) and a looper moth (*Macaria pallidata*) (Julien et al., 2004). However, widespread control has not been achieved. Supplementary control methods such as herbicides and fire are required to break up the stands (Julien et al., 2004).

As the detection of mimosa in Western Australia represents a new incursion of an invasive alien weed species, a single planning body (DAFWA's invasive species programme) is responsible for oversight. However, in other cases where the species concerned is established elsewhere in the state, the central planning body may be a local shire council or potentially an Industry Funding Scheme (IFS) formed under the Biosecurity and Agriculture Management Act 2007. IFSs are intended to direct funding to activities that benefit specific industries, rather than multiple industry groups; therefore, they are better suited to address host-specific insect pests, for example, as opposed to weeds or highly polyphagous insect pests.

As part of the emergency response to mimosa in Western Australia, an economic impact assessment was initiated by DAFWA to determine what, if any, funding should be allocated to eradicate mimosa. This chapter describes the approach used for the assessment. The economic impact assessment uses a stochastic bioeconomic model to simulate the likely effects of the weed over 20 years under two scenarios: eradication and nil management. The difference between these two scenarios indicates the likely financial benefits of eradication, taking into account the dynamics of spread and impact. The model accounts only for costs of livestock displacement and reduced grazing capacity caused by mimosa, which are relatively easy to quantify.

Models and Methods

The methodology used in this analysis assumes that the current infestation of mimosa in the area of Lake Argyle is eradicated and concentrates on events that might subsequently transpire. In this context, eradication simply describes the removal of above-ground plant biomass. Weeds specialists may apply 'eradication' differently due to the existence of seed banks that may persist over long periods of time and are extremely difficult to remove. Hence, in the analysis to follow it is assumed that the probability of re-establishment following the removal of above-ground plant biomass is relatively high.

Local eradication of future incursions is treated as an investment alternative to nil management of mimosa. We assume that a single planning body (i.e. DAFWA) determines appropriate weed control strategies. Predicted investment paths are defined as a function of expected yield and input cost changes in affected agricultural industries (and hence their profitability) as a result of investing in mimosa eradication relative to a nil management approach. We assume that the planning body will choose to invest in mimosa eradication in region i in time step (e.g. year) t if it is expected to reduce grower losses by a greater amount than additional costs. The dichotomous adoption variable, α_t, which takes the value of 1 if the central planner invests in eradication across n regions in year t and 0 otherwise, is defined as:

$$\alpha_t = \begin{cases} 1 \text{ if } \sum_{i=1}^{n} d_{it} \geq \sum_{i=1}^{n} c_{it} \\ 0 \text{ if } \sum_{i=1}^{n} d_{it} < \sum_{i=1}^{n} c_{it} \end{cases} \quad (10.1)$$

where d_{it} is the total difference in the present value of predicted production costs induced by mimosa between the eradication and nil management policy options in region i in time t and c_{it} is the present value of total costs of eradication in region i in time t. Because the value of $\sum_{i=1}^{n} d_{it}$ can be estimated, a 'break-even' analysis can be used. That is, the estimate of $\sum_{i=1}^{n} d_{it}$ indicates how large $\sum_{i=1}^{n} c_{it}$ would need to be before α_t assumes a value of zero (i.e. the central planner does not adopt the eradication policy). This approach is warranted given that the size of the investment to be made in mimosa eradication has yet to be negotiated. It presents decision makers with a cost ceiling beyond which further expenditure on eradication will result in a net loss in social welfare (i.e. the costs of eradication will exceed the benefits).

If, following eradication, an incursion is detected in a subsequent time step and occurs early enough in the weed's growth cycle, there may be a strong likelihood of local eradication. So, the value of d_{it} is influenced by local eradication costs and probability of eradication success. This probability of success is assumed to decline exponentially at an average rate of $e^{-0.15A_{it}}$,

where A_{it} is the area infected with mimosa in region i year t weighted by the probability of infestation and the density of infestation.

The rate of decline in eradication success is a difficult parameter to estimate, so we use expert elicitation to help us (see Box 10.1). Expert elicitation is a structured approach to systematically consult experts on uncertain issues and is typically used to quantify ranges for poorly known parameters (Knol et al., 2010). We asked three weed scientists from the DAFWA to provide estimates for the minimum, most likely and maximum exponential rates of decline in the probability of eradication success as the area of infestation at the time of detection increases. We used graphical representations to clarify the implications of their answers. These probabilities were combined using a PERT distribution – a type of beta distribution used extensively for modelling expert estimates, where we have the experts' minimum, most likely (i.e. skewness) and maximum guesses (Vose, 2008). The resultant distribution had a minimum value of 10%, a most likely value of 15% and a maximum value of 20% (see 'Exponential rate of decline for eradication success probability with respect to area affected' in Table 10.1). This relationship between eradication success and infestation area appears broadly consistent with other anecdotal evidence in the literature that suggests eradication of weed species is seldom achieved unless detection occurs soon after introduction and often requires sustained efforts over long periods (Dodd, 1990; Panetta and Lawes, 2005; Panetta, 2007).

If the incursion is not detected early enough, local eradication within the area affected may be aborted. Assume that the decision on when to abort is based purely on the affected area and that a threshold exists beyond which local eradication is technically infeasible. When this threshold is reached, management effort switches to a nil management strategy. In such cases, the eradication option fails. Note that, under nil management, we allow for the possibility that some modest efforts to control the weed will be exerted by leaseholders. For

> **Box 10.1.** Expert elicitation
>
> Specifying the details of a model to predict what might happen in the event of an invasive alien species incursion is complex and extremely uncertain. Empirical evidence to develop model parameter estimates and predictions of biosecurity management outcomes may not be available at the time a decision must be made. When this is the case, direct or indirect expert elicitation approaches to fill the data gaps are becoming common (Burgman, 2005; Low Choy et al., 2009; Kuhnert et al., 2010; Martin et al., 2012). Direct elicitation requires experts on a particular species to express their knowledge in terms of quantities required by analysts. For instance, experts may be asked to provide statistical summaries (e.g. a lower bound, a best estimate and an upper bound) or a full parametric probability distribution. In contrast, indirect elicitation requires experts to answer questions that relate to their experiences that are then encoded into the quantities needed. For example, the expert may be asked about expected economic cost given different efforts invested in eradication, which the analyst then translates into a corresponding probability distribution for a model parameter. This indirect approach tends to be more comfortable for the experts, but not necessarily for the analyst (Allan et al., 2010).
>
> If an elicitation process involves multiple experts, information can be elicited independently and then combined, or a group consensus opinion can be sought (Martin et al., 2012). In either case, the process is subject to 'heuristics and biases', as extensively documented in psychological research investigating the assessment of uncertain information (Kynn, 2008). In a recent study, assessing the potential impacts of marine invasive species, researchers showed that individual experts use both a spatial proximity heuristic (e.g. they attribute relatively more damage to invasive species in their home country than elsewhere) and an effect heuristic (e.g. tangible economic impacts are regarded as more severe than equivalent environmental impacts) (Dahlstrom Davidson et al., 2013). In a group elicitation setting, there are added concerns introduced via dominant group members and groupthink (Burgman, 2005). To overcome these limitations, structured approaches such as the Delphi method can help in allowing experts to adjust their own estimates in light of the answers of other group members while maintaining anonymity (de França Doria et al., 2009).

example, infrequent chemical applications may be applied to mimosa thickets, but they are likely to be insufficient to eradicate the weed or to significantly slow dispersal over time.

Algebraically, d_{it} is expressed as:

$$d_{it} = \begin{cases} E_{it} A_{it} & \text{if } A_{it} \leq A_{it}^{\text{erad}} \\ Y_{it} P_t A_{it} + V_{it} A_{it} & \text{if } A_{it} > A_{it}^{\text{erad}} \end{cases}$$

(10.2)

where E_{it} is the cost of eradication per hectare in region i in year t; A_{it}, as stated above, is the area infested with mimosa in region i in year t weighted by the probability of incursion and density of infestation; A_{it}^{erad} is the maximum technically feasible area of eradication in region i in year t; Y_{it} is the mean change in livestock yield resulting from mimosa becoming established across region i in year t; P_t is the prevailing domestic price for cattle in year $t - 1$; and V_{it} is the increase in variable cost of production per hectare induced by mimosa management methods in region i in year t.

Arrival

A_{it} is inclusive of mimosa entry and establishment probability in region i, z_i, and therefore represents the area predicted to be in need of additional management effort due to the weed's presence. If it is not present at the outset of the analysis, the possible transition from a *without*-mimosa to a *with*-mimosa situation must be modelled in subsequent time periods. This transition can be represented as a Markov process.

Table 10.1. Parameters for a model to assess invasion and impacts of mimosa in Western Australia under two management strategies.

Description	Nil management	Eradication
Area currently affected, A_{min} (m^2)[a]	PERT($9.0 \times 10^6, 1.2 \times 10^7, 1.5 \times 10^7$)	PERT($9.0 \times 10^6, 1.2 \times 10^7, 1.5 \times 10^7$)
Cost of eradication, E (AUS$/ha)[b]	0.0	PERT(15,25,35)
Demand elasticity[c]	Uniform(−1.5,−1.3)	Uniform(−1.5,−1.3)
Exponential rate of decline for eradication success probability with respect to area affected	PERT(−0.1,−0.15,−0.2)	PERT(−0.1,−0.15,−0.2)
Gross value of production of A_{max} divided by 100, G (AUS$)[d]	1.4×10^5	1.4×10^5
Increased herbicide and application cost if eradication fails, V (AUS$/ha)[e]	Discrete({0,120,140}, {1,0.5,0.25})	Discrete({0,120,140}, {1,0.5 0.25})
Intrinsic rate of infestation and density increase, r (/year)[f]	PERT(1,2,3)	PERT(1,2,3)
Intrinsic rate of satellite generation per unit area of infestation, μ (#/m^2)[f]	PERT($5.0 \times 10^{-2}, 7.5 \times 10^{-2}, 1.0 \times 10^{-1}$)	PERT($5.0 \times 10^{-2}, 7.5 \times 10^{-2}, 1.0 \times 10^{-1}$)
Maximum area affected, A_{max} (m^2)[g]	4.0×10^9	4.0×10^9
Maximum area considered for eradication, A_{erad} (m^2)	0.0	2.5×10^7
Maximum infestation density, K (#/m^2)[f]	PERT($1.0 \times 10^4, 5.5 \times 10^4, 1.0 \times 10^5$)	PERT($1.0 \times 10^4, 5.5 \times 10^4, 1.0 \times 10^5$)
Maximum number of satellite sites generated in a single time step, s_{max} (#)[f]	PERT(30,40,50)	PERT(30,40,50)
Minimum infestation density, N_{min} (#/m^2)	1.0×10^{-4}	1.0×10^{-4}
Minimum number of satellite sites generated in a single time step, s_{min} (#)	1.0	1.0
Population diffusion coefficient, D (m^2/year)[f]	PERT($0.0, 1.5 \times 10^{-3}, 2.0 \times 10^{-3}$)	PERT($0.0, 1.5 \times 10^{-3}, 2.0 \times 10^{-3}$)
Prevailing beef price (young steers for export) in the first time step, P_0 (AUS$/kg)[h]	1.75	1.75
Probability of re-entry and establishment in the first time step, z	1.0	Uniform(0.2,0.7)
Re-infestation detection probability[i]	0.0	Binomial(1.0,0.4)
Value of yield reduction despite control, Y (AUS$/ha)[j]	PERT(0.0,0.8,1.7)	PERT(0.0,0.1,0.3)

[a]Currently, the precise area of infestation is not yet known.
[b]Leavold *et al.* (2007).
[c]Ulubasoglu *et al.* (2011).
[d]ABS (2011).
[e]Assumes: (i) labour costs of AUS$100/ha (i.e. one application, 2 h/ha × AUS$50/h); (ii) Roundup™ Biactive applied at 100–150 ml per 15 l (i.e. approximately AUS$20/ha); and (iii) it is two times more likely that no herbicide treatments will be applied to an affected hectare of land by private land managers than one treatment in the absence of an eradication campaign, and four times more likely than two treatments.
[f]Specified with reference to Waage *et al.* (2005).
[g]Specified with reference to Walden *et al.* (2004). Note 1 ha = 10,000 m^2.
[h]Curtis and McCormick (2012).
[i]When a re-infestation event takes place the probability of it being detected is generated in the model as a binomial distribution. The probability of this distribution returning a 1 (i.e. detection) is given by a PERT distribution, PERT(20%,40%,60%).
[j]Assumes a yield loss of 0–5% and 0–1% in the nil management and status quo scenarios, respectively, and a beef gross margin of approximately AUS$35/ha (McCosker *et al.*, 2010).

Consider the probability of mimosa's arrival (call it event x) in a time period, $t + 1$, conditional on its absence (call it event y) in time period t, as z_{xy}. There is also a probability attached to event x occurring in both time periods, denoted z_{xx}. All possible outcomes for time period $t + 1$ are arranged in a transition matrix,

$$Z = \begin{pmatrix} z_{xx} & z_{xy} \\ z_{yx} & z_{yy} \end{pmatrix}$$

The elements in the matrix are conditional probabilities indicating the probability of experiencing event x in a time period given that a certain state (i.e. either y or x) was observed in the previous time period.

We use expert elicitation combined with the methodology of Waage et al. (2005) to estimate each of the transition probabilities. Three DAFWA weed scientists were asked to choose from a list of uniform (i.e. rectangular) distributions taken from Biosecurity Australia (2001) to represent the probabilities of re-entry and establishment in the eradication scenario. Their agreed choice – a uniform distribution with a minimum value of 30% and a maximum value of 70%, which can be written as uniform(30%,70%) – gives us the transitional probability z_{xy}. Similarly, z_{xx} is given by the experts' preferred choice for the probability distribution of establishment, uniform(70%,100%). The remaining transitional probabilities are calculated as $z_{yx} = (1 - z_{xx})$ and $z_{yy} = (1 - z_{xy})$.

If the probabilities of the events y and x occurring in any time period t are denoted $z_y(t)$ and $z_x(t)$, respectively, and $\mathbf{z}(t)$ is a column vector with elements $z_y(t)$ and $z_x(t)$ (where $z_y(t) + z_x(t) = 1$), the transition matrix can be used to calculate the probability of x occurring in $t + 1$ as $z_x(t + 1) = \mathbf{Zz}(t)$. The vector $\mathbf{z}(t)$ will converge to a unique vector as t increases (Moran, 1984; Hinchy and Fisher, 1991) and the probabilities of event y or x occurring in any given time period reduce to constant values after several time steps. Since event x (i.e. arrival) is of primary concern in a growing region i, $z_x(t)$ is denoted z_{it} hereafter.

Deriving estimates of z_{it} from risk maps is difficult. A simpler operation may involve using expert interpretations of map-based tools to assess areas according to their susceptibility to invasion (see Box 10.1). An estimate of the number of arrival events expected in a particular location over a given timeframe can then be used to generate an actual whole number from a Poisson distribution in each model iteration (Harwood et al., 2009).

Dispersal

To describe the dispersal of mimosa across multiple regions after its arrival, a stratified diffusion model that combines both short- and long-distance dispersal processes can be used (Hengeveld, 1989). It is derived from the reaction-diffusion models originally developed by Fisher (1937) which have been shown to provide a reasonable approximation of the spread of a diverse range of organisms (Dwyer, 1992; Holmes, 1993; McCann et al., 2000; Okubo and Levin, 2002). These models assert that an invasion diffusing from a point source will eventually reach a constant asymptotic radial spread rate of $2\sqrt{r_i D_{ij}}$ in all directions, where r_i describes a growth factor for mimosa per year in region i; growth is assumed constant over all infested sites (Hengeveld, 1989; Lewis, 1997; Shigesada and Kawasaki, 1997). D_{ij} is a diffusion coefficient for an infested site with an age index j in region i; diffusion is assumed constant over time in this analysis for simplicity. Hence, assume that the original infestation takes place in a homogeneous environment in region i and expands by a diffusive process such that area infested at time t, a_{ijt}, can be predicted by:

$$a_{ijt} = z_{it}\left[\pi\left(2t\sqrt{r_i D_{ij}}\right)^2\right] = z_{it}\left(4D_{ij}\pi r_i t^2\right)$$

(10.3)

Although demographic stochasticity is ignored here by assuming D_{ij} is constant across all sites with age index j, environmental heterogeneity can be incorporated by the inclusion of habitat layers derived from maps of the land area concerned. Incorporation of heterogeneity could be as simple as a binary susceptible (1) or not susceptible (0) multiplier for D_{ij}, or a more complicated weighting system could be used for each region i rating susceptibility to invasion by soil type, precipitation, elevation and so forth.

The density of mimosa within a_{ijt} influences the control measures required to counter the effects of an infestation and thus partially determines the value of A_{it}. Assume that in each site with age index j in region i affected, the infestation density, N_{ijt}, grows over time following a logistic growth curve until the carrying capacity of the environment, K_{ij}, is reached:

$$N_{ijt} = \frac{K_{ij} N_{ij}^{min} e^{r_i t}}{K_{ij} + N_{ij}^{min}(e^{r_i t} - 1)} \quad (10.4)$$

Here, N_{ij}^{min} is the size of the original influx at a site with age index j in region i and r_i is the rate of density increase in region i. Again, this may be informed by spatial information derived from risk maps in much the same way as the diffusion parameter using, for instance, a weighting system for each region i where r_i is partially determined by various physiological criteria.

Nascent foci

In addition to a_{ijt} and N_{ijt}, the size of A_{it} depends on the number of nascent foci in year t, s_{it}, which can take a maximum value of s_i^{max} in any year. Moody and Mack (1988) refer to these nascent foci as *satellite* infestation sites. These sites result from events external to the outbreak itself, such as weather phenomena, animal or human behaviour, which periodically jump the expanding infestation beyond the invasion front (Cook et al., 2011). Here, nascent foci are explicitly modelled since the species of concern is established in the environment and can potentially spread to any susceptible unit of land. In cases where a pest or disease is transported via trade pathways (e.g. international trade, inter- and intra-regional nursery, timber products), it may be more appropriate to use techniques like network theory (Paini, 2012). Where an invasive species has the potential to become established in both a trade network and the wider environment, both systems become important and may need to be modelled separately (Harwood et al., 2009). Here, a logistic equation is used to generate changes in s_{it} as an outbreak continues:

$$s_{it} = \frac{s_i^{max} s_i^{min} e^{\mu_i t}}{s_i^{max} + s_i^{min}(e^{\mu_i t} - 1)} \quad (10.5)$$

where μ_i is the intrinsic rate of new foci generation in region i and is assumed constant over time, and s_i^{min} is the minimum number of satellite sites generated in region i.

Total investment benefit

Total investment benefit is a function of the area, cattle price, yield reduction, control cost and gross value of production. Given the predicted area affected by mimosa in sites that have had mimosa for different lengths of time (i.e. given by Eqn 10.3), the density of these infestations (Eqn 10.4) and the number of satellite sites predicted to have been created (Eqn 10.5), the total area, A_{it}, is calculated across m sites as:

$$A_{it} = s_{it} \sum_{j=1}^{m} \left(a_{ijt} N_{ijt} \right) \quad \text{where } 0 \leq A_{it} \leq A_i^{max} \quad (10.6)$$

For the model, an estimate of cattle price, P_t, is also given for the first time step (i.e. P_0). This estimate is the initial price of beef, since only the costs of livestock displacement caused by mimosa are included in the assessment. Given that the demand for beef is elastic (i.e. price increases with relative scarcity and vice versa), the beef price in subsequent time periods will be partially influenced by mimosa's impact on beef production. In the case of mimosa, the pressure exerted on price is negligible. However, in other cases where output is severely affected, substantial price increases may accompany pest spread. This outcome is only true to the extent that the market is closed to competition. If the market is open to suppliers from other regions and the home region/economy contributes a relatively small amount to global production, the domestic price will remain unchanged in response to a pest incursion. Moreover, the probability of a pest incursion is also increased under this scenario, which may lead to countries protecting themselves through the use of phytosanitary measures. For a full discussion, see Cook and Fraser (2008).

For mimosa, we encounter a problem because a truncated analytical tool like partial budgeting makes linking changes in output to pest abundance difficult. But, the growers' production loss, $Y_{it}A_{it}$, can be used as a proxy for the reduction in output[1] and the lagged beef price, P_{t-1}, to calculate $P_t = P_{t-1}\left[1-(Y_{it}A_{it}/G_{it}\eta)\right]$. Here, G_{it} is the gross value of production divided by 100 and η is the elasticity of demand for beef (i.e. the ratio of percentage change in quantity demanded over the percentage change in price).

Values of the cost/loss parameters E_{it} (i.e. the cost of eradication), Y_{it} (i.e. the value of yield reduction) and V_{it} (i.e. herbicide costs) as they relate to mimosa appear in Table 10.1.

The total benefit to the central planner of adopting an eradication policy for mimosa in year t, B_t, across n regions can be expressed as:

$$B_t = \sum_{i=1}^{n} d_{it}\alpha_t \quad (10.7)$$

In the following section, $\sum_{i=1}^{n} d_{it}$ is estimated using multiple mimosa re-entry and spread scenarios for Western Australia over a 20-year period. From Eqn 10.2, d_{it} includes parameters which cannot be given definitive values. These uncertain parameters are specified within the model as distributions and a Latin hypercube sampling algorithm is used to sample from each distribution using the @Risk™ software package (Palisade Software, Ithaca, New York). In each of 10,000 model iterations, one value is sampled from the cumulative distribution function so that sampled parameter values are weighted according to their probability of occurrence. Model calculations use the sampled set of parameters.

A list of model parameters and their distributions appears in Table 10.1 (the i, j and t subscripts are omitted). Types of distributions used in the model include: (i) PERT, a type of beta distribution specified using minimum, most likely (i.e. skewness) and maximum values; (ii) uniform, a rectangular distribution bounded by minimum and maximum values; (iii) binomial, returning a 0 (i.e. failure) or 1 (i.e. success) based on a number of trials and the probability of a success; and (iv) discrete, a distribution in which several discrete outcomes and their probabilities of occurrence are specified. The process of estimating these distributions in the absence of data can sometimes be complex and a full discussion of the process is beyond the scope of this chapter. But, a brief introduction to the use of expert elicitation in forming parameter estimates is given in Box 10.1.

Results

This analysis assumes that the current mimosa outbreak was eradicated at time $t = 0$. We assume re-establishment is likely to occur at some point or multiple points over the estimation period (i.e. $z_{it} > 0$). Therefore, some spread is expected under the eradication and nil management scenarios. However, the extent of expected spread under an eradication programme is substantially below that of a nil management policy over the 20 years simulated in the model (Fig. 10.1). This mitigated spread is broadly consistent with observations of changes in mimosa stand area reported in Cook et al. (1996).

Translating the difference in prevalence between the two scenarios into economic impacts, Fig. 10.2 shows the present (i.e. discounted) value of benefits accruing from the eradication of mimosa from Western Australia over time[2]. This somewhat chaotic picture of possible future incursion scenarios highlights the large amount of uncertainty involved in forecasting impacts into the future. This significant uncertainty is particularly true for the eradication scenario, the costs of which depend on the size of future re-infestation when detected and the probability of eradication success. The variance in predicted costs under the eradication scenario falls after 10 to 12 years. Looking at the corresponding time periods in Fig. 10.1, the majority of model iterations show eradication failure by this time and the weed population grows thereafter.

Annualizing the difference between these two scenarios provides a summary of the complex information contained in Fig. 10.2. Over the first 20 time steps of the model, the average annual advantage of the eradication scenario in terms of expected damage over the nil management scenario is AUS$2.95 million per year (i.e. $\sum_{i=1}^{n} d_{it} = \text{AUS}\$2.95 \times 10^{6}$). This represents the threshold level of $\sum_{i=1}^{n} d_{it}$ beyond which the central planning body will choose not to invest in the eradication and eradication strategy as an alternative to a nil management strategy (i.e. $\alpha_t = 0$). The standard deviation of the distribution of

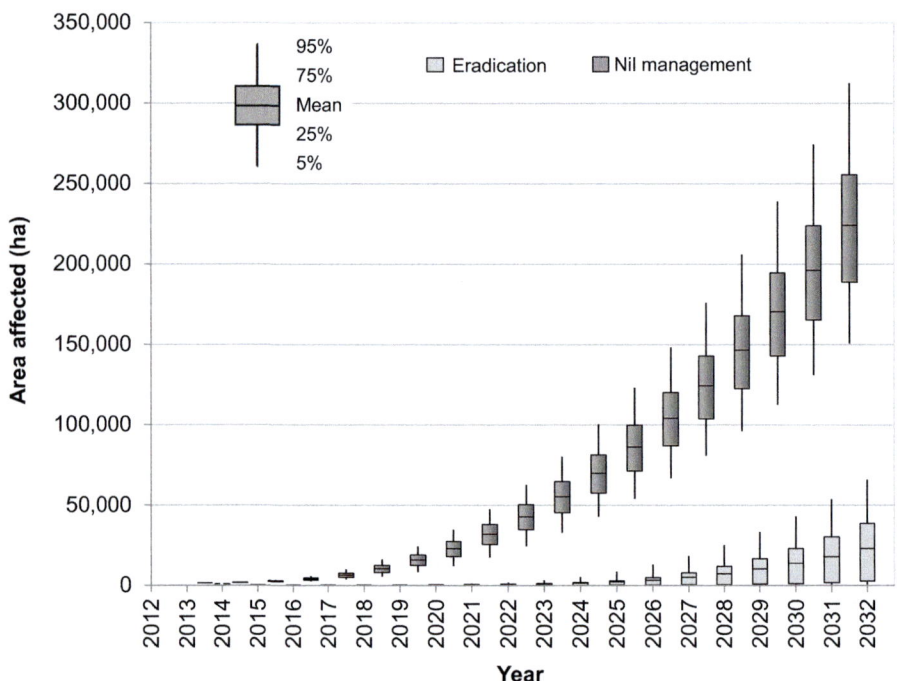

Fig. 10.1. Expected area affected by mimosa in Western Australia under the eradication and nil management scenarios. The box-and-whisker plot shows the extent of uncertainty in the model predictions, in turn dictated by the relative uncertainty in parameter specification. The box-and-whisker plot shows the 25th percentile, the median (i.e. the 50th percentile), the 75th percentile, and remaining values up to and including the 5th and 95th percentiles. The boundaries of the boxes (shaded grey) indicate the 25th percentiles and the 75th percentiles. The length of the box indicates the variability (i.e. the larger the box, the greater the spread of model predictions and the more uncertain the results). The horizontal lines inside the boxes represent the medians. If a median is not in the centre of a box, the distribution is skewed. Values lying between the 5th and 25th percentiles and between the 75th and 95th percentiles are denoted by the lines extending from the bottom and top of each box, respectively. These lines are called whiskers. If the upper whisker is longer than the lower whisker, it implies a positive skewness and vice versa.

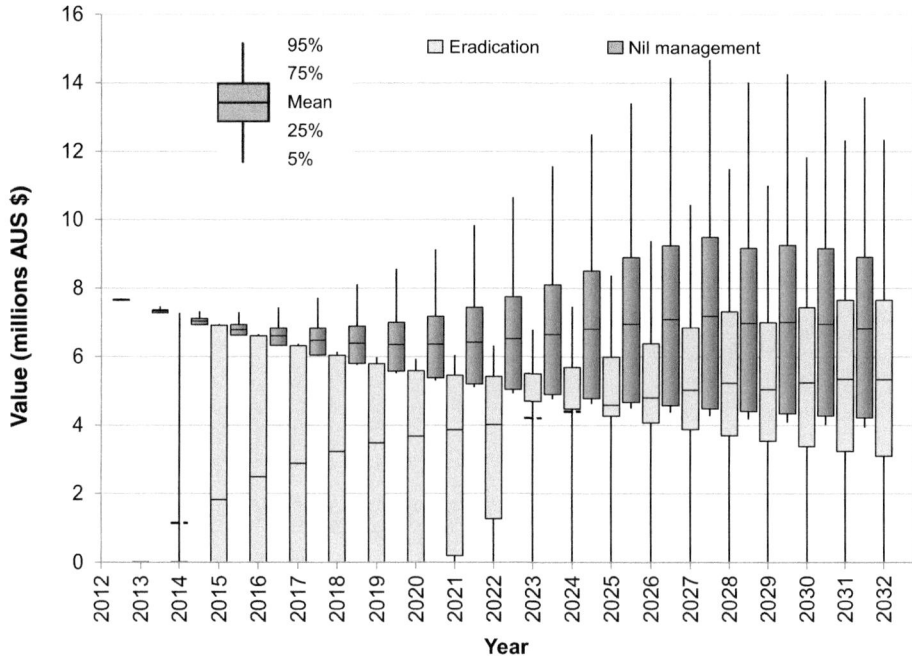

Fig. 10.2. Present value of annual losses from mimosa in Western Australia predicted under the eradication and nil management scenarios. The details of the box-and-whisker plot are described in Fig. 10.1.

average annual biosecurity benefits is AUS$1.69 million and skewness 0.86 (i.e. the distribution is skewed right such that the right tail is long compared with the left tail).

From the standard deviation, decision makers are able to determine the variation from the mean in the distribution of benefits, or how wrong their prediction of investment returns could be. The skewness indicates if their decision is more likely to produce higher or lower benefits than the mean suggests. So, in this case, the relatively high standard deviation and positive skewness of the output distribution might give the planning body reason for optimism in that the distribution suggests they may exceed their expected (i.e. mean) investment returns by quite some margin. But, the decision makers' interpretation of the information largely depends on their attitudes to risk and uncertainty.

While average eradication benefits are large, Fig. 10.3 illustrates how annual eradication benefits are expected to change over 20 years – in effect, the uncertainties in mimosa management over time. The figure is unusual in that the variance decreases with time. As stated previously, this result is largely attributable to the stochasticity in the arrival process with an eradication strategy in place.

In view of the large amount of uncertainty surrounding the parameters used to describe the mimosa (re-)infestation and spread process, it is prudent to test the sensitivity of the change in expected eradication benefits to the key assumptions of the model to gauge the robustness of the predictions. Parameters were sampled from a uniform distribution with a maximum (minimum) of +50% (−50%) of the original values in the model using Monte Carlo simulation. The Spearman's rank correlation

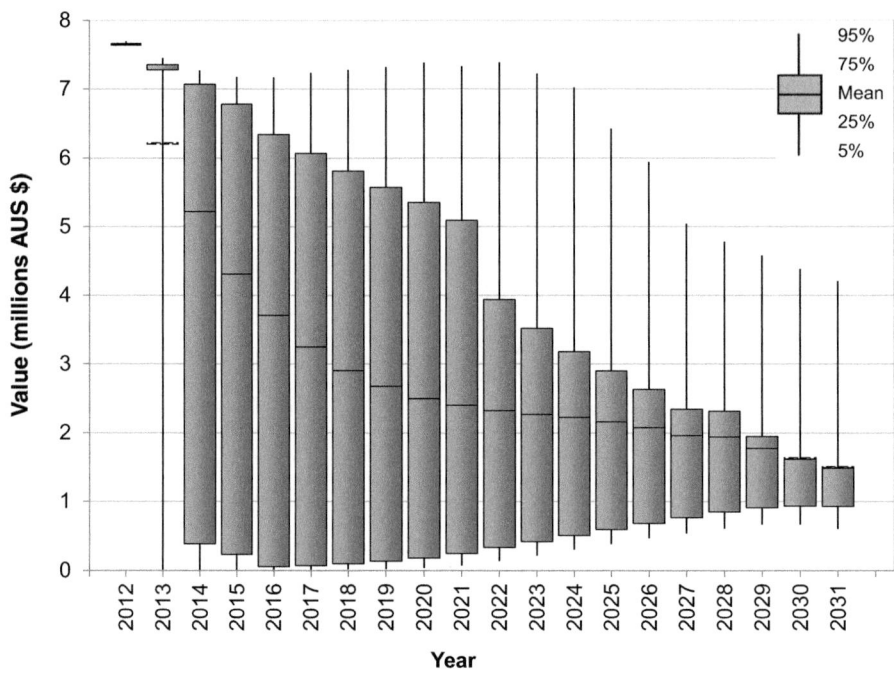

Fig. 10.3. Average industry benefit from mimosa eradication from Western Australia over 20 years. The details of the box-and-whisker plot are described in Fig. 10.1.

coefficients relating the sampled model parameter values and the change in $\sum_{i=1}^{n} d_{it}$ were then calculated.

The numerical value of the Spearman's rank correlation coefficient is denoted ρ, where $-1.0 \leq \rho \leq 1.0$. If ρ is positive it implies that $\sum_{i=1}^{n} d_{it}$ has a tendency to increase when the parameter increases and vice versa. A ρ value of 0 indicates that there is no tendency for $\sum_{i=1}^{n} d_{it}$ to either increase or decrease when the parameter values change, while a ρ of 1 indicates that they are perfect monotone functions of each other. Parameters and their ρ values are presented in Table 10.2.

The sensitivity tests indicate that the model is relatively responsive to changes in three of the parameters listed in Table 10.1 (11 of which are shown in Table 10.2). These parameters and their correlation with predicted $\sum_{i=1}^{n} d_{it}$ are the increased herbicide and application cost if eradication fails (ρ = 0.38), re-infestation detection probability (ρ = 0.24) and probability of re-entry and establishment (ρ = −0.10). All of these parameters can, to some extent, be influenced by policy and so are of interest to the decision-making body. For instance, if the decision makers were to consider subsidizing the price of herbicides in the event of mimosa's re-emergence post-eradication, the sensitivity analysis suggests that this would actually reduce the benefits of the initial eradication. However, the opposite effect would be achieved if they were to invest in surveillance activities that increase the re-infestation detection probability or phytosanitary measures that reduce the probability of re-entry and establishment.

Discussion

We have described a method to measure some agricultural impacts from a single

Table 10.2. Results of sensitivity analysis for model parameters. Herbicide use, re-infestation detection probability and re-entry probability have the greatest impact on model results.

Parameter	Correlation coefficient[a]
Increased herbicide and application cost if eradication fails	0.38
Re-infestation detection probability	0.24
Probability of re-entry and establishment	−0.10
Intrinsic rate of satellite generation per unit area of infestation	0.06
Population diffusion coefficient	0.03
Maximum number of satellite sites generated in a single time step	0.02
Exponential rate of decline for eradication success probability with respect to area affected	−0.02
Intrinsic rate of infestation and density increase	0.01
Demand elasticity	−0.01
Maximum area considered for eradication	0.01

[a]Spearman's rank correlation coefficient, varies from −1 to +1

invasive species, but it is important to recognize that other impacts may occur that are more difficult to evaluate. The invasion of natural communities by introduced plants, in particular, is considered one of the most serious threats to biodiversity and woody legumes like mimosa are some of the world's most important invasive weeds (Paynter and Flanagan, 2004). Mimosa has been identified as one of the top 20 weeds of national significance in Australia (Thorp and Lynch, 2000). The spread of this weed across northern parts of Australia has been one of the most serious examples of the impact of plant invasions on biodiversity (Braithwaite et al., 1989; Lonsdale et al., 1989; Lonsdale, 1993). Mimosa displaces native shrub and grass species, creating a dense tall scrubland where floodplains previously existed and shifting the wetland fauna to one with 'edge-habitat' affinities (Braithwaite et al., 1989). Such harm is extremely difficult to express in economic terms with any degree of accuracy, but raises the average damage cost per hectare.

Even if environmental and agricultural impacts are known with certainty, problems still remain that can severely hamper control efforts. Creating and maintaining a cooperative control relationship between many different parties is seldom straightforward due to externalities created in the provision of management services. Once a problem species like mimosa has been identified, the failure of one individual to control it within his or her area of influence creates a negative flow-on effect for all neighbouring areas due to the increased likelihood of transference. This outcome is termed a negative externality, which occurs when the welfare of one individual is adversely affected by the actions of another.

In the case of many invasive alien pest species where dispersal often takes place over large distances, a number of potential management parties and jurisdictions may be involved in control. While a negative externality is created through inaction (i.e. non-compliance or a lack of control) by any of these parties or jurisdictions, a positive externality is created by control. If one manager undertakes control measures, he/she is unable to exclude neighbouring areas from enjoying a portion of the benefits in the form of the reduced likelihood of weed transference. These neighbouring managers who did not contribute to the control activities are known as free riders.

While the free rider problem might not prevent the private sector from providing some level of mimosa control effort, the level provided will almost certainly be lower than a socially desirable amount. The diversity of stakeholders involved in control and the inherent uncertainty and variability within biological systems make it difficult to establish a management system that allows all parties to contribute appropriate knowledge, expertise and opinions in setting control targets. Even if these issues are overcome and a collective amount of weed control can be agreed upon as a standard, the amount of individual control effort required to maintain this standard will vary. The problem then becomes one of providing group members with sufficient information on weed risks and the consequences of inaction to encourage them to act in accordance with the collective good (Herb et al., 2002; Dietz et al., 2003). If this is not done, the continuity of control is put in jeopardy by parties defecting from the spirit of cooperation.

For government to correct the situation, some form of intervention is required, perhaps the payment of subsidies to private land managers or cooperatives to increase their level of control effort or assessment of penalties on non-compliant property managers to encourage them to control the weed on their properties. In both cases, establishing an amount of weed control that is deemed environmentally and economically acceptable in a given area will be difficult and may necessitate a degree of policy experimentation (Cook et al., 2010). Certainly, such experimentation would benefit greatly from economic impact assessments of the kind demonstrated here, but the political implications of policy failures make this approach an uncertain one for accountable decision makers.

Conclusion

This chapter has described a method of estimating the likely returns to biosecurity investments by simulating economic effects over time. The simulation uses a relatively simple representation of pest ecology and economic behaviour but can be developed quickly for use in pest and disease emergency responses. Analyses of this type can be of tremendous benefit to decision makers, especially early in an incursion when eradication remains a viable response.

Although the extent of the environmental threat this species poses to Western Australia wetlands is difficult to estimate in monetary terms, the analysis shows that the likely impact of mimosa on cattle grazing enterprises is sufficiently large to warrant an eradication investment of up to AUS$2.95 million per year. If eradication can be achieved for less than this amount, a net benefit will have been generated for the Western Australian economy. However, the model projections come with a large amount of uncertainty. Sensitivity analysis can indicate the benefit of reducing uncertainty in the future, but even without this, decision makers gain an understanding of what current knowledge implies about possible results of their decisions.

Acknowledgements

We would like to thank Andrew Reeves, Jon Dodd, Viv Read and Simon Merewether from the Department of Agriculture and Food, Western Australia's Biosecurity and Regulation Directorate for their valuable contributions to the analysis. We would also like to thank the editor and two anonymous reviewers for their helpful comments and suggestions.

Notes

[1] This is a practical solution to a non-trivial problem. The yield loss is only part of the story. To truly gauge a producer's output response requires information about their cost curves (Cook and Fraser, 2002).

[2] We discounted benefits because a dollar available for investment in the present is more valuable than a dollar that will not become available until the future. The future dollar has an opportunity cost associated with it (i.e. investment opportunities we have to forgo while we wait for it to become available for spending). Identifying exactly what return could have been earned on those forgone investments is a challenge. In the absence of definitive information on opportunity costs relevant to a specific project like mimosa control, we simply refer to Australian government guidelines, which recommend a standard discount rate of 5% (Commonwealth of Australia, 2006).

References

ABS (2011) *Agricultural Commodities, Australia, 2010–11*. Australian Bureau of Statistics, Canberra.

Allan, J., Low Choy, S. and Mengersen, K. (2010) Elicitator: an expert elicitation tool for regression in ecology. *Environmental Modelling & Software* 25, 129–145.

Biosecurity Australia (2001) *Guidelines for Import Risk Analysis*. Agriculture, Fisheries and Forestry Australia/Biosecurity Australia, Canberra.

Braithwaite, R.W., Lonsdale, W.M. and Estbergs, J.A. (1989) Alien vegetation and native biota in tropical Australia: the impact of *Mimosa pigra*. *Biological Conservation* 48, 189–210.

Burgman, M. (2005) *Risks and Decisions for Conservation and Environmental Management*. Cambridge University Press, New York.

Cacho, J.O., Spring, D., Pheloung, P. and Hester, S. (2006) Evaluating the feasibility of eradicating an invasion. *Biological Invasions* 8, 903–917.

Commonwealth of Australia (2006) *Handbook of Cost–Benefit Analysis*. Department of Finance and Administration, Canberra.

Cook, D.C. and Fraser, R.W. (2002) Exploring the regional implications of interstate quarantine policies in Western Australia. *Food Policy* 27, 143–157.

Cook, D.C. and Fraser, R.W. (2008) Trade and invasive species risk mitigation: reconciling WTO compliance with maximising the gains from trade. *Food Policy* 33, 176–184.

Cook, D.C., Liu, S., Murphy, B. and Lonsdale, W.M. (2010) Adaptive approaches to biosecurity governance. *Risk Analysis* 30, 1303–1314.

Cook, D.C., Fraser, R.W., Paini, D.R., Warden, A.C., Lonsdale, W.M. and De Barro, P.J. (2011) Biosecurity and yield improvement technologies are strategic complements in the fight against food insecurity. *PLoS One* 6, e26084.

Cook, G.D., Setterfield, S.A. and Maddison, J.P. (1996) Shrub invasion of a tropical wetland: implications for weed management. *Ecological Applications* 6, 531–537.

Curtis, K. and McCormick, B. (2012) *Western Australian Beef Commentary*. Department of Agriculture and Food Western Australia, Bunbury, Australia. Available at: http://www.agric.wa.gov.au/objtwr/imported_assets/content/aap/bc/2012q1%20beef%20commentary%2004.pdf (accessed 29 December 2013).

Dahlstrom Davidson, A., Campbell, M.L. and Hewitt, C.L. (2013) The role of uncertainty and subjective influences on consequence assessment by aquatic biosecurity experts. *Journal of Environmental Management* 127, 103–113.

de França Doria, M., Boyd, E., Tompkins, E.L. and Adger, W.N. (2009) Using expert elicitation to define successful adaptation to climate change. *Environmental Science & Policy* 12, 810–819.

Dietz, T., Ostrom, E. and Stern, P.C. (2003) The struggle to govern the commons. *Science* 302, 1907–1912.

Dodd, J. (1990) The role of ecological studies in assessing weed eradication programs. In: Heap, J.W. (ed.) *Proceedings of the 9th Australian Weeds Conference*. Crop Science Society of South Australia, Inc., Glen Osmond, Australia, pp. 416–426.

Dwyer, G. (1992) On the spatial spread of insect pathogens – theory and experiment. *Ecology* 73, 479–494.

Fisher, R.A. (1937) The wave of advance of advantageous genes. *Annals of Eugenics* 7, 353–369.

Harwood, T.D., Xu, X., Pautasso, M., Jeger, M.J. and Shaw, M.W. (2009) Epidemiological risk assessment using linked network and grid based modelling: *Phytophthora ramorum* and *Phytophthora kernoviae* in the UK. *Ecological Modelling* 220, 3353–3361.

Hengeveld, B. (1989) *Dynamics of Biological Invasions*. Chapman and Hall, London.

Herb, J., Helms, S. and Jensen, M.J. (2002) Harnessing the 'power of information': environmental right to know as a driver of sound environmental policy. In: Deitz, T. and Stern, P. (eds) *New Tools for Environmental Protection: Education, Information, and Voluntary Measures*. National Academy Press, Washington, DC, pp. 253–262.

Hinchy, M.D. and Fisher, B.S. (1991) *A Cost–Benefit Analysis of Quarantine*. Bureau of Agricultural and Resource Economics, Canberra.

Holmes, E.E. (1993) Are diffusion-models too simple – a comparison with telegraph models of invasion. *The American Naturalist* 142, 779–795.

Julien, M., Flanagan, G., Heard, T., Hennecke, B., Paynter, Q. and Wilson, C. (2004) Research and management of *Mimosa pigra*. Papers Presented at the 3rd International Symposium on the Management of *Mimosa pigra*. CSIRO Entomology, Canberra.

Knol, A., Slottje, P., van der Sluijs, J. and Lebret, E. (2010) The use of expert elicitation in environmental health impact assessment: a seven step procedure. *Environmental Health* 9, 19.

Kuhnert, P.M., Martin, T.G. and Griffiths, S.P. (2010) A guide to eliciting and using expert knowledge in Bayesian ecological models. *Ecology Letters* 13, 900–914.

Kynn, M. (2008) The 'heuristics and biases' bias in expert elicitation. *Journal of the Royal Statistical Society: Series A (Statistics in Society)* 171, 239–264.

Leavold, V., Lloyd, L. and Lepetit, J. (2007) *Development of Case Studies on the Economic Impacts of Invasive Species in Africa:* Mimosa pigra. Global Invasive Species Programme, Melbourne, Australia.

Lewis, M.A. (1997) Variability, patchiness, and jump dispersal in the spread of an invading population. In: Tilman, D. and Kareiva, P. (eds) *Spatial Ecology: The Role of Space in Population Dynamics and Interspecific Interactions.* Princeton University Press, Princeton, New Jersey, pp. 46–74.

Lloyd, S.G. and Vinnicombe, T.L. (2010) *Mimosa pigra* L. – a new incursion into Western Australia. In: Zydenbos, S.M. (ed.) *17th Australasian Weeds Conference.* New Zealand Plant Protection Society, Christchurch, New Zealand, pp. 180–181.

Lonsdale, W.M. (1993) Rates of spread of an invading species – *Mimosa pigra* in northern Australia. *Journal of Ecology* 81, 513–521.

Lonsdale, W.M., Miller, I.L. and Forno, I.W. (1989) The biology of Australian weeds. 20. *Mimosa pigra. Plant Protection Quarterly* 4, 119–131.

Low Choy, S., O'Leary, R. and Mengersen, K. (2009) Elicitation by design in ecology: using expert opinion to inform priors for Bayesian statistical models. *Ecology* 90, 265–277.

Martin, T.G., Burgman, M.A., Fidler, F., Kuhnert, P.M., Low-Choy, S., McBride, M. and Mengersen, K. (2012) Eliciting expert knowledge in conservation science. *Conservation Biology* 26, 29–38.

McCann, K., Hastings, A., Harrison, S. and Wilson, W. (2000) Population outbreaks in a discrete world. *Theoretical Population Biology* 57, 97–108.

McCosker, T., McLean, D. and Holmes, P. (2010) *Northern Beef Situation Analysis 2009.* Meat and Livestock Australia, Sydney, Australia. Available at: http://www.mla.com.au/News-and-resources/Publication-details?pubid=5029 (accessed 27 February 2015).

Moody, M.E. and Mack, R.N. (1988) Controlling the spread of plant invasions: the importance of nascent foci. *Journal of Applied Ecology* 25, 1009–1021.

Moran, P. (1984) *An Introduction to Probability Theory.* Oxford University Press, New York.

Myers, J.H., Simberloff, D., Kuris, A.M. and Carey, J.R. (2000) Eradication revisited: dealing with exotic species. *Trends in Ecology & Evolution* 15, 316–320.

Okubo, A. and Levin, S.A. (2002) *Diffusion and Ecological Problems: Modern Perspectives.* Springer, New York.

Paini, D.R. (2012) *Six Degrees of Preparation. Final Report CRC10161 Network Theory and Invasive Species.* Cooperative Research Centre for National Plant Biosecurity, Canberra.

Panetta, F.D. (2007) Evaluation of weed eradication programs: containment and extirpation. *Diversity and Distributions* 13, 33–41.

Panetta, F.D. and Lawes, R. (2005) Evaluation of weed eradication programs: the delimitation of extent. *Diversity and Distributions* 11, 435–442.

Parsons, W.T. and Cuthbertson, E.G. (1992) *Noxious Weeds of Australia.* Inkata Press, Melbourne, Australia.

Paynter, Q. (2005) Evaluating the impact of a biological control agent *Carmenta mimosa* on the woody wetland weed *Mimosa pigra* in Australia. *Journal of Applied Ecology* 42, 1054–1062.

Paynter, Q. and Flanagan, G.J. (2004) Integrating herbicide and mechanical control treatments with fire and biological control to manage an invasive wetland shrub, *Mimosa pigra. Journal of Applied Ecology* 41, 615–629.

Pitt, J.P.W., Worner, S.P. and Suarez, A.V. (2009) Predicting Argentine ant spread over the heterogeneous landscape using a spatially explicit stochastic model. *Ecological Applications* 19, 1176–1186.

Rejmánek, M. and Pitcairn, M.J. (2002) When is eradication of exotic pest plants a realistic goal? In: Veitch, C.R. and Clout, M.N. (eds) *Turning the Tide: The Eradication of Invasive Species. Proceedings of the International Conference on Eradication of Island Invasives.* IUCN SSC Invasive Species Specialist Group, University of Auckland, Auckland, New Zealand, pp. 249–253.

Shigesada, N. and Kawasaki, K. (1997) *Biological Invasions: Theory and Practice.* Oxford University Press, New York.

Shigesada, N., Kawasaki, K. and Takeda, Y. (1995) Modeling stratified diffusion in biological invasions. *The American Naturalist* 146, 229–251.

Simberloff, D. (2003) Eradication – preventing invasions at the outset. *Weed Science* 51, 247–253.

Thorp, J.R. and Lynch, R. (2000) *The Determination of Weeds of National Significance.* National Weeds Strategy Executive Committee, Launceston, Australia.

Ulubasoglu, M., Mallick, D., Wadud, M., Hone, P. and Haszler, H. (2011) *How Price Affects the Demand for Food in Australia – An Analysis of*

Domestic Demand Elasticities for Rural Marketing and Policy. Rural Industries Research and Development Corporation, Canberra.

Vose, D. (2008) *Risk Analysis: A Quantitative Guide*. Wiley, Chichester, UK.

Waage, J.K., Fraser, R.W., Mumford, J.D., Cook, D.C. and Wilby, A. (2005) *A New Agenda for Biosecurity*. Department for Environment Food and Rural Affairs, London.

Walden, D., van Dam, R., Finlayson, M., Storrs, M., Lowry, J. and Kriticos, D. (2004) *A Risk Assessment of the Tropical Wetland Weed Mimosa pigra in Northern Australia*. CSIRO Entomology, Canberra.

11 Spatial Modelling Approaches for Understanding and Predicting the Impacts of Invasive Alien Species on Native Species and Ecosystems

Craig R. Allen,[1]* Daniel R. Uden,[2] Alan R. Johnson[3] and David G. Angeler[4]

[1]*US Geological Survey – Nebraska Cooperative Fish and Wildlife Research Unit, University of Nebraska–Lincoln, Lincoln, Nebraska, USA;* [2]*Nebraska Cooperative Fish and Wildlife Research Unit, School of Natural Resources, University of Nebraska–Lincoln, Lincoln, Nebraska, USA;* [3]*School of Agricultural, Forest, and Environmental Sciences, Clemson University, Clemson, South Carolina, USA;* [4]*Department of Aquatic Sciences and Assessment, Swedish University of Agricultural Sciences, Uppsala, Sweden*

Abstract

Biological invasions threaten native species and ecosystems worldwide. Estimating the level of risk that an invasive alien species poses to native species across landscapes is important for prioritizing mitigation efforts. We describe a risk assessment approach that incorporates spatial heterogeneity in effects and illustrate this method by considering the risk that the red imported fire ant (*Solenopsis invicta*) presents to two native birds. The common ground-dove (*Columbina passerina*), an oviparous, ground-nesting species with altricial young that prefers open habitats, is more susceptible to impacts from fire ants than the swallow-tailed kite (*Elanoides forficatus*), which occupies closed-canopy forests, nests high in trees, is oviparous and has altricial young. Risk approaches that consider landscapes and that are spatially explicit are of particular relevance as remaining undeveloped lands become increasingly uncommon, disjointed and more important for the management and recovery of native species and ecosystems.

Introduction

Invasive alien species rank among the greatest threats to native biodiversity and ecosystems worldwide (Mack *et al.*, 2000; Didham *et al.*, 2007; Hulme, 2009) and are responsible for substantial costs to human societies by diminishing ecosystem goods and services (Pejchar and Mooney, 2009). The effects of invasive alien species on native species and ecosystems are known to be region- and context-specific (Vila *et al.*, 2011; Pysek *et al.*, 2012). Methodologies are needed to incorporate spatial analysis into risk assessments for invasive alien species

*Corresponding author. E-mail: allencr@unl.edu

and to evaluate a large number of potentially affected native species.

Modelling the potential overlap between invasive alien species and native species distributions within a spatial risk assessment framework can help determine which native species and/or ecosystems may be most influenced by the establishment and spread of alien invasive species (Allen et al., 2006). Here, we use Allen et al.'s (2006) application of a spatial risk assessment framework to an invasive alien species of particular concern in the south-eastern USA: the red imported fire ant (*Solenopsis invicta*; Wojcik et al., 2001).

Risk assessment progresses through the phases of: (i) problem formulation; (ii) assessment of exposure and effects; and (iii) risk characterization (EPA, 1998). Probabilistic risk assessments have been used extensively by engineers and mathematicians (Seife, 2003). Toxicologists have extended probabilistic methods to the assessment of risks posed by toxicants to humans and wildlife, and considerable literature addresses potential impacts of chemical stressors on animals and humans. Technological advances in spatial assessments and theoretical advances in landscape ecology now enable explicit consideration of the spatial aspects of many stressors, both chemical and non-chemical (Hope, 2005; Bradley and Mustard, 2006; Lahr et al., 2010). The application of risk assessments to invasive alien species is an example of this widening scope of analysis (Andersen et al., 2004; Hulme, 2013; Kriticos et al., 2013). Invasive species distribution models (iSDMs) can incorporate measures of landscape change and climate change to facilitate comparisons of alternative restoration and policy interventions and support adaptive management of invasive alien species (Hulme, 2013; Kriticos et al., 2013).

Spatial risk analysis, as we envisage its application to impacts of invasive alien species on native species, should proceed through two levels of analysis. In the initial phase, the goal is to determine the probability of spatial co-occurrence of stressors and a target native species. We use the term 'target' to refer to taxa under consideration. Clearly, for invasive alien species to have direct, deleterious impacts on a native species, their geographic distributions must overlap in both space and time. However, the demonstration of overlapping distributions is not, in itself, sufficient evidence to conclude that the target species will be impacted. Thus, this initial level of analysis eliminates combinations of target invasive alien species and native species that fail to co-occur and combinations that do overlap are subjected to the next tier of analysis.

The second tier of analysis attempts to gauge the potential impact of an invasive alien species on the local abundance, regional occurrence or geographic range of target native species and/or the degree of functional connectivity among species in the landscape. These end points are of clear ecological relevance and are likely to be important to managers interested in conservation of one or more target native species. Given sufficient time and knowledge, it is possible to quantitatively estimate stressor-induced changes in abundance for some species, for example, through population viability analyses (Akcakaya, 2004). However, simple methods are needed to sort rapidly through a large number of potentially impacted species and identify those most at risk. We have developed a semi-quantitative index of effects based on categorical ratings of direct and indirect invasion impacts – similar to the rank-based approach to regional risk assessment developed by Landis and Wiegers (1997) and proposed for application to invasive alien species by Landis (2004) – for estimating the relative risks posed by invasive alien species.

Co-occurrence of target invasive alien species and native species is assessed on the basis of respective, field-calibrated species distribution models for native and invasive alien species (Elith et al., 2006; De Marco et al., 2008; Elith and Leathwick, 2009; Gallien et al., 2012). Spatial co-occurrence (Allen et al., 2001) and relative abundances of invasive alien species and native species can be estimated by overlaying forecasted distributions for both species within a geographic information system (Forys et al., 2002).

Frequently, it is not possible to derive precise quantitative estimates of the effects of an invasive alien species; instead, a 'hazard index' may be assigned to each invasive alien species/native species combination. This hazard index is developed based on conceptual models of the modes of interaction between invasive alien species and native species. For each invasive alien species/native species pair, each possible mode of interaction – whether direct or indirect – is assessed as likely (+) or unlikely (–). Scores should be informed with information from peer-reviewed publications whenever possible, with expert opinion being utilized only when reliable local information is unavailable. In our modelling approach, a likely interaction is assumed to have negative outcomes for the native species (i.e. population declines), rather than positive outcomes. Scores can be conveniently represented in a matrix format, analogous to the matrix approach discussed by Foran and Ferenc (1999). The hazard index is calculated by dividing the number of interaction modes under which a target is vulnerable by the total number of interaction modes. Although this index is not directly interpretable in terms of predicted population-level consequences, it provides a comparable set of values that allow for ranking of target native species in terms of the likely magnitude of impact. Such qualitative approaches in ecological risk assessment are well established (Foran and Ferenc, 1999) and have been applied to predicting the risk of spread of invasive alien species (USDA, 1993; USFWS/NOAA, 1996; Landis and Wiegers, 1997; Pheloung et al., 1999).

The Case of the Red Imported Fire Ant

We illustrate our methods with an example from Allen et al. (2006), using South Carolina, USA, as our study area, an invasive alien ant as our environmental stressor and two declining native vertebrates as our target native species. This methodology was developed to sift among a large number of potentially impacted species to identify those species most at risk, thereby enabling scarce resources to be more effectively targeted towards the analysis of potentially deleterious stressor effects on those species most at risk. Thus, this method is a preliminary 'coarse-filter' approach.

Problem formulation

The problem formulation phase of a risk assessment involves careful delineation of the stressor or stressors to be considered, the ecological receptors that might be exposed and the spatial and temporal scale of the analysis. At a minimum, the problem formulation should specify the assessment end points that are going to be used as measures or indicators of effects and provide a conceptual model of how the stressor(s) may interact with organisms or ecosystems to produce those effects (EPA, 1998).

For our example we are focusing on a single stressor, the red imported fire ant. The red imported fire ant is an established, aggressive invasive alien species in the south-eastern USA that is native to the Paraguay and Parana Rivers of South America. Since being introduced to the port of Mobile, Alabama, in the 1930s, fire ants have become the dominant ant species throughout much of the south-eastern USA, having outcompeted or displaced many other ant species. Fire ants have also invaded California (USA), numerous Caribbean Islands, Australia, Taiwan and China (Callcott and Collins, 1996).

Several native species can be affected by fire ants. Fire ants may affect entire ecosystems by altering or eliminating ecological processes like seed dispersal (Zettler et al., 2001). Allen et al. (2004) documented the potential negative impacts of red imported fire ants on wildlife species. Many species of wild animals are susceptible to both direct and indirect fire ant impacts. Ground-nesting species that lay eggs or produce altricial young may be especially vulnerable to direct impacts, which can include predation of newly hatching young and reduced weight gain and survival of non-lethally attacked offspring (Allen *et al.*, 1995). Clearly, however, it is neither cost- nor conservation-efficient to treat all species as equally vulnerable or to conduct experiments to assess the vulnerability of each. Therefore, we decided to develop and apply a relative risk assessment method to focus research and conservation efforts on wildlife species at greatest risk from fire ants now and in the future.

The assessment end point, broadly speaking, is biodiversity, but our analysis focuses on two native bird species. The conceptual model involves a two-step process: if fire ants are to cause impacts on native biodiversity, they must first spatially co-occur with native biota, and then cause negative effects on the native species with life-history characteristics detailed in Table 11.1.

Our example applies to South Carolina, USA, where fire ants are widespread and well established. We focus on assessing their potential adverse impacts.

Assessment of exposure

Fire ants were sampled throughout South Carolina. Sampling was stratified by ecoregion (e.g. sandhills, coastal plain, piedmont and mountains) and by land-cover types as described by the South Carolina gap analysis, derived from 30 m resolution Landsat Imagery (http://www.dnr.sc.gov/GIS/gap/mapping.html). Approximately ten replicates of each land-cover type in each ecoregion were sampled by establishing a linear transect of sample points through a patch. Sample points consisted of bait attractants and pitfall traps.

We modelled the presence/absence of fire ants, as the dependent variable. Independent variables included habitat, soils, aspect and landscape metrics such as patch size, shape, the Euclidean distance between the survey location and nearest development, the area of development at various buffer distances around each survey location and the Euclidean distance between the survey location and nearest paved road. Some variables were recorded in the field at sample locations, while others were derived from digital soil and land-cover maps. After we evaluated and eliminated strongly collinear variables (i.e. those with co-efficients of determination >0.75), the landscape variables were entered into a stepwise logistic analysis to derive a multivariate model that predicted the presence or absence of red imported fire ants. The resulting models were evaluated using goodness-of-fit tests based on maximum likelihood estimates and/or Akaike's Information Criterion (Akaike, 1969; Burnham and Anderson, 1998).

We used two approaches to create maps of the distribution of fire ants. The first approach was based on the outputs of logistic regression models and the second approach was based simply on densities of fire ants sampled in the field. Maps based on logistic regression presence/absence models considered fire ants as 'present' when there was a predicted probability of presence >90%, based on statistical models. The second approach was to build simpler, empirical (i.e. sample-based) models. The percentage of field samples from each land-cover type that detected fire ants was used

Table 11.1. Elements of the ecologically based hazard index to assess the safety/vulnerability of birds with different life-history characteristics to fire ant predation.

Life-history characteristic	Safe	Vulnerable
Reproduction	Viviparous	Oviparous
Eggs	Hard-shelled	Soft-shelled
Nests	In trees	On ground
Foraging	Aerial/arboreal	On ground
Young	Precocious	Altricial
Reproductive timing	Autumn and winter	Spring and summer

as an estimate of the likelihood of fire ants being present in that cover type. In our example, only habitat variables were significant in logistic models. Both approaches provide continuous estimates of the probability of fire ant presence. By applying thresholds to the probability of fire ants being present, we have converted the output from the logistic model into a binary map (Fig. 11.1) of fire ant presence. We used our estimates of the probability of fire ants being present in different land-cover types to produce a continuous map (Fig. 11.2; see colour plate section) that is consistent with varying fire ant densities across land-cover classes. We characterized the exposure of native species to fire ants with a probability of co-occurrence model.

Assessment of effects

We focus our example on two declining species identified as at risk by the state of South Carolina. At-risk species represent animals with small or declining populations, for which additional stressors may be a proximate cause of extinction. The common ground-dove (*Columbina passerina*) is associated with open habitats such as grasslands and savannahs and has been declining across much of its range, including South Carolina, where it is listed as threatened. Much of this decline has been attributed to habitat loss, but fire ants might also present a threat (Cely, 2000). The swallow-tailed kite (*Elanoides forficatus*) is associated with forested wetlands. Its regional population may be stable, but the species is generally rare and has been proposed for listing as an endangered species at federal and state levels (Meyer, 2004). The predominant identified threat is the loss of suitable nesting habitat. We chose these two species as examples because they are of conservation concern and have different life histories which may affect their vulnerability to fire ant effects.

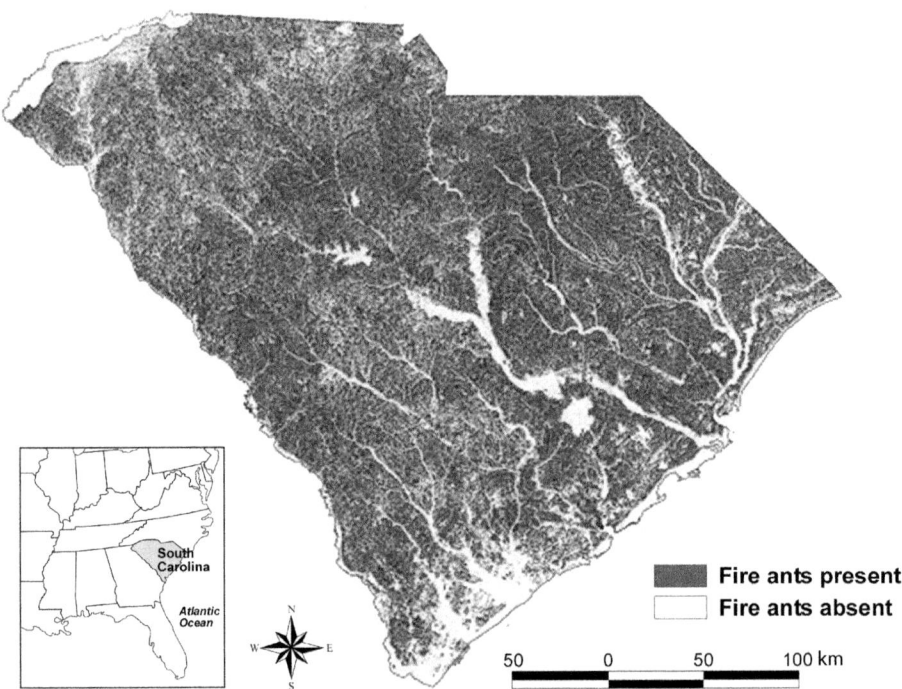

Fig. 11.1. Predicted fire ant distribution map (90% probability) for the state of South Carolina, USA, based on stepwise logistic regression analysis.

When fire ants encounter these two bird species, the outcomes are expected to be different. The common ground-dove has altricial young, prefers open habitats, nests on the ground and is oviparous, all of which makes it more susceptible to impacts from fire ants than the swallow-tailed kite. The swallow-tailed kite has altricial young, occupies closed-canopy forests, nests high in trees and is oviparous (Table 11.2). Thus, the common ground-dove receives higher scores for the overall hazard index (Table 11.2). This form of the index assumes additivity among the potential vulnerabilities.

We used the general species–habitat models developed by the South Carolina Gap Analysis Project (http://www.dnr.sc.gov/GIS/gap/mapping.html) to characterize the potential geographic distributions of the two birds. Gap analysis models create county-range maps for native species based on occurrence records, a habitat matrix based on known species–habitat associations and land-cover classification maps (Scott et al., 1993). Gap-analysis species models are peer reviewed and made freely available in the public domain. Different states release models at different resolutions. South Carolina species models are available at a 30 m resolution (http://www.dnr.sc.gov/GIS/gap/mapping.html).

When invasive alien species have a severe impact or have been present for an extended period, they may have already affected the distribution of a target native species. Field samples might suggest the presence of fire ants but not the presence of our target bird species. Care must be taken to avoid the conclusion based on these data that ants and birds occupy different habitats. While this might be true, it is also possible that ants have already locally extirpated a target species. In our example, bird distribution models are based on potential habitat and, thus, are not influenced by fire ant distributions.

Risk characterization

The first spatial level of risk assessment is accomplished by overlaying the results of the predictive models of fire ant distribution with habitat-relationship models of the two birds (Figs 11.3 and 11.4; see colour plate section). In doing so, we have characterized the probable co-occurrences of fire ants and each of the bird species. The biggest difference in the two models of fire ant occurrence is that in the logistic models, all modelled habitat is either occupied by fire ants or unoccupied (i.e. binary), whereas with the empirically based fire ant distribution models, there is a probability of occurrence associated with each habitat type. These different types of models affect risk characterization. For example, consider upland pine habitats. With logistic models, upland pine is occupied by fire ants and if another species were restricted only to this habitat its co-occurrence with fire ants would be 100% (i.e. 100% spatial overlap); whereas empirically based models attribute upland pine with a 45% occurrence of fire ants (i.e. probability of encounter) and if another species were restricted only to this habitat its co-occurrence with fires ants would be 45%. The risk characterization phase combines co-occurrence and effects into an integrated estimate of risk. Risk is

Table 11.2. Calculation of hazard indices based on six life-history traits that potentially increase species' vulnerability to fire ant impacts. A '+' indicates vulnerability. Hazard indices are produced by dividing the number of vulnerabilities for each target by the total number of potential vulnerabilities.

Species	Life-history trait						Hazard index
	Oviparity	Soft-shelled	Ground-nesting	Ground-foraging	Altricial young	Spring/summer young	
Common ground-dove	+	−	+	+	+	+	0.83
Swallow-tailed kite	+	−	−	−	+	+	0.50

estimated as the product of the hazard index (Table 11.2) and probability of co-occurrence (Table 11.3). Although this risk estimate is not directly interpretable in terms of predicted population-level consequences, it provides a set of values that permits ranking of native species based on the magnitude of likely impact.

Conclusions

Our procedures provide managers an efficient and flexible tool for identifying native species at risk by calculating relative risk for multiple species, determining the spatial distribution of risk and exploring landscape manipulations for reducing risk. This procedure can be applied rapidly and yields a spatially and ecologically based risk assessment. The explicit spatial nature of the assessment could allow one to develop a general framework for evaluating how functional connectivity of habitat and other aspects of landscape patterns affect the risk faced by wildlife species from various stressors. This in turn could allow for the determination of impacts of alternative land-use/land-cover change scenarios on the distribution of stressors and the risk faced by wildlife species and for the exploration of alternative landscape remedial and/or proactive actions to reduce that risk.

Our fire ant distribution model was developed for South Carolina, where local variables were important for predicting fire ant presence/absence. In larger areas, or areas with steep environmental gradients, climatic variables may also be important for modelling invasive alien species' distributions. Sufficient a priori consideration must be given to the appropriate spatial scale(s) for an analysis and the appropriate predictor variables for inclusion in distribution models.

Our methods allow for the incorporation of scenarios of potential land-use or land-cover change. Land-use or land-cover changes affect the distribution of both invasive alien species and native species. In our example system, decreasing closed-canopy forest area and increasing open and disturbed habitats likely would lead to increased distributions and abundances of fire ants. These same changes differentially affect target native species. Closed-canopy specialists, such as the swallow-tailed kite, will have decreased distributions and less core area, and vulnerability to fire ant impacts will increase. Open habitat specialists, such as the ground-dove, will have an increase in overall distribution, but the total area and proportion of range overlapping with fire ants will also increase, as will risk.

In addition to their applicability for a broad class of invasive alien species and native species, ecological risk assessments can be adapted for use at a broad range of spatial scales. 'Coarse-filter' assessments are most appropriate at regional scales. These methods are potentially useful for risk managers in both the private and public sectors at federal, state and local scales. There is clear applicability for screening risks at broad scales, for example, to determine risks faced by native wildlife across a landscape with the introduction of new diseases. There is also utility for managers of conservation areas, who may wish to determine the potential impacts of a new invasive alien species or wish to explore the

Table 11.3. Probability of spatial co-occurrence between bird species and fire ants in South Carolina. Co-occurrence is based on results of a logistic regression model and an empirical model of fire ant distribution. Overall risk index, shown in parentheses, is calculated as the product of the probability of co-occurrence and the hazard index.

Species	Probability of co-occurrence with fire ants	
	Logistic regression model	Empirical model
Common ground-dove	0.83 (risk index = 0.69)	0.64 (risk index = 0.53)
Swallow-tailed kite	0.50 (risk index = 0.34)	0.40 (risk index = 0.20)

effect of habitat manipulations on it. When multiple stressors create management challenges, these methods will allow for the identification of the native species most at risk, the stressors with the greatest impact, and habitat and landscape manipulations that may reduce impacts.

Our approach leads to a landscape spatial risk assessment of the potential impacts of invasive alien species on native wildlife. After native species potentially at risk because of spatial overlap are identified, the level of risk faced by that native species is determined by incorporating ecological attributes of the target and invasive alien species. Species judged to be facing significant risk may then be further investigated, allowing scarce resources to be spent only on those native species identified as being most at risk. Further investigation may include experiments conducted in the laboratory or controlled field experiments, which may identify the most effective means of risk reduction for a particular native species or stressor. Such experiments will also be useful for identifying whether multiple stressors interact in synergistic or antagonistic ways.

The approach to assessing impacts from invasive alien species that we have described focuses on landscapes and is explicitly spatial. Such models will be important for the conservation and restoration of species and ecological systems as undeveloped lands become increasingly restricted and isolated. Effective ecosystem management includes the control of invasive alien species and other stressors with large impacts, understanding where those impacts may be the most severe and implementing management strategies to reduce impacts.

References

Akaike, H. (1969) Fitting autoregressive models for prediction. *Annals of the Institute of Statistical Mathematics* 21, 243–247.

Akcakaya, H.R. (2004) Using models for species conservation and management: an introduction. In: Akcakaya, H.R., Burgman, M.A., Kindvall, O., Wood, C.C., Sjogren-Gulve, P., Hatfield, J.S. and McCarthy, M.A. (eds) *Species Conservation and Management: Case Studies*. Oxford University Press, New York, pp. 2–14.

Allen, C.R., Lutz, R.S. and Demarais, S. (1995) Red imported fire ant impacts on northern bobwhite populations. *Ecological Applications* 5, 632–638.

Allen, C.R., Pearlstine, L.G., Wojcik, D.P. and Kitchens, W.M. (2001) The spatial distribution of diversity between disparate taxa: spatial correspondence between mammals and ants. *Landscape Ecology* 16, 453–464.

Allen, C.R., Epperson, D. and Garmestani, A. (2004) The impacts of fire ants on wildlife: a decade of research. *American Midland Naturalist* 152, 88–103.

Allen, C.R., Johnson, A.R. and Parris, L. (2006) A framework for spatial risk assessments: potential impacts of nonindigenous invasive species on native species. *Ecology and Society* 11, 39.

Andersen, M.C., Adams, H., Hope, B. and Powell, M. (2004) Risk assessment for invasive species. *Risk Analysis* 24, 787–793.

Bradley, B.A. and Mustard, J.F. (2006) Characterizing the landscape dynamics of an invasive plant and risk of invasion using remote sensing. *Ecological Applications* 16, 1132–1147.

Burnham, K.P. and Anderson, D.R. (1998) *Model Selection and Inference: A Practical Information-Theoretic Approach*. Springer, New York.

Callcott, A.-M. and Collins, H.L. (1996) Invasion and range expansion of imported fire ants (Hymenoptera: Formicidae) in North America from 1918–1995. *Florida Entomologist* 79, 240–251.

Cely, J.E. (2000) Status and distribution of the common ground-dove in South Carolina. *The Chat* 64, 37–46.

De Marco, P. Jr, Diniz-Filho, J.A.F. and Bini, L.M. (2008) Spatial analysis improves species distribution modelling during range expansion. *Biology Letters* 4, 577–580.

Didham, R.K., Tylianakis, J.M., Gemmell, J.M., Rand, T.A. and Ewers, R.M. (2007) Interactive effects of habitat modification and species invasion on native species decline. *Trends in Ecology and Evolution* 22, 489–496.

Elith, J. and Leathwick, J.R. (2009) Conservation prioritization using species distribution models. In: Moilenan, A., Wilson, K.A. and Possingham, H.P. (eds) *Spatial Conservation Prioritization: Quantitative Methods and Computational Tools*. Oxford University Press, Oxford, pp. 70–93.

Elith, J., Graham, C.H., Anderson, R.P., Dudik, M., Ferrier, S., Guisan, A., Hijmans, R.J.,

Huettmann, F., Leathwick, J.R., Lehmann, A., Lucia, J.L., Lohmann, G., Loiselle, B.A., Manion, G., Moritz, C., Nakamura, M., Nakazawa, Y., Overton, J.M., Peterson, A.T., Phillips, S.J., Richardson, K., Scachetti-Pereira, R., Schapire, R.E., Soberon, J. Williams, S., Wisz, M.S. and Zimmermann, N.E. (2006) Novel methods improve prediction of species' distributions from occurrence data. *Ecography* 29, 129–151.

EPA (1998) *Guidelines for Ecological Risk Assessment. EPA/630/R-95/002F*. Environmental Protection Agency, Washington, DC.

Foran, J.A. and Ferenc, S.A. (1999) *Multiple Stressors in Ecological Risk and Impact Assessment*. SETAC Press, Pensacola, Florida.

Forys, E.A., Allen, C.R. and Wojcik. D.P. (2002) Influence of proximity and amount of human development and roads on the occurrence of the red imported fire ant in the Lower Florida Keys. *Biological Conservation* 108, 27–33.

Gallien, L., Douzet, R., Pratte, S., Zimmerman, N.E. and Thuiller, W. (2012) Invasive species distribution models – how violating the equilibrium assumption can create new insights. *Global Ecology and Biogeography* 21, 1126–1136.

Hope, B.K. (2005) Performing spatially and temporally explicit ecological exposure assessments involving multiple stressors. *Human and Ecological Risk Assessment* 11, 539–565.

Hulme, P.E. (2009) Trade, transport and trouble: managing invasive species pathways in an era of globalization. *Journal of Applied Ecology* 46, 10–18.

Hulme, P.E. (2013) Weed risk assessment: a way forward or a waste of time? *Journal of Applied Ecology* 49, 10–19.

Kriticos, D.J., Venette, R.C., Baker, R.H.A., Brunel, S., Koch, F.H., Rafoss, T., van der Werf, W. and Worner, S.P. (2013) Invasive alien species in the food chain: advancing risk assessment models to address climate change, economics and uncertainty. *NeoBiota* 18, 1–7.

Lahr, J., Münier, B., De Lange, H.J., Faber, J.F. and Sørensen, P.B. (2010) Wildlife vulnerability and risk maps for combined pollutants. *Science of the Total Environment* 408, 3891–3898.

Landis, W.G. (2004) Ecological risk assessment conceptual model formulation for nonindigenous species. *Risk Analysis* 24, 847–858.

Landis, W.G. and Wiegers, J.A. (1997) Design considerations and a suggested approach for regional and comparative ecological risk assessment. *Human and Ecological Risk Assessment* 3, 287–297.

Mack, R.N., Simberloff, D., Lonsdale, W.M., Evans, H., Clout, M. and Bazzaz, F.A. (2000) Biotic invasions: causes, epidemiology, global consequences, and control. *Ecological Applications* 10, 689–710.

Meyer, K.D. (2004) *Demography, Dispersal, and Migration of the Swallow-Tailed Kite: Final Report*. Florida Fish and Wildlife Conservation Commission, Tallahassee, Florida.

Pejchar, L. and Mooney, H.A. (2009) Invasive species, ecosystem services and human well-being. *Trends in Ecology and Evolution* 24, 497–504.

Pheloung, P.C., Williams, P.A. and Halloy, S.R. (1999) A weed risk assessment model for use as a biosecurity tool evaluating plant introductions. *Biological Conservation* 57, 239–251.

Pysek, P., Jarosik, V., Hulme, P.E., Pergl, J., Hejda, M., Schaffner, U. and Vila, M. (2012) A global assessment of invasive plant impacts on resident species, communities and ecosystems: the interaction of impact measures, invading species' traits and environment. *Global Change Biology* 18, 1725–1737.

Scott, J.M., Davis, F., Csuti, B., Noss, R., Butterfield, B., Groves, C., Anderson, H., Caicco, S., Derchia, F., Edwards, T.C. Jr, Ulliman, J. and Wright, R.G. (1993) Gap analysis: a geographical approach to protection of biological diversity. *Wildlife Monographs* 123, 3–41.

Seife, C. (2003) The Columbia disaster underscores the risky nature of risk analysis. *Science* 299, 1001–1002.

USDA (1993) *Pest Risk Assessment of the Importation of* Pinus radiata, Nothofagus dombeyi, *and* Laurelia philippiana *Logs from Chile. Forest Service Miscellaneous Publication No. 1517*. US Department of Agriculture, Forest Service, Washington DC.

USFWS/NOAA (1996) *Generic Nonindigenous Aquatic Organisms Risk Analysis Review Process. Report to the Aquatic Nuisance Species Task Force*. US Fish and Wildlife Service/National Oceanic and Atmospheric Administration, Washington, DC.

Vila, M., Espinar, J.L. Hejda, M., Hulme, P.E., Jarosik, V., Maron, J.L., Pergl, J., Schaffner, U., Sun, Y. and Pysek, P. (2011) Ecological impacts of invasive alien plants: a meta-analysis of their effects on species, communities and ecosystems. *Ecology Letters* 14, 702–708.

Wojcik, D.P., Allen, C.R., Brenner, R.J., Forys, E.A., Jouvenaz, D.P. and Lutz, R.S. (2001) Red imported fire ants: impact on biodiversity. *American Entomologist* 47, 16–23.

Zettler, J.A., Spira, T.P. and Allen, C.R. (2001) Ant–seed mutualisms: can fire ants sour the relationship? *Biological Conservation* 101, 249–253.

12 Process-based Pest Risk Mapping using Bayesian Networks and GIS

Rieks D. van Klinken,[1]* Justine V. Murray[1] and Carl Smith[2]

[1]*CSIRO Biosecurity Flagship, Brisbane, Queensland, Australia;*
[2]*School of Agriculture and Food Sciences, The University of Queensland, St Lucia, Queensland, Australia.*

Abstract

Predicting the potential distribution of invasive alien pests (i.e. habitat suitability modelling) and their potential spread from existing populations (i.e. habitat susceptibility modelling) is critical to guide management responses at local, regional and national scales. We use the management of Chilean needle grass (*Nassella neesiana*) invasion in a 260,791 km^2 part of eastern Australia as an example to describe a process-based approach for making such predictions with publicly available software (e.g. NETICA and ESRI products). The approach is deductive, with causal relationships captured in a Bayesian network and represented spatially at fine resolution using a geographic information system (GIS). Pest risk responses to changing environments, such as land-use change, climate change or altered flood regimes, and to management interventions can be tested through scenario analysis. Predictive risk mapping of invasive aliens is often knowledge-constrained; therefore, our approach seeks to capture the best available knowledge from often disparate sources in a transparent and explicit manner. For Chilean needle grass, we elicited process understanding from experts through a participatory approach, integrated an existing bioclimatic model and obtained our own field data. Our model, thereby, represents a hypothesis of what determines the distribution, abundance and spread of Chilean needle grass in the modelled region. Specifically, the model forecasts the likelihood of the weed reaching a threshold density (e.g. in this case, >30% ground cover) as defined by the experts. This approach to likelihood estimation contrasts with the presence/absence predictions of most other models. Modelling was done at a sufficiently fine spatial resolution (i.e. 30 m) to capture relevant invasion dynamics. Finally, we illustrate how validation can be used to give end users confidence in model predictions and to identify important knowledge gaps and uncertainties. We demonstrate how the resulting pest risk maps for Chilean needle grass can guide management decisions.

Introduction

Pest risk modelling aims to help decision makers identify and quantify risks of establishment and spread of invasive alien

* Corresponding author. E-mail: Rieks.VanKlinken@csiro.au

© CAB International 2015. *Pest Risk Modelling and Mapping for Invasive Alien Species* (ed. R.C. Venette)

species and to craft and implement management responses to best manage those risks. Regional-scale pest risk modelling requires several features for it to be of most value to decision makers. First, it needs to be foretelling as it can rarely be assumed that the pest has already reached all possible environments within the region (Sutherst and Bourne, 2009); and even if it has, end users often want to know how pest abundance will respond under novel environmental or management conditions. Second, fine-spatial-resolution processes are likely to be important in determining pest risk and therefore need to be incorporated when possible. For example, risk of invasion by a riparian weed may depend on the interaction of weed inundation patterns with suitable soil, the herbaceous layer and disturbance history, each of which can vary considerably at fine spatial resolutions. Third, a general need exists to forecast where a pest will reach densities that cause serious impact, rather than just presence and absence. Finally, assessments of risk ideally should be probabilistic to reflect the inherent uncertainty in risk analysis and risk management.

This chapter describes a recently developed causal (i.e. process-based) modelling approach that combines Bayesian networks and geographic information systems (GISs) to meet requirements for regional-scale forecasts of species' invasions. The approach aims to capture causes of invasion in order to forecast where the organism can reach high densities if it arrives (i.e. habitat suitability) and where invasion of suitable habitat from existing populations is most imminent (i.e. habitat susceptibility). It also aims to characterize how habitat suitability and susceptibility might be affected through management alternatives or environmental changes (e.g. global climate, land use). Down-scaling forecasts in this way can be constrained by data availability. Our approach, therefore, draws upon all available information sources, including distributional data, expert knowledge and other models, to identify causal processes that may affect the future course of an invasion.

This approach to regional-scale pest risk modelling has been used to predict habitat suitability and susceptibility of rabbits (J.V. Murray, Queensland, 2012, personal communication) and several weed species in Australia (van Klinken and Murray, 2011; Murray et al., 2012; Smith et al., 2012). The regional foci for these models have been large, including the Murray Darling Basin in south-eastern Australia (1,061,469 km^2), or a part thereof (i.e. Queensland Murray Darling Basin, 260,791 km^2), and the Desert Channels Region in north-eastern Australia (509,933 km^2). In each case, modelling has been in close collaboration with relevant regionally focused natural resource management groups. Models were constructed to examine the effect of a wide range of different management scenarios and the potential effects of changes in land management or climate.

This approach should be applicable to any spatial extent (i.e. local, regional, national, continental or global) or pest organism. The process is currently time-consuming, so it is most applicable to serious invaders that pose significant threats or may require substantial management effort. Time requirements will change as the process is streamlined. Spatial extent is not a computational impediment. The main limitation is the availability of requisite environmental data, although useful results can still be generated in relatively data-poor regions (Smith et al., 2012). Models aim to capture the current state of qualitative and quantitative knowledge regarding the circumstances under which an invasive alien organism can reach high densities and under which it can spread. Therefore, the more there is known of the species, the greater the predictive confidence. In many cases, modelling serves an important role in identifying the key knowledge gaps for less-understood species and regions.

In this chapter, we present a broad overview of the causal modelling approach with specific details provided for one case study. Because the approach can be applied with a range of readily available software, we focus on the principles behind the approach rather than prescriptive instructions.

Instructions would vary depending on the software of choice. This modelling approach has been successful, in part, because we emphasize verification, validation and quantification of uncertainty, so these topics receive special attention in this chapter. We illustrate our approach using Chilean needle grass (*Nassella neesiana*) in the Queensland Murray Darling Basin (QMDB) in eastern Australia. Chilean needle grass is a temperate perennial from South America. It established at Clifton, a small town in the QMDB, in the early 1970s and is slowly spreading into pastures and grasslands in eastern Australia. This weed threatens pastoral production and the environment, and, as such, is currently an eradication target in the QMDB. The modelling effort aimed to identify the area at risk and to provide guidance on the best management options and extension messages for Chilean needle grass across the QMDB.

Overview of the Modelling Approach

The modelling approach is described graphically in Fig. 12.1. Modelling starts with a question, which depends on the problems expressed by end users. In this example, end users wanted to know what threat Chilean needle grass posed to the QMDB; how best to manage that threat across the QMDB; whether natural barriers exist that might limit spread; who and what was at risk if eradication failed; and whether land managers could minimize potential impacts through preparations. Answers to

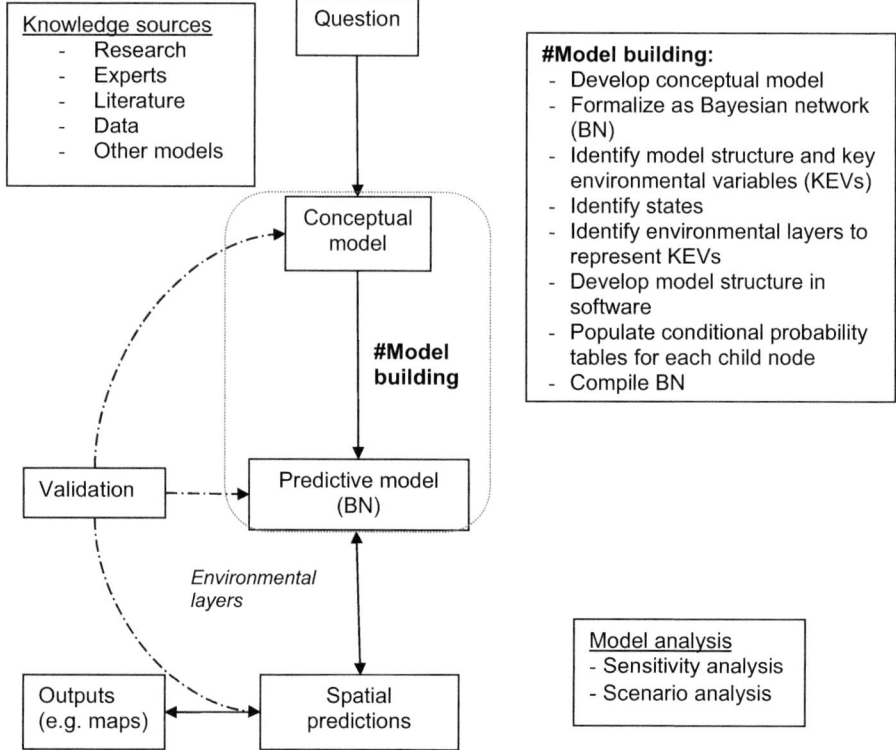

Fig. 12.1. Overview of the modelling approach and an outline of the key methodological steps in building the model and generating spatial risk predictions. See the online supplement to Chapter 12 for details on making spatial predictions.

these questions involve: (i) identifying the area potentially at risk from Chilean needle grass in the QMDB (i.e. suitable habitat); (ii) identifying the area at most immediate threat of invasion based on dispersal pathways and current distribution (i.e. susceptible habitat); (iii) determining how suitability and susceptibility could be reduced, such as by managing dispersal pathways and altering land-management practices; and (iv) determining the likelihood of Chilean needle grass reaching harmful densities (i.e. the response variable of interest in our model). A conceptual model is built to capture the main invasion processes (i.e. introduction, establishment and persistence of populations) and the key environmental variables that drive these processes (Fig. 12.1). The conceptual model is formalized as a predictive Bayesian network (Figs 12.1 and 12.2). Forecasts are made spatially explicit by linking key environmental variables to available environmental data (see the online supplement to Chapter 12) and are presented in the form of risk maps and summary statistics. Several knowledge sources can be consulted when building the model (Fig. 12.1). Validation can target different components of the model framework (Fig. 12.1) and is important in gaining support. An important role of validation is testing how well the predictive model captures current understanding and identifying gaps or limitations in that understanding which may affect spatial predictions (Fig. 12.1).

Information Sources to Build Models

Once the purpose of the model and the response variable have been determined (see section on 'Overview of the Modelling Approach' above), different data sources can be used to develop the model. For example, the model theoretically could rely entirely on distributional data to identify the key environmental drivers for habitat suitability and susceptibility; however, in our experience, sufficient data are rarely, if ever, available. Existing distribution records are often incomplete, lack the associated environmental information necessary to infer responses to land management or other factors and are available at insufficient spatial resolution to infer process understanding. For example, distributional data from which effects of disturbance history can be inferred are rare. Some of these

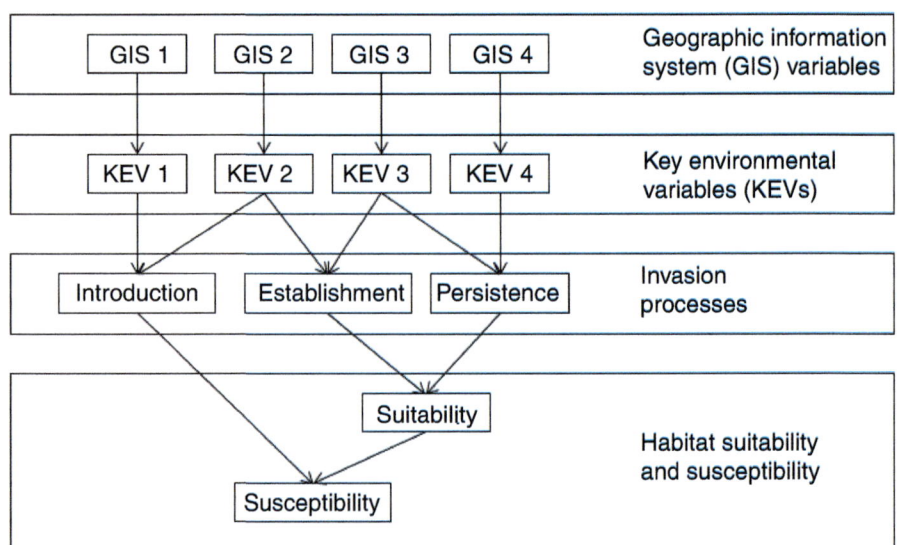

Fig. 12.2. Conceptual framework for modelling habitat suitability and habitat susceptibility to Chilean-needle-grass invasion.

problems can be addressed through stratified field sampling (Murray et al., 2012). Often only a subsample of the environmental space can be sampled, so relationships still must be inferred from other data sources. Also, site-level drivers of pest density such as historical disturbance regimes may not be obvious at the time of sampling.

After we had reviewed available information for Chilean needle grass, we concluded that available published information was inadequate for our purposes. In response, we applied participatory modelling techniques that sought to capture relevant information from experts and supplemented those opinions with additional data sources such as bioclimatic models and distributional data.

Example of a Participatory Modelling Process

A 1.5-day workshop was held in the QMDB to identify consensus about the most likely drivers of Chilean-needle-grass abundance and spread among local experts. This consensus represents a working hypothesis for how Chilean needle grass will respond to the environment and management within the QMDB. This approach is akin to collaborative scientific enquiry where diverse scientists work together to develop a deeper understanding of a system. It is quite different from elicitation processes that seek the range of individual opinions or judgements (Allan et al., 2010; Low Choy et al., 2012) from representative and independent experts (Martin et al., 2012). Rather, it is a facilitated process to generate a predictive model (Fig. 12.1) that captures current process understanding derived from shared and tested (i.e. debated) experiences, observations and interpretations from individuals with first-hand knowledge.

Selection of experts is important. Our workshop had carefully selected experts with direct knowledge of diverse aspects of Chilean needle grass in the QMDB. The experts also had a keen interest in better understanding the species and its management. Our workshop participants included a soil scientist, environmental scientists and other researchers with knowledge of the organism from laboratory studies or field tests; farmers and landholders with observational knowledge that often spanned decades; and natural resource managers who had a regional perspective on Chilean needle grass from extensive discussions with local landholders and had first-hand experience controlling the spread of this weed. These experts were among the primary end users who were interested in the management implications of the modelling. Therefore, the participatory approach has the added benefit of establishing engagement, ownership and trust of the modelling process and outcomes among end users from the outset.

The workshop accessed knowledge from experts by using various elicitation techniques, including small breakout groups to encourage individual input and entire-group participation for feedback and group consensus (Murray et al., 2012; Smith et al., 2012). For example, experts were asked to identify individually the most important spread pathways by drawing on their first-hand experiences of Chilean-needle-grass spread and knowledge of its biology. These hypotheses were tested against the combined experiences of the group. Alternative models, representing different hypotheses, could be built if consensus could not be reached. For the engagement of experts to succeed, the facilitator must have a good understanding of pest risk processes and the ability to stimulate hypothesis generation/testing among the group, but not first-hand knowledge of the organism or its invasion threat.

Our goal during the Chilean-needle-grass workshop was to develop an initial Bayesian network on the first day and to solicit feedback on the model structure, populate the conditional probability tables and gather information to develop spatial predictions on the second day. A follow-up workshop, which included most of the same participants, was held to obtain feedback on model performance (see section on 'Validation and Quality Control' below).

Developing the conceptual framework

First, objectives for the Chilean-needle-grass model and their implications for model structure were agreed upon. The workshop attendees agreed that the main objectives of the model were to determine: (i) how quickly and by what means Chilean needle grass will spread from existing locations; (ii) where Chilean needle grass might reach harmful densities; and (iii) how sensitive spread and habitat suitability were to management. Therefore, we decided that the model needed to: (i) compare the relative effects of managing grazing, Chilean needle grass itself or dispersal pathways; (ii) include a model of spread over 10 years, a timeframe relevant for management; (iii) include all of the QMDB; and (iv) determine the likelihood of Chilean needle grass reaching or exceeding densities (e.g. 30% cover) that were relevant to management. With these objectives and goals in mind, we then framed a conceptual model (Fig. 12.1) that included obtaining agreement on the invasion processes and identifying and prioritizing the key environmental variables thought to be most important in determining distribution, abundance and spread in the modelled region (Fig. 12.2).

Suitable habitat occurs where conditions allow populations to establish and persist after they arrive (Fig. 12.2). This definition of habitat suitability differs slightly from other uses in this book. The initial establishment (i.e. founding) of a population of Chilean needle grass may require ephemeral conditions for germination and recruitment, whereas a different set of conditions may be needed for populations to persist (Murray et al., 2012; Smith et al., 2012). These two phases are influenced by key environmental variables, which for Chilean needle grass include physical conditions (e.g. climate, soil moisture and soil type), biotic conditions (e.g. competition with ground cover and disturbance regime) and pest- or land-management practices (e.g. cultivation, grazing and Chilean-needle-grass management; Fig. 12.3). We model each phase separately.

Susceptible habitat is where propagules from existing source populations can reach suitable habitat (Fig. 12.2). Most pests can disperse through multiple pathways and the challenge is to identify and include the most important ones in the model. For Chilean needle grass, we identified several important modes of dispersal: (i) local (i.e. farm-scale) dispersal by animals, humans and grass-cutting machinery (i.e. 'slashers' in Australia); and (ii) long-distance dispersal in contaminated vehicles, farm goods and flood waters (Fig. 12.3). These modes of dispersal operate at different spatial scales and are captured in the model.

Formalizing the model as a Bayesian network

A Bayesian network is a directed acyclic graph that describes probabilistic relationships among variables, represented as boxes (henceforth, 'nodes'), and is used to convert a conceptual model into a quantitative model. In this case, our Bayesian network is used to calculate an invasion probability for any set of environmental variables (Fig. 12.3). Nodes can represent continuous data or have two or more mutually exclusive categories (henceforth, 'states'). Causal relationships within the network are represented by directional links (i.e. arrows) from parent nodes to child nodes (Marcot et al., 2001; Marcot, 2006, McCann et al., 2006). Behind each child node, a conditional probability table (CPT) specifies the probability of the node being in each of its states given the state of parent nodes that link into it (Fig. 12.3). Nodes without parents (i.e. inputs) also have probability tables; however, these simply store the prior unconditional probabilities for each of their states. Prior probabilities for the input nodes can come from data, scenario analysis or, as in our case, from spatial data layers when they are read into the model.

Several software packages are available for constructing Bayesian networks. For details, see Korb and Nicholson (2004) and Uusitalo (2007). For Chilean needle grass,

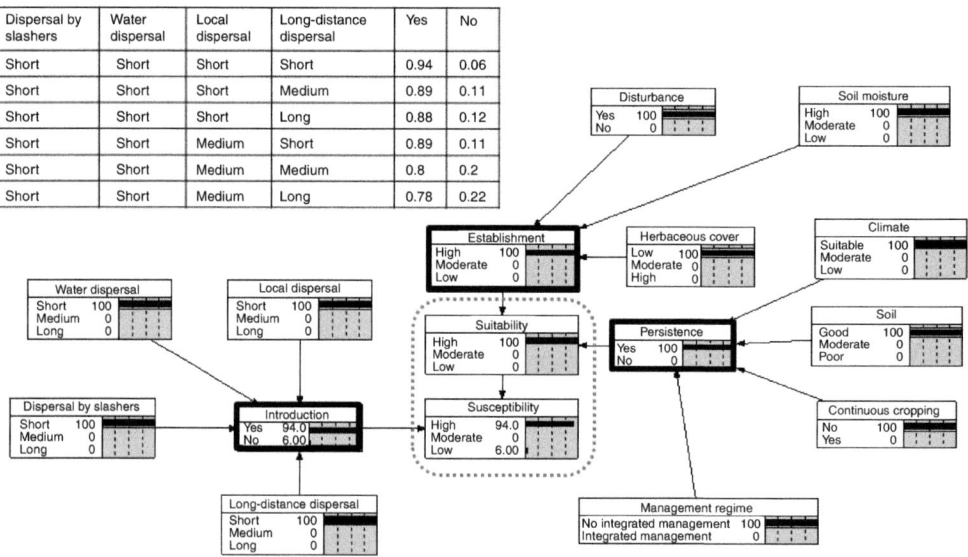

Fig. 12.3. Bayesian network for Chilean-needle-grass habitat suitability and susceptibility developed with experts in Queensland. In this scenario, likelihoods for each parent environmental variable are set to 100% for the most favourable state to give the best possible conditions for habitat suitability and susceptibility. An example of the conditional probability tables (CPTs) behind each node is also given for Introduction. Key environmental variables drive invasion processes (outlined in bold), as described in Fig. 12.2.

we used NETICA™ version 4.12 (Norsys Software Corporation, 2009) which has the advantages of being relatively inexpensive, easy to use and adaptable to a workshop setting. However, it lacks sophisticated analytical tools and other features (e.g. ability to handle continuous variables without discretizing, Monte Carlo simulation methods and object-orientated Bayesian networks) of programs such as HUGIN, BAYESIA and WINBUGS that would be useful for more advanced applications and validation approaches.

The conceptual framework for Chilean needle grass was formalized into a Bayesian network during the model-building workshop using the following steps:

1. Habitat suitability, habitat susceptibility, invasion processes and key environmental variables were defined (Fig. 12.2). The processes and key environmental variables were each divided into discrete states (Fig. 12.3) and each state was defined. We were careful to ensure that definitions could be assessed in the field. Pragmatically, the current approach is constrained to three or four environmental variables and two or three states for each variable because CPTs (Fig. 12.3) can quickly become large and cumbersome to populate as the number of nodes and states increases. For example, we constrained the states for soil-type requirements for Chilean-needle-grass persistence as good (i.e. soil types that maintain vigorous populations when all other conditions are met), moderate (i.e. vigorous populations are possible, but plants rarely become dominant and may be sensitive to other conditions being suboptimal) and poor (i.e. only sufficient to sustain isolated plants at best, even when all other conditions are ideal).

2. In our study, experts populated CPTs behind each node (Fig. 12.3) by using a structured probability elicitation technique from a CPT calculator (Cain, 2001). The CPT calculator works by reducing the number of scenarios in a CPT to those in which only one parent node is not in its most favourable state (i.e. most favourable for introduction, establishment or persistence in this case) and to best and worst case scenarios. Probabilities are then elicited for these scenarios from experts and the calculator interpolates probabilities linearly for all scenarios in the full CPT. Subsequently each probability can be adjusted if required, for example, to capture non-linear relationships. The CPT calculator reduces the number of probabilities that need to be elicited from experts and also maintains logical consistency in the probability elicitation process. Nonetheless, the size of CPTs should be constrained to simplify checking and validation. When using the CPT calculator, the first state of each node must be the 'best case' for the invasive alien species (e.g. requirements necessary for the highest chance of establishment) and the last state is the 'worst case'.

The model structure and definitions were presented back to participants at the workshop to ensure that the structure and definitions captured their understanding. The presentation included running intermediate condition tests (i.e. likely scenarios) and extreme condition tests (i.e. special or unusual scenario s where factors are very favourable or very unfavourable) to explore model behaviour. Alterations were made to the model when experts felt that their understanding was not adequately captured. In the case of Chilean needle grass, we only needed to adjust a small number of CPT values.

Workshop participants might disagree about the importance of some processes, environmental variables or model structure. Divergent views do not need to be resolved. In fact, alternative explanations should be actively sought and evaluated during the workshop process. When options can be well argued and supported, alternative models can be developed, ultimately to test how important the disagreement is in terms of risk forecasts. For example, experts might disagree about how Chilean-needle-grass persistence is affected by shading. If they did disagree, forest cover could be added to the model to test how spatial risk predictions are influenced by this assumption. However, when building the Queensland Chilean-needle-grass model, all experts agreed.

The habitat susceptibility model requires estimates of the dispersal capacity of Chilean-needle-grass propagules. Dispersal might vary from hundreds of metres to hundreds of kilometres. Useful empirical data are rare and expert elicitation is particularly difficult because experts are asked to estimate probabilities of propagule movement through different pathways. Such questions are common to most regional-scale dispersal modelling efforts (Crossman et al., 2011; Frid et al., 2013). Our focus has been on the relative importance of different pathways for moving Chilean-needle-grass propagules (Box 12.1). Qualitative estimates of importance are sufficient for predicting relative risk, but not for estimating absolute seed numbers which requires demographic approaches that are computationally more intensive (Hadincova et al., 2008; Coutts et al., 2011).

Making Spatial Predictions

The Bayesian network is used to generate spatial predictions by importing environmental variables for each pixel from a text file, calculating likelihoods (e.g. likelihood of being highly suitable) for each combination of environmental variable values and exporting the outcomes into GIS to produce maps of the likelihood of Chilean-needle-grass invasion (Fig. 12.1; online supplement to Chapter 12). We converted each environmental data layer to a grid using the finest spatial resolution available (~30 m^2). This scale is also used for a number of derived data layers from remote sensing data, most notably digital elevation maps and herbaceous cover maps (Table 12.1). This resolution is, therefore, sufficient to capture some spatial heterogeneity in some key environmental variables (e.g. ground cover and disturbance regime). Inaccuracies from

Box 12.1. Eliciting dispersal kernels

Empirical estimates for long-distance dispersal were unavailable for Chilean needle grass, so we used experts to generate probabilistic dispersal kernels (i.e. estimates of the probability that an individual will move a particular distance) for each pathway and asked them to characterize this probability over 10 years. This period was sufficient to allow experts to integrate temporal idiosyncrasies in dispersal distance, such as periodic flooding, into their estimates. This assumption ignores the possibility that spread over 10 years might be the result of annual step-wise dispersal.

Experts can find it difficult to estimate low-probability events. We therefore asked experts to estimate how far propagules will travel given certain likelihoods (Fig. 12.4), rather than estimate the probability of propagules travelling certain distances. Experts were asked to estimate distances that Chilean-needle-grass seeds would disperse from known sources for given likelihoods of occurrence, certain (100% probability), likely (50%), unlikely (10%) and exceptional (< 1%), together with a rationale or evidence for those estimates (Fig. 12.4). These estimates were then incorporated into the Bayesian network for each dispersal state. The result is a relativized assessment of dispersal risk for each pathway, which in our experience is sufficiently robust to provide the necessary guidance to end users.

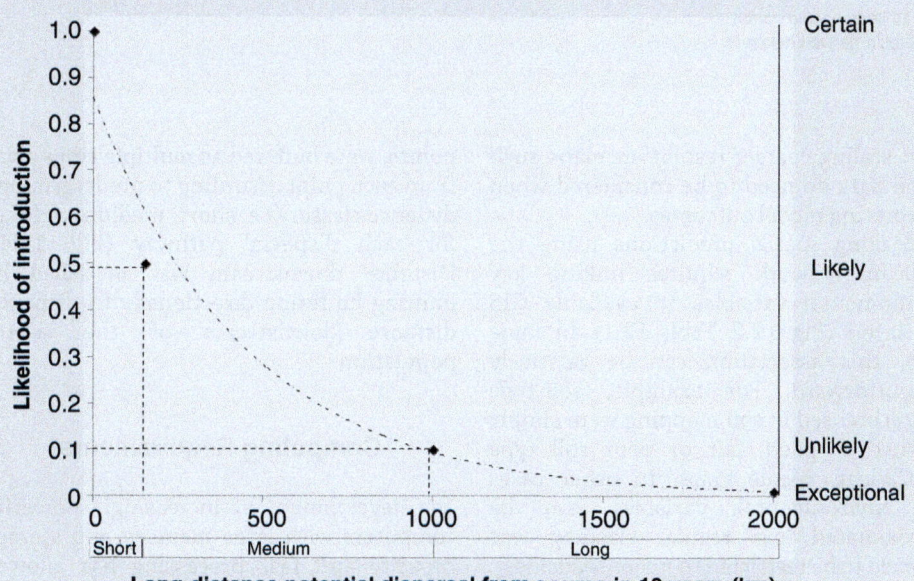

Fig. 12.4. Chilean-needle-grass dispersal distances for the long-distance dispersal pathway over 10 years estimated by experts for different likelihoods (right axis). Likelihood of introduction (left axis) was then conservatively calculated for each distance class as originally defined by experts, in this case 0–100 km, 100–1000 km, 1000–2000 km and >2000 km, for inclusion in the Bayesian network.

Table 12.1. A description of the GIS data layers used to represent key environmental variables in the Chilean-needle-grass model.

Key environmental variable	GIS data layer	Data type	Spatial resolution
Soil type[a]	Land systems data (DSITIA, 2013)	Polygon	1:100,000
Climate	Bioclimatic model output (Bourdôt et al., 2010)	Raster	0.01°
Soil moisture	AWAP data (Raupach et al., 2009, 2012)	Raster	0.05°
Herbaceous cover	Derived from Landsat imagery (Scarth et al., 2006)	Raster	30 m
Disturbance[a]	Inferred: roadways	Polygon	1:100,000
Continuous cropping	ACLUMP (DAFF, 2013)	Raster	0.01°
Local dispersal[a]	Distance function from known CNG sites (concentric buffer circles)	Polygon	1:100,000
Long-distance dispersal[a]	Distance function from known CNG sites (along road corridors)	Polygon	1:100,000
Water dispersal[a]	Downstream distance from known flood-prone CNG sites	Polygon	1:100,000
Dispersal by slashers[a]	Inferred: primary roadsides	Polygon	1:100,000

CNG, Chilean needle grass.
[a]Indicates a derived measure.

down-scaling coarser-resolution maps such as soil data do need to be considered when interpreting model outcomes.

Making spatial predictions using the Bayesian network requires linking key environmental variables to available GIS data layers (Fig. 12.2; Table 12.1). In some cases, this association can be relatively straightforward. For example, soil-type categories used in soil mapping were simply matched to good, fair or poor soil type for Chilean needle grass. In other cases key environmental variables can be approximated (e.g. regular slashing was assumed to be restricted to major roadsides); surrogates can be used (e.g. buffered roadways for disturbance); or they can be modelled using global scenarios when GIS variables are simply not available (e.g. to test the role of integrated Chilean-needle-grass management; Fig. 12.3). New environmental data layers with improved accuracy and spatial resolution are continually becoming available. Because many pest models require similar environmental data layers, successive models become easier to create.

Current distribution data were used as input into the susceptibility model (Table 12.1). The distribution data, generally as points, were buffered as multiple rings away from each point according to predetermined distance classes (i.e. short, medium or long) for each dispersal pathway (Fig. 12.4). Distance downstream was modelled by limiting buffering directionally to dispersal distance downstream of the source population.

Computing Requirements

The development of increasingly powerful computers with large memory and storage capacity and fast processing has allowed modellers to take advantage of the increasingly finer-scale data and sophisticated software packages. A number of GIS options exist including free and open-source software (e.g. DIVA-GIS, QGIS, GRASS), popular commercial packages (e.g. ESRI products, MAPINFO, ERDAS IMAGINE) and GIS-linked databases (PostgreSQL and Microsoft SQL) that could be used for mapping. We used ESRI ARCGIS 10. Nonetheless, there is a limit to the ability of a computer to handle high-resolution data over large spatial extents (i.e. areas of interest). An average personal computer (e.g. 2–4 GB RAM, dual core 2–3 GHz processor) can easily handle small extents and coarser-resolution spatial data, but a large-capacity, high-speed computer, such as a server in a high-performance computer facility, may be necessary for large extents at fine scales (e.g. state and national level). We ran the Chilean-needle-grass model on a high-capability computer server (Dell Poweredge R910 with 64-bit core Xeon processor and 512 GB RAM). The most time-consuming step was combining the spatial layers initially and joining the Bayesian-network results back to the attribute table of the combined layer and exporting it to a new data layer (see Stages 1–3 in the online supplement to Chapter 12). Run times will depend on the resolution and extent of the study region and the number of environmental variables. Running this step for the QMDB at 30 m resolution took about 20 min. Each data output layer was approximately 25 MB.

Generating Pest Risk Maps

Risk maps can be presented in many ways. The challenge is to present them in ways that serve the needs of end users. Maps must be visually decipherable and they should be described in plain language. Our models have largely been aimed at identifying areas where pest impacts will be greatest. Risk is presented as the likelihood that the pest will reach 'high densities' as defined by experts during model development (i.e. Chilean needle grass exceeding 30% cover). Setting and articulating likelihood thresholds when presenting maps can have a large bearing on how risk is perceived and therefore needs to be done transparently. Our approach has been to calibrate likelihood thresholds where possible with observed densities in the field so that areas where the weed has already reached high densities are clearly identifiable on the map. Thresholds should also clearly contrast situations where habitat suitability is high but where the risk of invasion may be low. Thresholds are then maintained across all scenarios.

Validation and Quality Control

All risk models should be validated to provide confidence in the model's predictions, but validation is especially critical for models developed using participatory methods as these models contain inherent subjectivity. In our approach, the predictive model (Fig. 12.1) attempts to capture the current understanding of what determines habitat suitability and habitat susceptibility (Fig. 12.3). Validation allows the workshop-generated hypotheses to be tested using a range of approaches. Validation also helps to identify any gaps in important information and current misunderstandings of the ecology of this weed. The Bayesian network can be iteratively updated as understanding of the system is improved.

We take a broad view on validation to provide confidence in the model and its ability to make useful spatial predictions by testing the most likely sources of error. Various papers address specific sources of error (Kuhnert and Hayes, 2009; Marcot, 2012). Nonetheless, a comprehensive review of model error and validation approaches is urgently needed for process-based models but is beyond the scope of this chapter. Here, we limit discussion to major classes of error that we have identified as relevant to our modelling approach and provide a brief description of validation approaches that we have used (Table 12.2). Some of our validation procedures may be described by others as quality assurance and quality control measures. In each case, validation measures have been specifically designed to test the sources of error that are likely to be most influential on model outputs. Those sources of error can then be addressed or accommodated. Most of these errors are not unique to the use of experts. Broadly the predictive model may correctly capture the factors determining distribution, abundance and spread, but inaccuracies in the distributional and environmental data used to generate spatial predictions may create errors in spatial predictions. Alternatively, the experts' knowledge about factors that govern invasion dynamics or

Table 12.2. Approaches to test major sources of error and uncertainty in our modelling approach.

Types and sources of error	Validation and quality control methods
Model is correct, but spatial data are wrong	
Distribution or abundance data used to determine habitat susceptibility or validate the model	Verify that distributional data are correctly georeferenced, have been collected at sufficiently fine spatial resolution and reflect real densities
	Verify identity of mapped populations
Environmental data used for generating or validating spatial predictions	Validate GIS data layers in the field at spatial resolution being used in generating spatial predictions
	Identify any important discrepancies between GIS data layer and field-collected environmental data by comparing Bayesian-network-model results for pixels where both are available
Knowledge is correct, but model is wrong	
Model does not behave as experts expected	Revise model structure, definitions and probabilities with experts
	Check the relative importance of key environmental variables in the model (e.g. 'Sensitivity to findings analysis' within NETICA) against expert expectations
	Perform scenario analysis using a non-spatial Bayesian network, including actual scenarios and locations that are familiar to experts
	Test extreme and intermediate conditions
Knowledge is wrong	
General	Compare independently constructed models
	Compare against density data obtained through stratified sampling at sites where key environmental variables can be recorded and the pest has had an opportunity to invade
Model structure	Structure field survey or experimental work to demonstrate that invasion processes and key environmental variables are correctly captured
Categorization of environmental variables	Test categorizations of key environmental variables against field data to ensure that they match a priori definitions for each state
Parameterization	Structure field survey or experimental work to test the relative strength between key environmental variables against what is described by probability tables in the Bayesian network
	Test sensitivity to parameter estimates

responses to management may be correct but not adequately captured in the model. Finally, the experts' knowledge or working hypotheses may be incorrect.

We applied a range of error-checking approaches, some of which are described below, for the Chilean-needle-grass model. As is typically the case, validating the habitat suitability forecasts was relatively straightforward. However, validation did require us to obtain new field data to minimize the possibility that any inaccuracies were the result of problems with the validation data rather than the model itself. In contrast, validating the spread model was exceptionally difficult as independent data sets were not available. This situation is common in dispersal modelling.

Model is correct, but spatial data are wrong

The most common validation approach with ecological statistical models is to compare the fit between model predictions and independent data using a metric such as receiver-operating characteristic curves and area under the curve (Peterson *et al.*, 2008; Elith and Graham, 2009; Robinson *et al.*, 2010). There are, however, several reasons why a model may correctly capture invasion processes despite discrepancies between model predictions and observations. First, the value of these metrics has been questioned and the metrics are not always amenable to modelling invasive organisms where high omission errors can indicate an invasion in progress rather than error (Peterson et al., 2008; Robinson et al., 2010). The current distribution of Chilean needle grass in Queensland is known to be restricted due to dispersal limitation and ongoing management. Second, distribution data rarely include sufficient information about abundance to validate models that attempt to forecast abundance. Third, distribution data themselves can be incorrect, as a result of errors in identification or inaccurate georeferencing (Smith et al., 2012), be at the wrong spatial resolution (e.g. national records of Chilean-needle-grass density in Australia are reported in 50 km × 50 km grid cells, which may be too coarse for some of our models), be auto-correlated or be biased towards unusual and outlier records. Finally, and perhaps most critically for models that seek to capture underlying mechanisms, these metrics fail to explain causes of error. We found that distribution data for Chilean needle grass in the QMDB were inadequate for validating the Chilean-needle-grass model, because density data were unavailable at the spatial resolution we required; therefore, we collected our own field validation data.

A related problem is that environmental layers used to characterize conditions in which the species does or does not occur can be incorrect and result in incorrect inferences. Under these circumstances, spatial predictions (Fig. 12.1) can be inaccurate despite the predictive model being correct. This problem becomes more common as spatial heterogeneity of environmental layers becomes increasingly fine-grained (e.g. soil type mapped on a 1:100,000 scale can capture broad soil categories but generally misses finer-scale differences such as on ridgelines). Likewise, available GIS layers rarely capture temporal dynamics that may be important in assessing risk, such as changes in land use from cultivation to pasture in mixed farming regions (Murray et al., 2012). Again, environmental data can be field-validated and the implications of any discrepancies tested by comparing Bayesian-network-model results (Table 12.2). In another study, field validation found the expert model to be correct, with discrepancies between observed and predicted weed densities primarily the result of site-scale inaccuracies in environmental data layers (Murray et al., 2012).

Knowledge is correct, but the predictive model is wrong

The model is developed to capture experts' understanding of the ecology and management of Chilean needle grass. A range of approaches could be employed to test whether the model behaves as experts would expect and to identify why differences might occur (Table 12.2). 'Play back' feedback to experts has the added advantage of helping them gain ownership of the result and to improve their understanding of the pest and the risks that it might pose. With the Chilean-needle-grass model, performance assessment was done during a half-day, follow-up workshop, which included at least some of the original participants. First, the model was described to experts in terms of definitions, states and model behaviour to ensure that their understanding of the weed and the QMDB was captured correctly. Often definitions and states need to be refined between workshops to make them work with available GIS layers and these changes require checking. Model behaviour can be tested by looking at aspatial scenarios within the Bayesian network, checking whether relative weightings of different environmental variables reflect expert expectations (Murray et al., 2012), and critically examining model forecasts under different scenarios in locations most familiar to experts. Explanations are sought when discrepancies are found. For example, discrepancies could result from problems with the classification of environmental variables into states. Alternatively, experts may provide a good rationale to address discrepancies by modifying the model. For example, a recent workshop on Chilean needle grass found that different types of basalt soils originally coded as 'fair' needed to be recorded as 'fair' and 'poor' depending on the sub-type. This reclassification was supported by experts' field observations of spread.

Knowledge is wrong

The predictive model can be incorrect because expert knowledge or other information sources are wrong. Experts may incorrectly identify key environmental variables or interactions among those variables. Environmental variables, although correctly identified, may not be properly defined and categorized into states (e.g. good, fair and poor soil). Finally, the relative importance of each environmental variable or their interactions may not be properly understood. Modelling that relies on expert elicitation is arguably particularly prone to error as the outcome can be subjectively dependent on 'the day', the facilitator, the facilitation process and the knowledge of the experts (Murray et al., 2009), as well as the social interactions/influences within the group.

We tested whether the Chilean-needle-grass predictive model would differ depending on the group of experts being used to develop it. This was done by independently developing two habitat suitability and susceptibility models through two independent expert panels, one in Queensland and one in Victoria, 11 months apart (J.V. Murray, Queensland, 2012, personal communication). We were encouraged to find that both panels identified the same key environmental drivers and generated similar parameter estimates for the models. However, there were also important differences. The primary difference was that the Queensland group thought persistence of Chilean needle grass was much more sensitive to cultivation, climate and soil type than the Victorian group. As a result, the Queensland model predicted the risk of invasion to be much less than that of the Victorian model. Field data specifically collected to test the two models supported the Queensland model, but even this model appeared to overestimate the ability of Chilean needle grass to outcompete herbaceous cover. In this case, differences among the two expert panels probably reflected differences in knowledge and experiences. Consequently, their models represent different working hypotheses and discrepancies highlight critical knowledge needs. When the Queensland group of experts was presented with the field observations that suggested Chilean needle grass was less competitive than they had proposed, the new results were accepted and the model was adjusted accordingly.

Field validation alone can also help test for different types of model error (Smith et al., 2007; Murray et al., 2012). Field observations on pest density should be collected across environmental gradients to validate the model. Our approach for Chilean needle grass has been to identify environmentally heterogeneous areas where the weed is known to have reached high densities and has had the opportunity to disperse at least locally. Stratified sampling is conducted across the available environmental gradients identified in the model.

Field validation for Chilean needle grass supported the experts' selection of key environmental variables, model structure, categories for environmental variables and parameter estimates. However, expert opinions differed, especially on the role of disturbance, particularly from slashing, in creating a suitable habitat for establishment by Chilean needle grass. Queensland experts believed that Chilean needle grass could readily invade intact pastures in the absence of disturbance. On investigation, we found this was an expectation that they had learned from their Victorian colleagues. In contrast, field data suggested that Chilean needle grass rarely invades and dominates intact pastures unless they are routinely slashed. Extensive slashing of pastures is widespread in Victoria, but not Queensland. The modelling and validation process successfully identified an important knowledge gap that required further investigation and led to the inclusion of slashing as another key environmental variable in the predictive model.

Implications of Model Results for Management of Chilean Needle Grass

The objective of risk modelling for Chilean needle grass in Queensland was to identify areas at risk and to provide guidance on how best to manage this species across the QMDB. Risk models for Chilean needle grass have found that highly suitable habitat is restricted to high-altitude locations in the eastern QMDB as a consequence of climatic requirements (Bourdôt et al., 2010), soil type and the absence of cultivation (Fig. 12.5a; see colour plate section). The areas at risk are spatially more restricted than originally thought. These results suggest that management responses could be more focused than they are currently, a useful finding for decision makers.

In Queensland, Chilean needle grass is still spreading from Clifton, where the original incursion occurred in the early 1970s (Fig. 12.5b; see colour plate section). Interestingly, the current distribution is still largely restricted to the original Clifton Shire Council. This supports the importance of slashers as a dispersal pathway for Chilean needle grass (Fig. 12.3) because roadside slashing within the shire was conducted out of Clifton. However, this administrative arrangement changed in 2008 with the amalgamation of the Clifton Shire with seven neighbouring shires. Managing the risks posed by a much less centralized roadside slashing programme is now a priority.

Chilean needle grass remains a target for eradication in Queensland. Our results suggest that suitable habitat is relatively contiguous throughout eastern QMDB (Fig. 12.5b). No natural barriers exist to help a containment or eradication programme. In addition, long-distance dispersal through human activity (Fig. 12.3) is probable, so most suitable habitat has at least a small chance of being invaded within a 10-year period (Fig. 12.5c; see colour plate section). Again, large areas of susceptible habitat are expected to make successful eradication or containment difficult to achieve.

Discussion: Future Improvements to the Approach

We have described a pest risk modelling approach that can draw upon all available information sources to model habitat suitability and susceptibility. The method we have developed involves using expert elicitation workshops, combined with Bayesian networks and GIS, to capture the essential drivers of species' distribution and abundance. However, this approach can be improved. For example, more quantitative elicitation methods (e.g. Elicitator; Low Choy et al., 2012) could be used to estimate parameters. New methods of network and pathway analysis could be used to model long-distance dispersal (see Colunga-Garcia and Haack, Chapter 3 in this volume; Koch and Yemshanov, Chapter 13 in this volume). New and improved environmental data layers are always welcomed. Distributional data might be analysed with environmental covariates in new ways to better infer causal relationships than is possible currently (Ready et al., 2010). Finally, the entire process might be streamlined to generate spatial predictions more quickly than we can now (Fig. 12.1).

The major development planned in the short term is to incorporate the consequences of invasion into the same modelling framework. Impacts will depend on how pest density and the threatened value intersect in time and space and on the strength of the relationship between pest density and impact. Preliminary work on *Parthenium hysterophorus* L. has shown that this dimension of pest risk can be readily incorporated through the same use of experts, Bayesian networks and GIS. Experts are responsible for identifying the values (e.g. pasture production) that are threatened by the pest and estimating the functional relationship between pest density and impact for each value. These relationships are then captured in the Bayesian network and impact maps generated in GIS.

We have so far applied our approach to relatively well-understood organisms and demonstrated that it is easily applied to diverse organisms and transferable to new regions (van Klinken and Murray, 2011; Murray et al., 2012; Smith et al., 2012). However, further streamlining of the modelling approach and methodological improvements may make it amenable to modelling many species efficiently, as is often required for pest prioritization. The approach should also yield useful results even if only relatively general information is known about the organism, as might be the case for invasive alien weeds not yet in the country. Even in those cases, it should be possible to determine, for example, whether invasive alien weeds require inundation or particular disturbance regimes such as fire to establish.

Our modelling approach relies on explicitly and transparently harnessing diverse information sources to develop prognostic risk models. The resulting models represent hypotheses based on the best available knowledge. This strategy is readily applicable to a wide range of modelling techniques. For example, global fish distribution models use existing distributional data to fit functional relationships between fish occurrence and environmental variables, and those relationships can then be modified by experts (Ready et al., 2010). However, our approach has the benefits of relying on publicly available software, being intuitive to experts who may not have been previously exposed to models and readily allowing the characterization of functional relationships between diverse environmental variables.

Acknowledgements

We thank Luis Laredo and Pierre Goulier for assistance with GIS work; the Rural Industries Research & Development Corporation (RIRDC) and the Queensland Murray Darling Committee (QMDC) for funding to conduct the work on Chilean needle grass; Vanessa MacDonald and Darren Marshall for their contribution to developing, testing and applying the approach to diverse pests; and Jens Froese and Carmel Pollino for their comments on an earlier draft.

This chapter is enhanced with supplementary resources.

To access the material please visit:

www.cabi.org/openresources/43946/

References

Allan, J., Low Choy, S. and Mengersen, K. (2010) Elicitator: an expert elicitation tool for regression in ecology. *Environmental Modelling & Software* 25, 129–145.

Bourdôt, G.W., Lamoureaux, S.L., Kriticos, D.J., Watt, M.S. and Brown, M. (2010) Current and potential distributions of *Nassella neesiana* (Chilean needle grass) in Australia and New Zealand. In: Zydenbos, S.M. (ed.) *Proceedings of the 17th Australasian Weeds Conference 2010*. New Zealand Plant Protection Society, Christchurch, New Zealand, pp. 424–427.

Cain, J. (2001) *Planning Improvements in Natural Resources Management*. Centre for Ecology and Hydrology, Wallingford, UK.

Coutts, S.R., van Klinken, R.D., Yokomizo, H. and Buckley, Y.M. (2011) What are the key drivers of spread in invasive plants: dispersal, demography or landscape: and how can we use this knowledge to aid management? *Biological Invasions* 13, 1649–1661.

Crossman, N.D., Bryan, B.A. and Cooke, D.A. (2011) An invasive plant and climate change threat index for weed risk management: integrating habitat distribution pattern and dispersal process. *Ecological Indicators* 11, 183–198.

DAFF (2013) *Land Use and Management Information for Australia*. Australian Government, Department of Agriculture, Fisheries, and Forestry, Canberra. Available at: http://www.daff.gov.au/abares/aclump (accessed 20 November 2013).

DSITIA (2013) *Land Systems Series*. Queensland Government, Department of Science, Information Technology, Innovation and the Arts, Brisbane, Australia. Available at: https://data.qld.gov.au/dataset/land-systems-series (accessed 20 November 2013).

Elith, J. and Graham, C.H. (2009) Do they? How do they? WHY do they differ? On finding reasons for differing performances of species distribution models. *Ecography* 32, 66–77.

Frid, L., Holcombe, T., Morisette, J.T., Olsson, A.D., Brigham, L., Bean, T.M., Betancourt, J.L. and Bryan, K. (2013) Using state-and-transition modelling to account for imperfect detection in invasive species management. *Invasive Plant Science and Management* 6, 36–47.

Hadincova, V., Munzbergova, Z., Wild, J., Sajtar, L. and Maresova, J. (2008) Plant invasions: human perception, ecological impacts and management. In: Tokarska-Guzik, B., Brock, J.H., Brundu, G., Child, L., Daehler, C.C. and Pyšek, P. (eds) *Dispersal of Invasive* Pinus strobus *in Sandstone Areas of the Czech Republic*. Backhuys Publishers, Leiden, The Netherlands, pp. 117–132.

Korb, K.B. and Nicholson, A.E. (2004) *Bayesian Artificial Intelligence*. Chapman & Hall/CRC Press, London.

Kuhnert, P.M. and Hayes, K.R. (2009) How believable is your BBN? *Proceedings of the 18th World IMACS Congress and MODSIM09 International Congress on Modelling and Simulation*. Modelling and Simulation Society of Australia and New Zealand and International Association for Mathematics and Computers in Simulation, Cairns, Australia, pp. 4319–4325.

Low Choy, S., James, A., Murray, J. and Mengersen, K. (2012) Elicitator: a user-friendly, interactive tool to support scenario-based elicitation of expert knowledge. In: Perera, A.H., Drew, C.A. and Johnson, C.J. (eds) *Expert Knowledge and Its Application in Landscape Ecology*. Springer, New York, pp. 39–67.

Marcot, B.G. (2006) Characterizing species at risk I: modeling rare species under the Northwest Forest Plan. *Ecology and Society* 11, 10. Available at: http://www.ecologyandsociety.org/vol11/iss2/art10/ (accessed 20 December 2013).

Marcot, B.G. (2012) Metrics for evaluating performance and uncertainty of Bayesian network models. *Ecological Modelling* 230, 50–62.

Marcot, B.G., Holthausen, R.S., Raphael, M.G., Rowland, M.M. and Wisdom, M.J. (2001) Using Bayesian belief networks to evaluate fish and wildlife population viability under land management alternatives from an environmental impact statement. *Forest Ecology and Management* 153, 29–42.

Martin, T.G., Burgman, M.A., Fidler, F., Kuhnert, P.M., Low Choy, S., McBride, M. and Mengersen, K. (2012) Eliciting expert knowledge in conservation science. *Conservation Biology*, 26, 29–38.

McCann, R.K., Marcot, B.G. and Ellis, R. (2006) Bayesian belief networks: applications in ecology and natural resource management. *Canadian Journal of Forest Research* 36, 3053–3062.

Murray, J.V., Goldizen, A.W., O'Leary, R.A., McAlpine, C.A., Possingham, H.P. and Low Choy, S. (2009) How useful is expert opinion for predicting the distribution of a species within and beyond the region of expertise? A case study using brush-tailed rock-wallabies Petrogale penicillata. *Journal of Applied Ecology* 46, 842–851.

Murray, J.V., Stokes, K.E. and van Klinken, R.D. (2012) Predicting the potential distribution of a riparian invasive plant: the effects of changing climate, flood regimes and land-use patterns. *Global Change Biology* 18, 1738–1753.

Norsys Software Corporation (2009) *Netica 4.12*. Norsys Software Corporation, Vancouver, Canada. Available at: http://www.norsys.com (accessed 21 December 2013).

Peterson, A.T., Papes, M. and Soberon, J. (2008) Rethinking receiver operating characteristic analysis applications in ecological niche modelling. *Ecological Modelling* 213, 63–72.

Raupach, M.R., Briggs, P.R., Haverd, V., King, E.A., Paget, M. and Trudinger, C.M. (2009) *Australian Water Availability Project (AWAP). Final Report for Phase 3. CAWCR Technical Report No. 013*. CSIRO Marine and Atmospheric Research Component, Canberra.

Raupach, M.R., Briggs, P.R., Haverd, V., King, E.A., Paget, M. and Trudinger, C.M. (2012) *Australian Water Availability Project*. CSIRO Marine and Atmospheric Research, Canberra. Available at: http://www.csiro.au/awap/ (accessed 29 July 2013).

Ready, J., Kaschner, K., South, A.B., Eastwood, P.D., Rees, T., Rius, J., Agbayani, E., Kullander, S. and Froese, R. (2010) Predicting the distributions of marine organisms at the global scale. *Ecological Modelling* 221, 467–478.

Robinson, T.P., van Klinken, R.D. and Metternicht, G.I. (2010) Comparison of alternative strategies for invasive species distribution modelling. *Ecological Modelling* 221, 2261–2269.

Scarth, P., Byrne, M., Danaher, T., Henry, B., Hassett, R., Carter, J. and Timmers, P. (2006) State of the paddock: monitoring condition and trend in ground cover across Queensland. *Proceedings of the 13th Australasian Remote Sensing and Photogrammetry Conference.* Spatial Sciences Institute, Canberra. Available at: http://files.figshare.com/96429/scarth.2006.state_of_the_paddock_monitoring_condition_and_trend_in_groundcover_across_queensland.pdf (accessed 13 October 214).

Smith, C.S., Howes, A.L., Price, B. and McAlpine, C.A. (2007) Using a Bayesian belief network to predict suitable habitat of an endangered mammal – the Julia Creek dunnart (*Sminthopsis douglasi*). *Biological Conservation* 139, 333–347.

Smith, C., van Klinken, R.D., Seabrook, L. and McAlpine, C. (2012) Estimating the influence of land management change on weed invasion potential using expert knowledge. *Diversity and Distributions* 18, 818–831.

Sutherst, R.W. and Bourne, A.S. (2009) Modelling non-equilibrium distributions of invasive species: a tale of two modelling paradigms. *Biological Invasions* 11, 1231–1237.

Uusitalo, L. (2007) Advantages and challenges of Bayesian networks in environmental modelling. *Ecological Modelling* 203, 312–318.

van Klinken, R.D. and Murray, J.V. (2011) Challenges, constraints and solutions for modelling regional-scale dispersal of invasive organisms: from practice to policy. In: Chan, F., Marinova, D. and Anderssen, R.S. (eds) *MODSIM2011, Proceedings of the 19th International Congress on Modelling and Simulation.* Modelling and Simulation Society of Australia and New Zealand, Perth, Australia. pp. 2570–2577. Available at: http://www.mssanz.org.au/modsim2011/E16/vanklinken.pdf (accessed 20 December 2013).

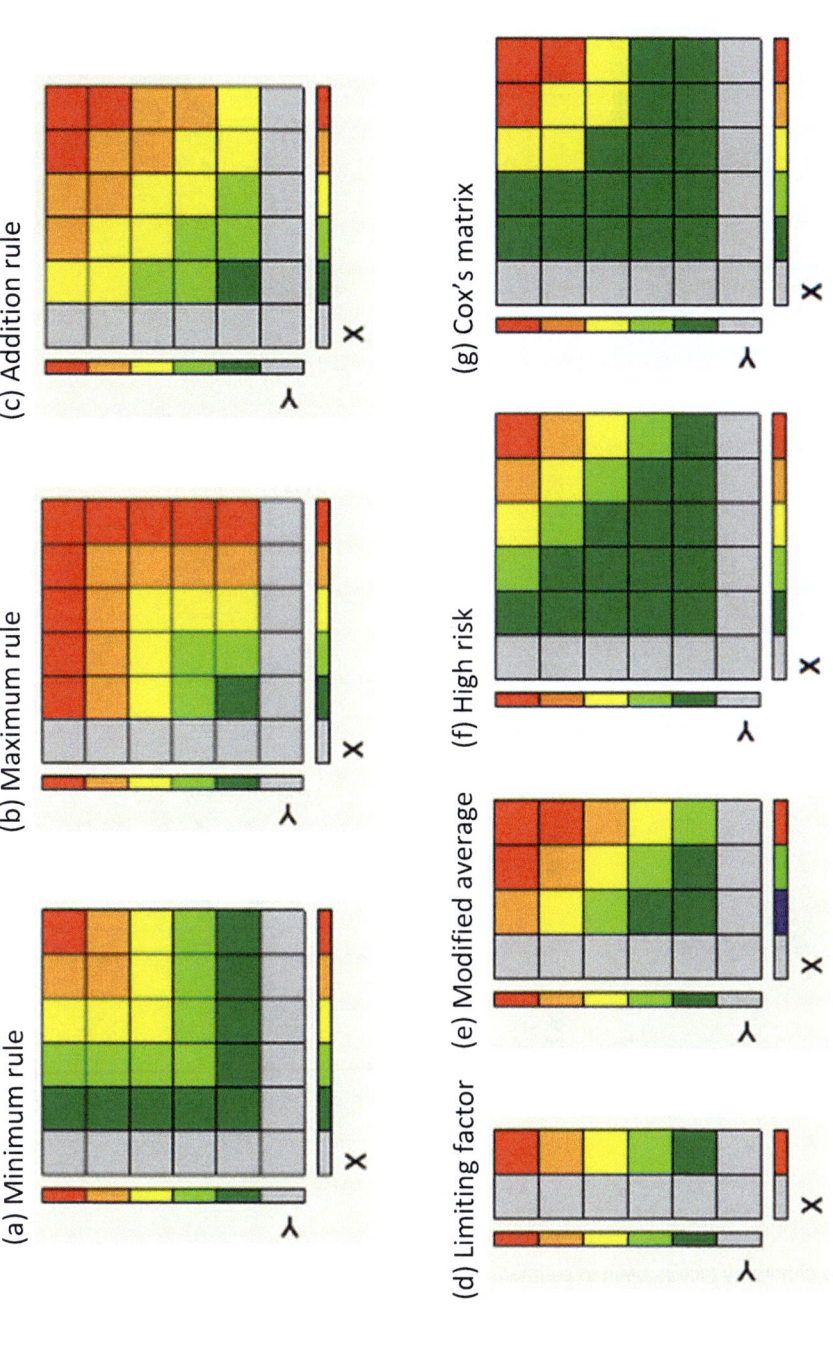

Fig. 2.2. Matrix rules to combine pest risk maps. X and Y represent different factors that affect overall pest risk. Each factor is classified on an ordinal scale from absent (grey) to very high/present (red). The central matrix illustrates the possible outcomes for a single grid cell on a map. The logic behind the rules continues to apply with different numbers of ordinal classifications for a factor.

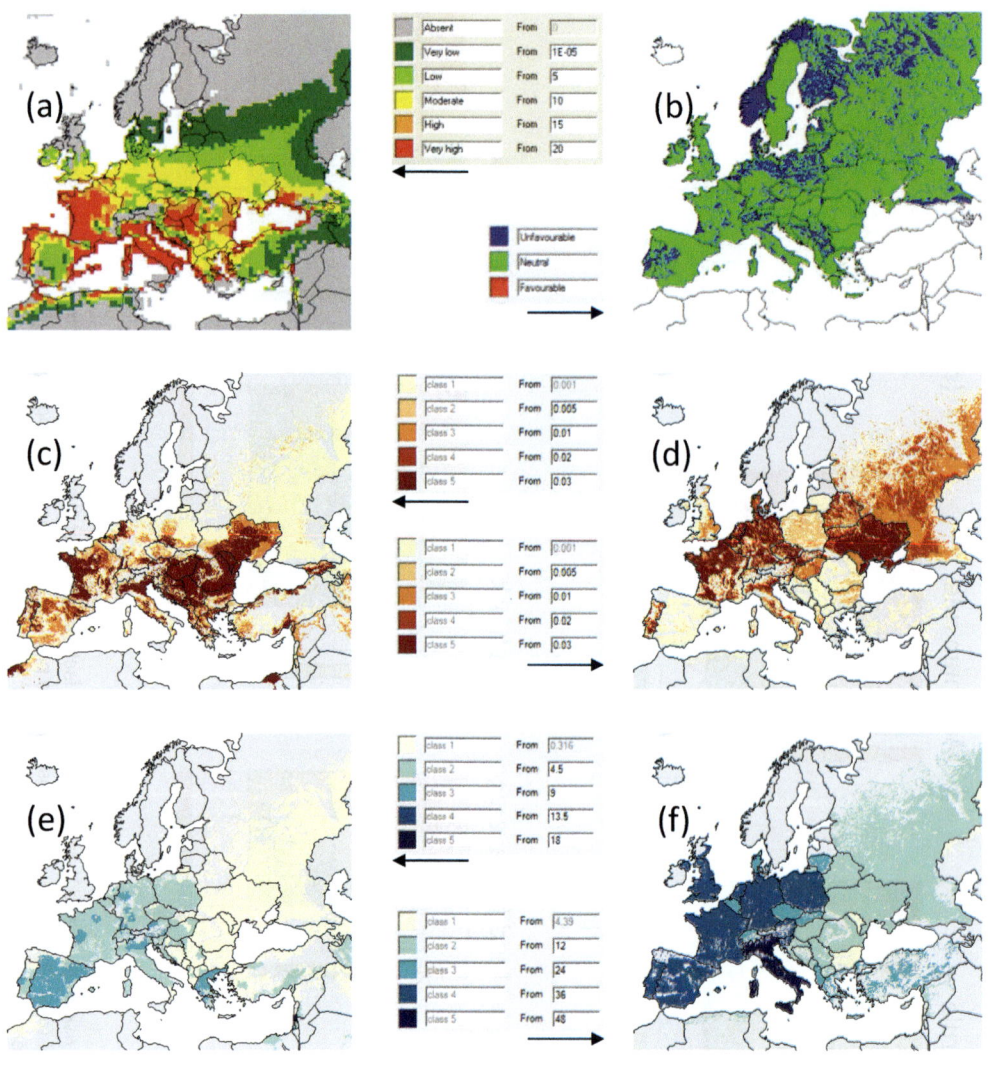

Fig. 2.3. Maps of primary factors used to estimate risks from *Diabrotica virgifera virgifera*: (a) climatic suitability measured by CLIMEX Ecoclimatic Index; (b) presence of sandy soils (<18% clay and >65% sand, blue; other soil textures, green) as described in the European Soil Database version 2 (JRC, 2010); (c) area of harvested grain maize (percentage of each 10 km × 10 km grid) based on data from Monfreda *et al.* (2008); (d) area of harvested forage maize (percentage of each 10 km × 10 km grid) based on data from Monfreda *et al.* (2008); (e) grain maize yield (tonnes per hectare) from Monfreda *et al.* (2008); and (f) forage maize yield (tonnes per hectare) from Monfreda *et al.* (2008).

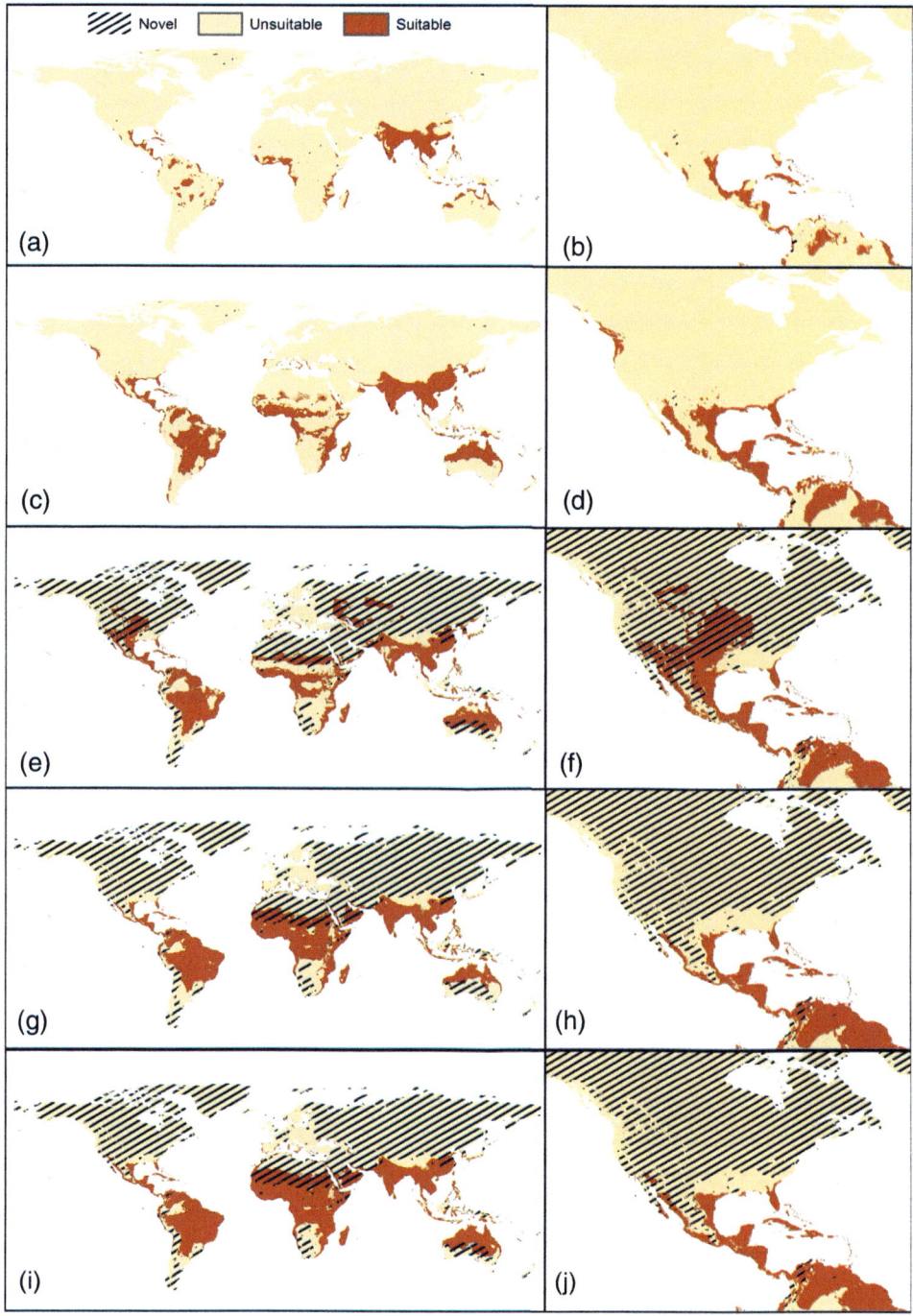

Fig. 5.1. MAXENT maps of climate suitability for Burmese pythons. Maps in the left column were produced using the minimum training presence (MTP) threshold for the globe and in the right column, using the MTP threshold for North America. Five models were developed: (a, b) model 'Default 90', run with default settings and 90 presence points; (c, d) model 'Default 86', run with default settings and corrected presence points; (e, f) model 'MCP 86', background data restricted to a minimum convex polygon around corrected presence points; (g, h) model 'MCP 90 reg3', background data restricted to a minimum convex polygon around 90 presence points and the regularization value set to 3; (i, j) model 'MCP 86 reg4', background data restricted to a minimum convex polygon around 86 presence points and the regularization value set to 4.

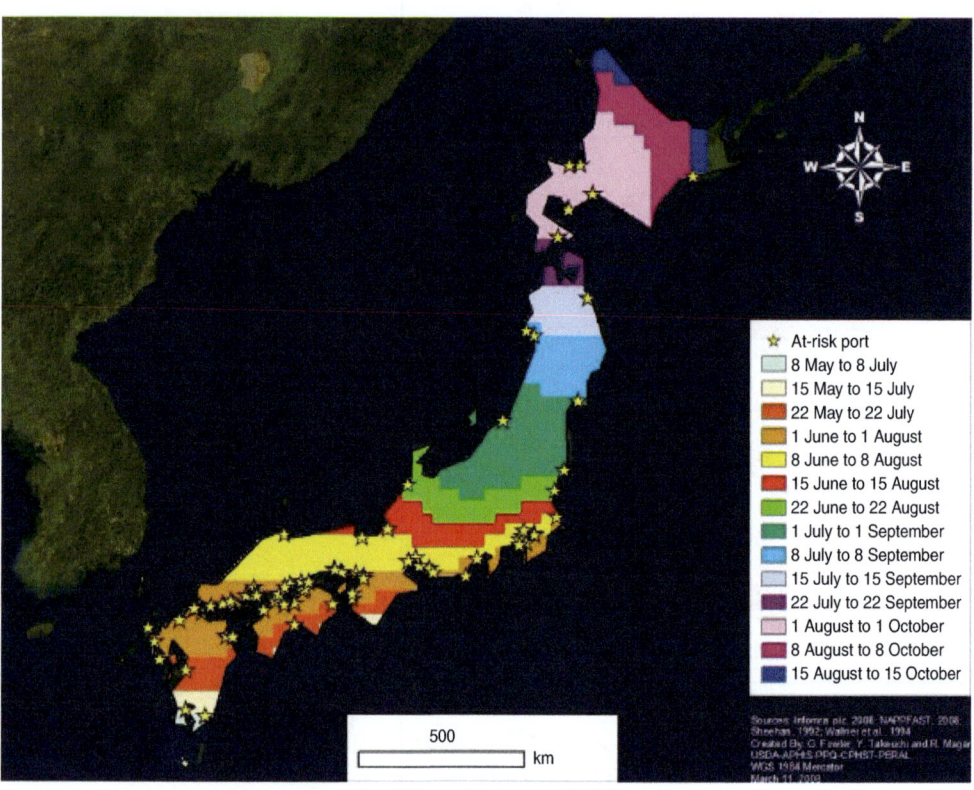

Fig. 6.1. Forecasted flight periods for Asian gypsy moths (*Lymantria dispar asiatica*) around at-risk ports.

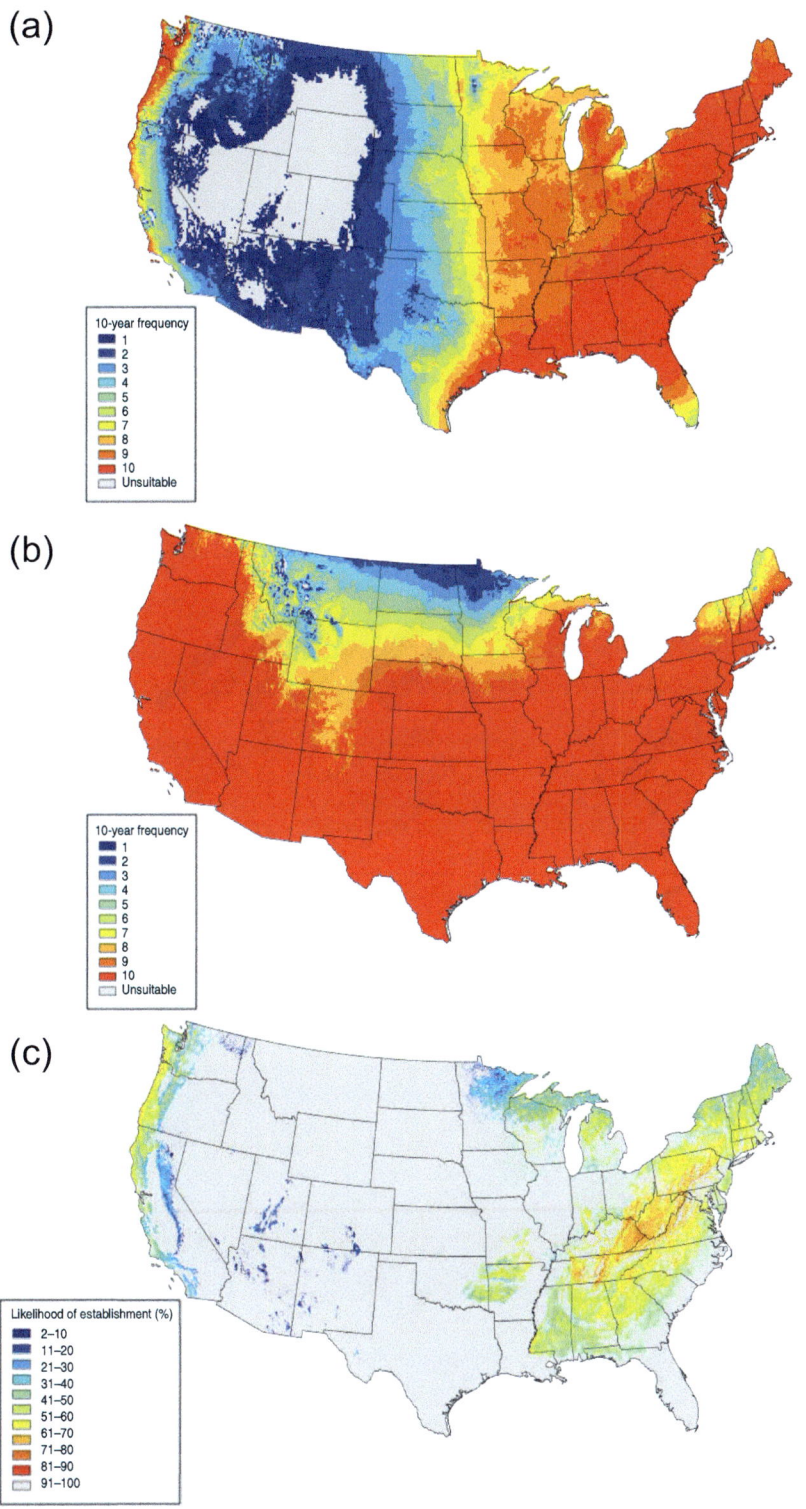

Fig. 6.3. *Phytophthora ramorum* risk maps for: (a) infection; (b) cold survival; and (c) likelihood of establishment on deciduous and understorey hosts.

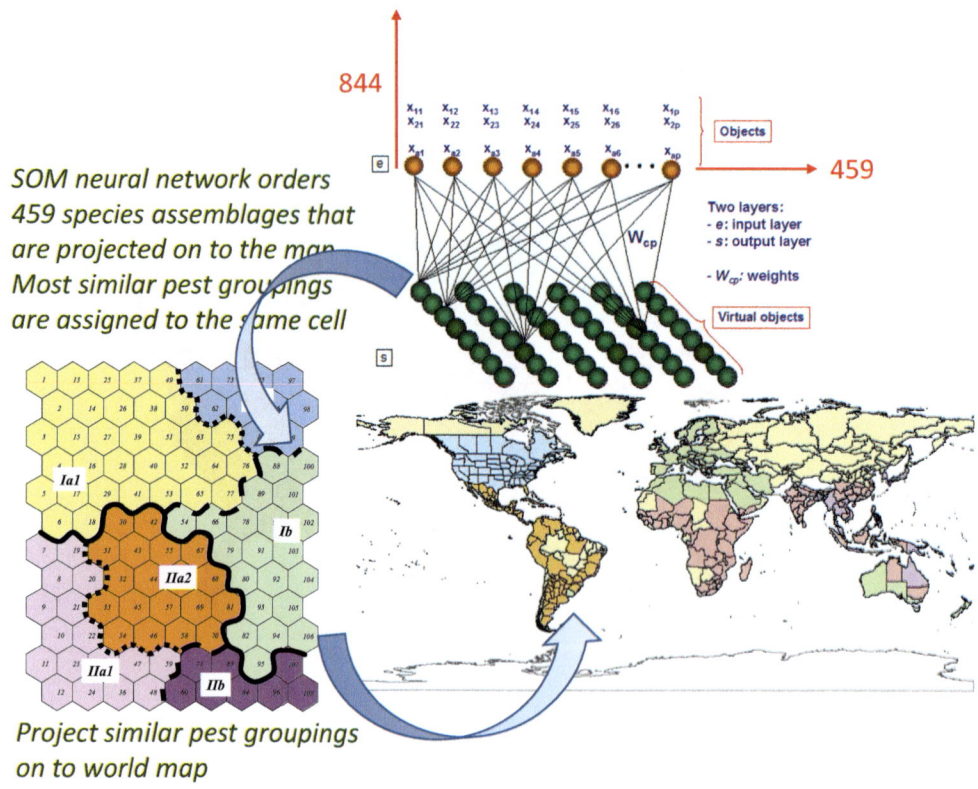

Fig. 7.2. A further representation of the output of a self-organizing map (SOM) illustrating the two-dimensional feature map that is the result of projecting similar regional (459) pest profiles comprising the presence or absence of 844 species into the same or nearby cells. Further cluster analysis gives crisp edges to the lower-resolution clusters (cells of the same colour) within the map that are projected on to a map of the world to illustrate potential donor and recipient areas. (Reproduced with permission from Worner et al., 2013.)

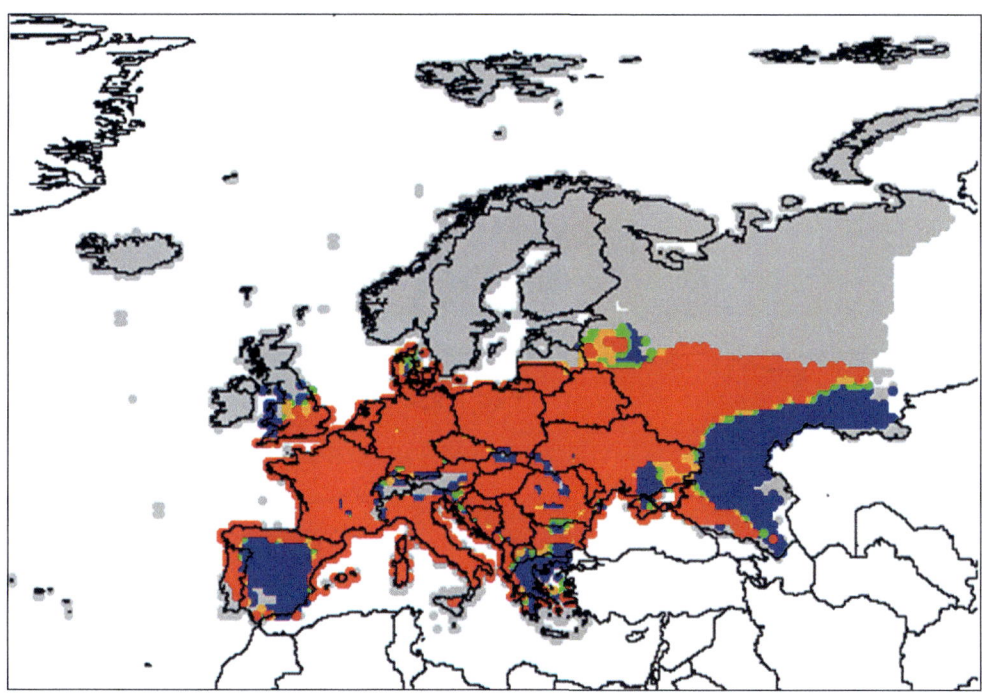

Fig. 8.3. Output of model C. Potential spread of the western corn rootworm at time $t = 20$ (year 2012). Grey dots indicate a population density of 0; blue dots indicate a population density in [0, 25%]; green dots indicate a population density in [25%, 50%]; orange dots indicate a population density in [50%, 75%]; red dots indicate a population density in [75%, 100%]. White means that data are missing (outside the study area).

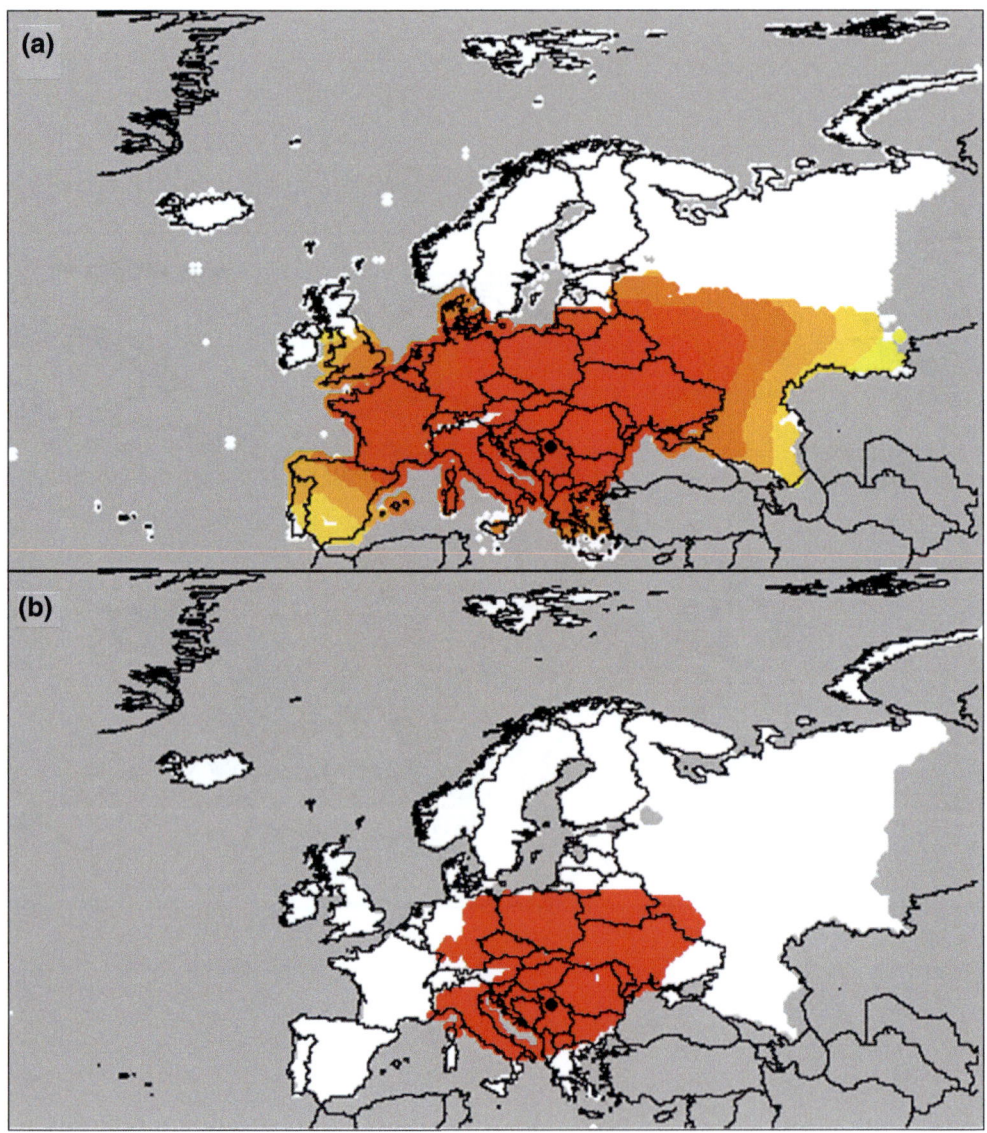

Fig. 8.4. Output of model D. Potential spread of the western corn rootworm at time $t = 20$ (year 2012). (a) Population density expressed as a percentage of the carrying capacity. The graduated colour indicates the population density, from white (when it is below 10^{-6}%) to yellow, orange and red (when it is above 10%). (b) Area in which the population density exceeds a given threshold. Red indicates that the population density exceeds 10% in this example.

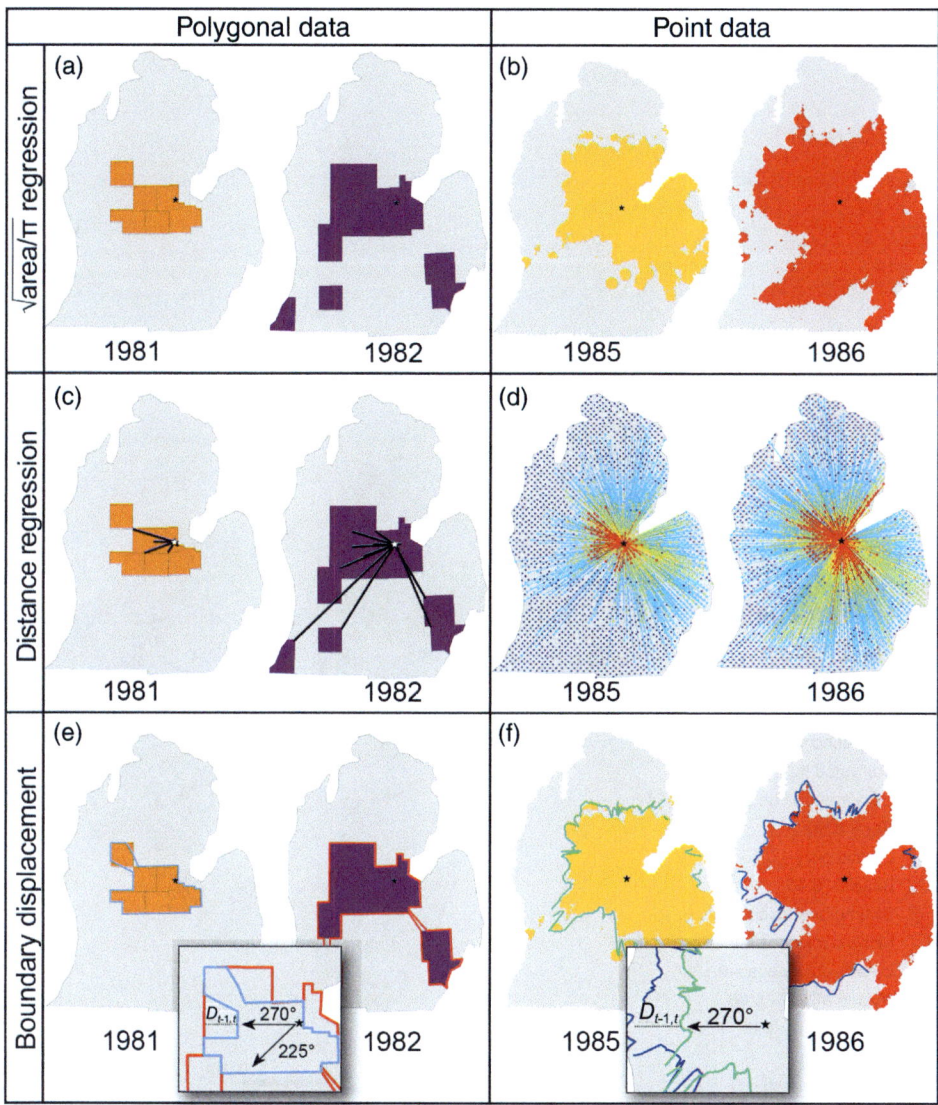

Fig. 9.3. Description of the square-root area method, the distance regression method and the boundary displacement method when applied to polygonal data (a, c and e, using 1981 and 1982 as example years) or point data (b, d and f, using 1985 and 1986 as example years). In the square-root area method, both data sources can be used to define the infested area in each year (a, b). In the distance regression method, the distance between Midland, Michigan and each quarantined county is measured (c); alternatively, the distance between Midland and traps recording at least 1 (pale blue), 10 (yellow) or 100 (red) moths can be measured if point data are available (d; traps recording 0 moths are indicated as dark blue dots). In the boundary displacement method, polygonal data can be used to interpolate a quarantine boundary (e) and then the boundaries can be compared by measuring the displacement at transects radiating from a fixed point, in this case, Midland. In the insert in (e), as an example, the displacement at a 270° transect is $D_{t-1,t}$, while at 225° the displacement between successive boundaries is 0. Displacement at all transects can be averaged to estimate an annual year-to-year rate of spread. This same method can be applied to point data (f); in this case, trapping data are interpolated using kriging (Isaaks and Srivastava, 1989), and then an optimization method (Sharov et al., 1995) can be used to estimate population threshold boundaries. Displacement between like thresholds (i.e. the 1-moth population boundary in 1985 and 1986) can be measured using radiating transects in the same manner as (e), and averaged to estimate year-to-year rates of spread.

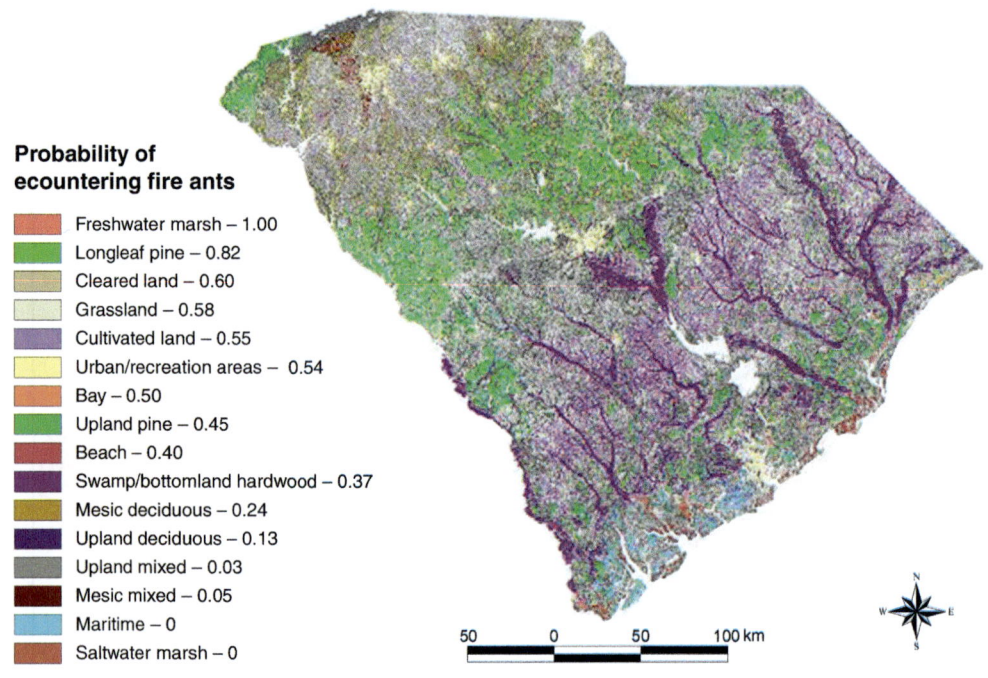

Fig. 11.2. Predicted fire ant distribution map for South Carolina, USA, based on empirical estimates of the probability of encountering fire ants during sampling, 1999–2000.

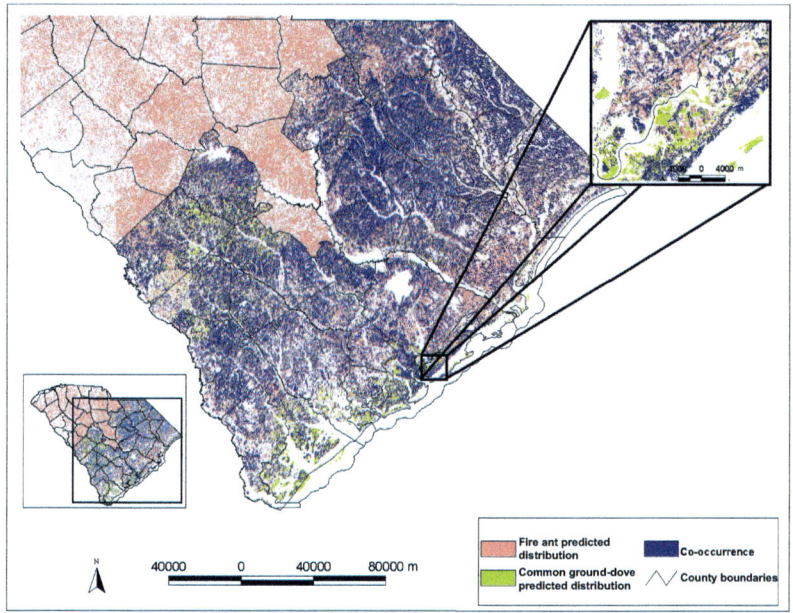

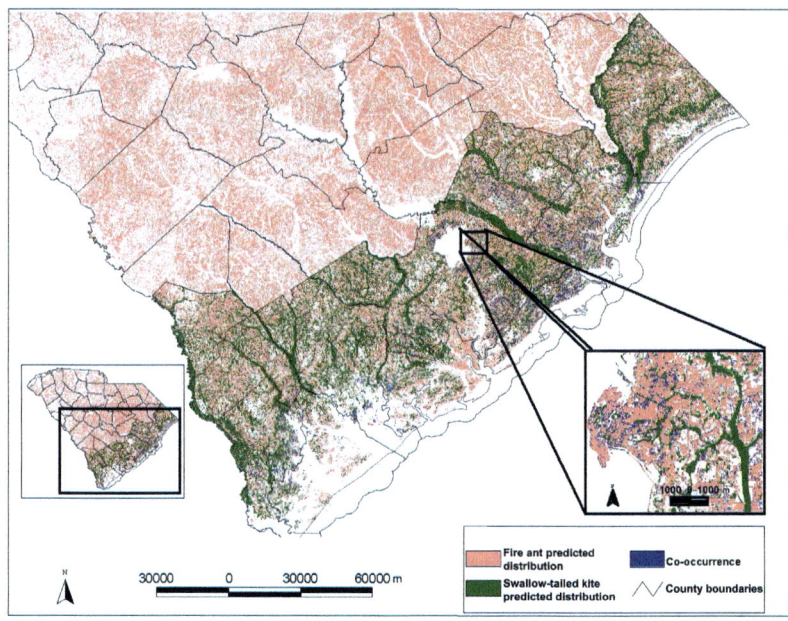

Fig. 11.3. Spatial correspondence of the distributions of common ground-doves and red imported fire ants in South Carolina, USA, at a 30 m resolution. The map of the common ground-dove is based on a predictive habitat model developed by the South Carolina Gap Analysis Program. Red imported fire ant predicted distribution based on logistic regression analysis of statewide presence/absence ant sampling. Eighty-three per cent of the common ground-dove's predicted distribution is shared with fire ants.

Fig. 11.4. Spatial correspondence of the distributions of swallow-tailed kites and red imported fire ants in South Carolina, USA, at a 30 m resolution. The map of the swallow-tailed kite is based on a predictive habitat model developed by the South Carolina Gap Analysis Program. Red imported fire ant predicted distribution based on logistic regression analysis of statewide presence/absence ant sampling. Seven per cent of the swallow-tailed kite's predicted distribution is shared with fire ants.

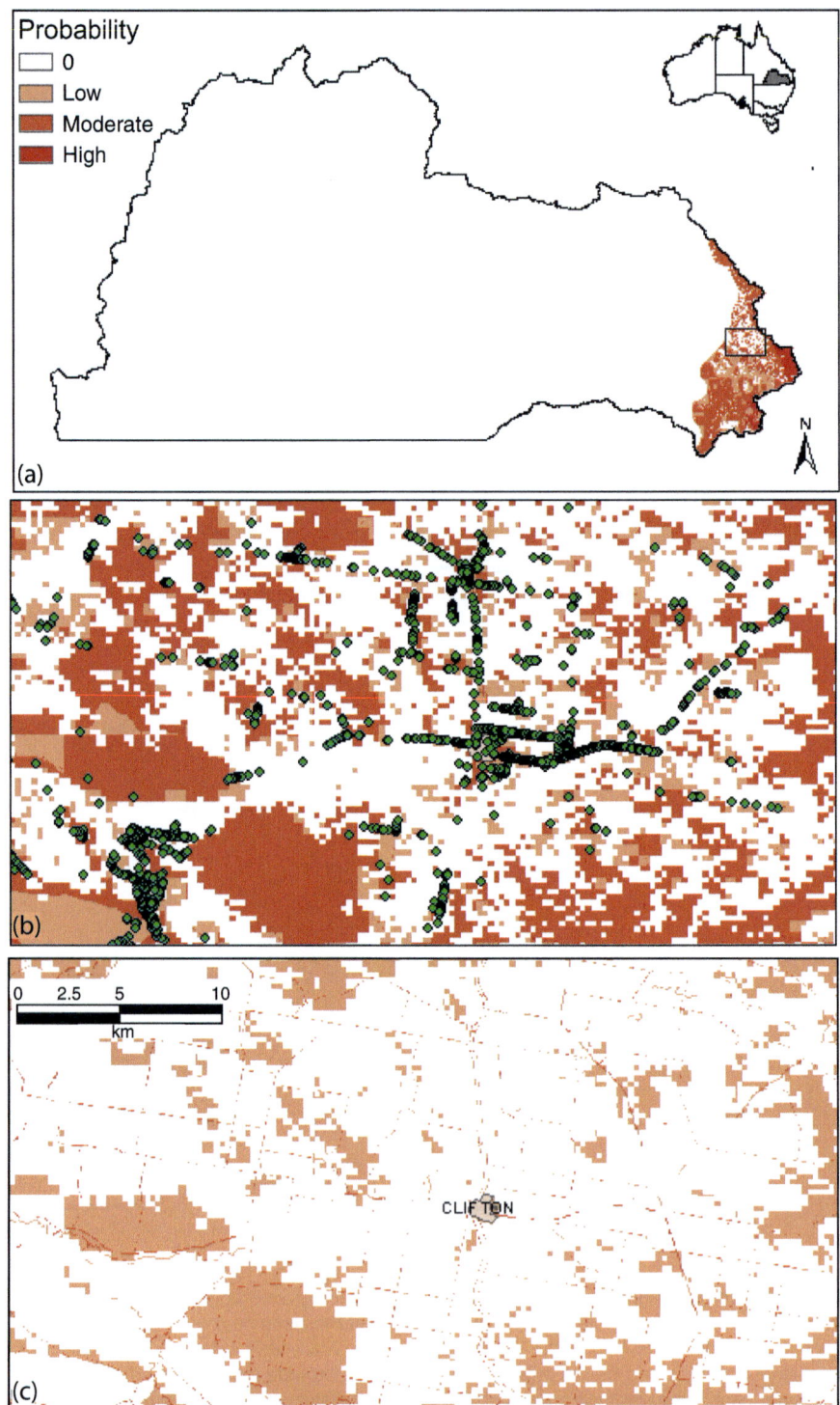

Fig. 12.5. Chilean-needle-grass maps: (a) probability of habitat being highly suitable in the Queensland Murray Darling Basin; (b) habitat suitability around the township of Clifton where the initial outbreak occurred; and (c) habitat susceptibility around Clifton where most Chilean-needle-grass records occur. Linear features apparent in (c) are roadways along which Chilean needle grass is spread by slashers.

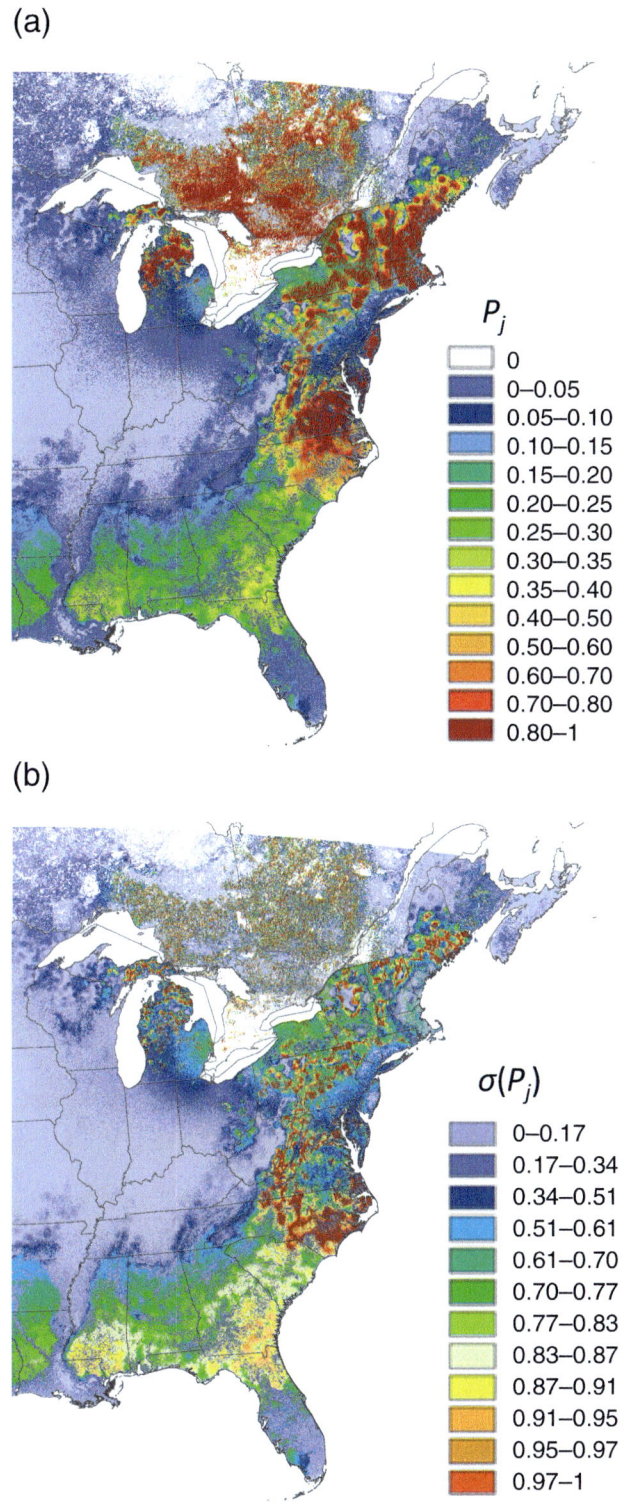

Fig. 13.1. *Sirex noctilio* risk maps for the baseline scenario: (a) invasion risk, P_j; (b) standard deviation of the risk values, $\sigma(P_j)$.

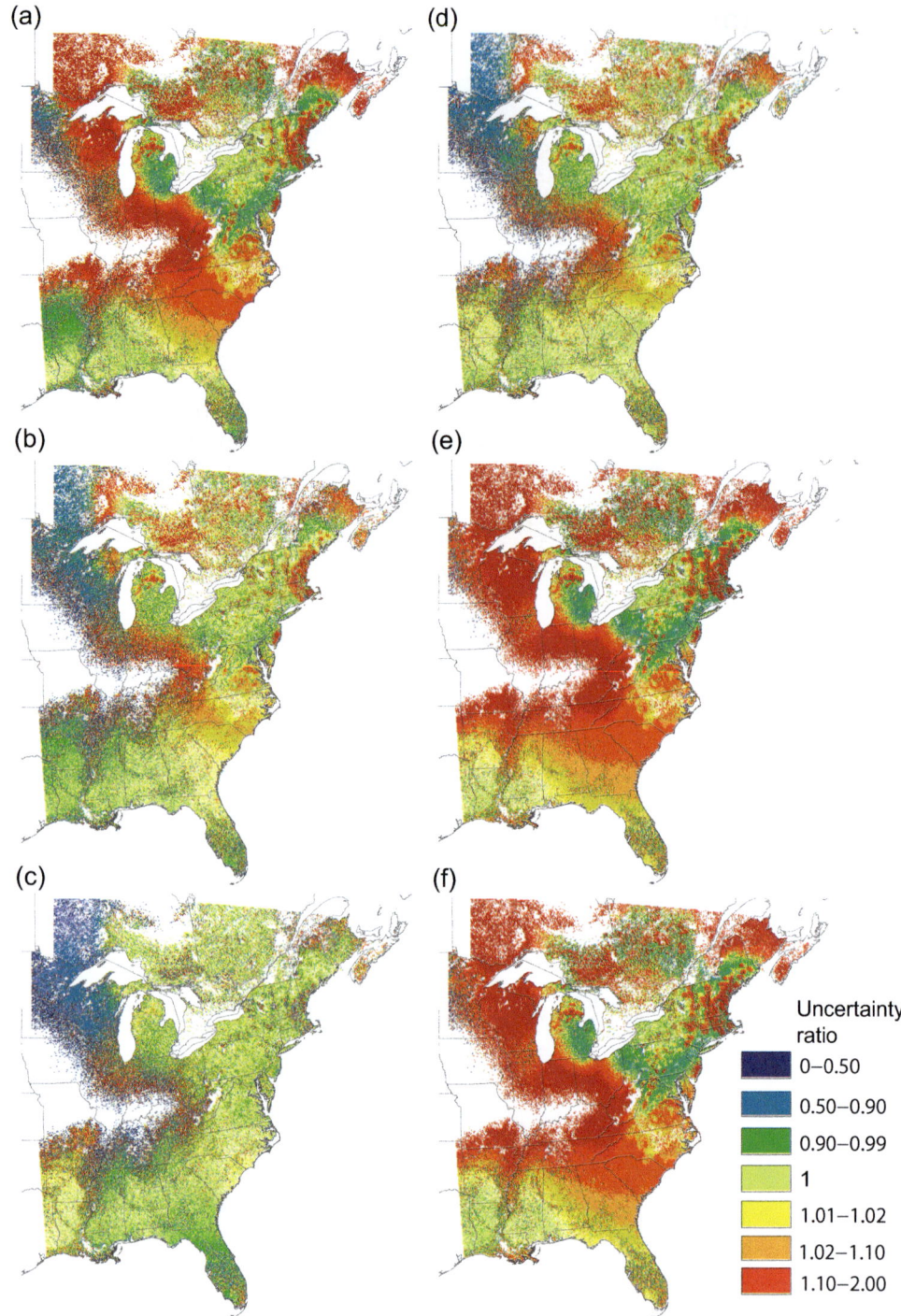

Fig. 13.3. Maps of uncertainty ratios generated through sensitivity analyses. One-parameter-at-a-time sensitivity analysis scenarios: (a) maximum annual spread distance, d_{max}; (b) local dispersal probability, p_0; (c) local probabilities of entry at marine ports, $W_{x(f)}$. All-but-one (i.e. single parameter left fixed) scenarios: (d) fixed d_{max}; (e) fixed population-carrying capacity, k; (f) fixed host growth rate, g_v. Scenarios at ±40% parametric uncertainty are shown. For any map cell, the uncertainty ratio is the $\sigma(P_j)$ value for the sensitivity scenario of interest divided by the $\sigma(P_j)$ value from the baseline scenario.

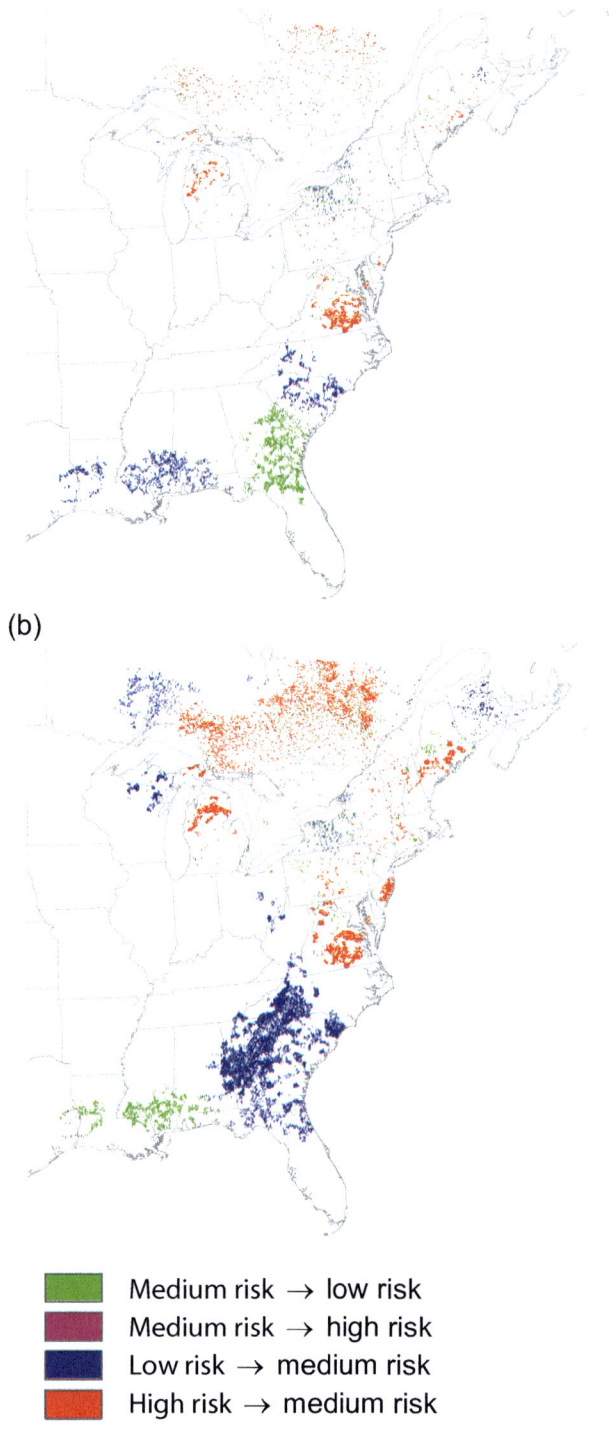

Fig. 13.5. Geographic distribution of the shifts in predicted risk classes due to increased uncertainty in the d_{max} parameter: (a) ±15% uncertainty in d_{max}; (b) ±50% uncertainty in d_{max}. White areas delineate no changes in risk class. Risk classes: low risk, $P_j < 0.25$; medium risk, $0.25 \leq P_j \leq 0.75$; high risk, $P_j > 0.75$.

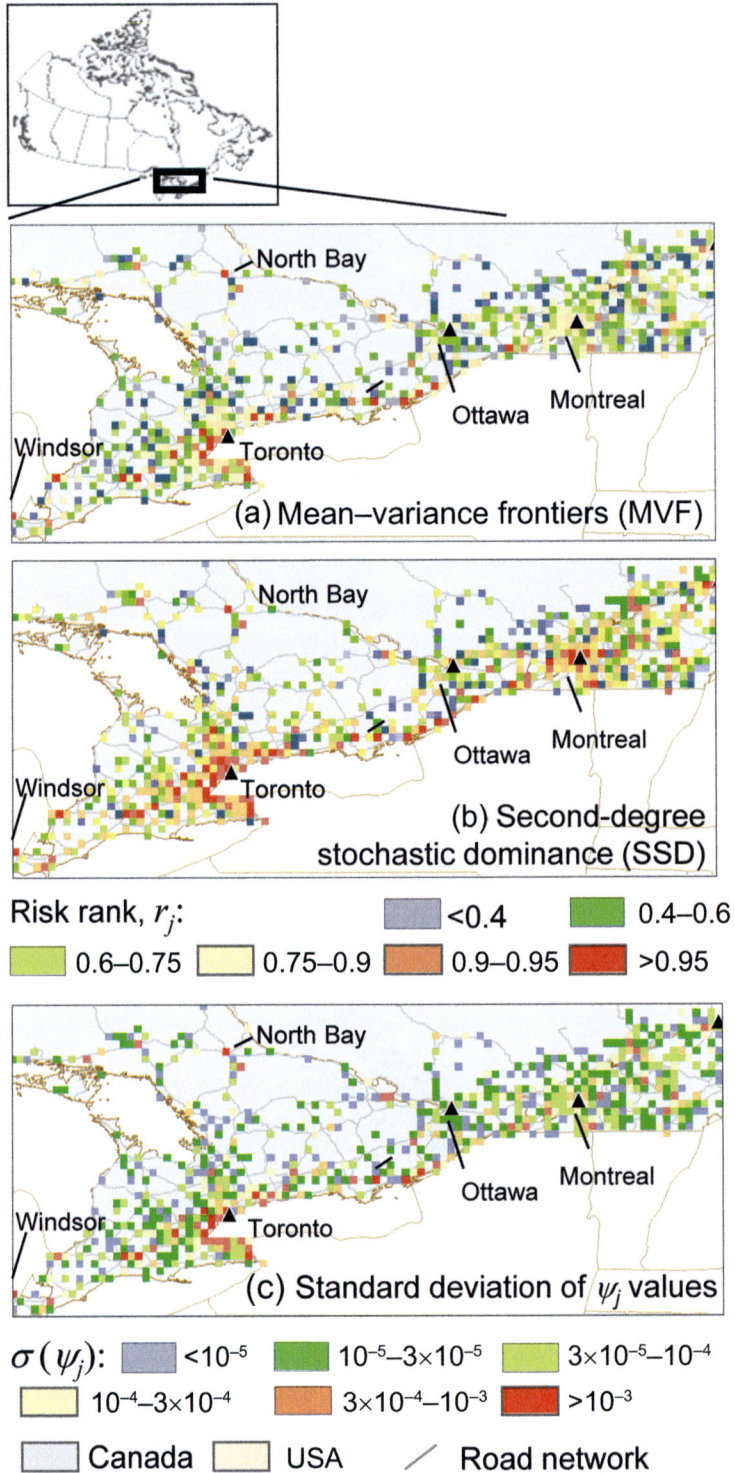

Fig. 14.4. A comparison of geographical risk delineations based on: (a) nested mean–variance frontiers (MVF); (b) the second-order stochastic dominance rule (SSD); and (c) the standard deviation of the pest arrival rate ($\sigma(\psi_j)$). The maps show a portion of eastern Canada with the highest overall risk estimates.

13 Identifying and Assessing Critical Uncertainty Thresholds in a Forest Pest Risk Model

Frank H. Koch[1]* and Denys Yemshanov[2]

[1]*USDA Forest Service, Southern Research Station, Eastern Forest Environmental Threat Assessment Center, Research Triangle Park, North Carolina, USA;* [2]*Natural Resources Canada, Canadian Forest Service, Great Lakes Forestry Centre, Sault Ste. Marie, Ontario, Canada*

Abstract

Pest risk maps can provide helpful decision support for invasive alien species management, but often fail to address adequately the uncertainty associated with their predicted risk values. This chapter explores how increased uncertainty in a risk model's numeric assumptions (i.e. its principal parameters) might affect the resulting risk map. We used a spatial stochastic model, integrating components for entry, establishment and spread, to estimate the risks of invasion and their variation across a two-dimensional gridded landscape for *Sirex noctilio*, a non-native woodwasp detected in eastern North America in 2004. Historically, *S. noctilio* has been a major pest of pine (*Pinus* spp.) plantations in the southern hemisphere. We present a sensitivity analysis of the mapped risk estimates to variation in six key model parameters: (i) the annual probabilities of new *S. noctilio* entries at US and Canadian ports; (ii) the *S. noctilio* population-carrying capacity at a given location; (iii) the maximum annual spread distance; (iv) the probability of local dispersal (i.e. at a distance of 1 km); (v) the susceptibility of the host resource; and (vi) the growth rate of the host trees. We used Monte Carlo simulation to sample values from symmetric uniform distributions defined by a series of nested variability bounds around each parameter's initial values (i.e. ±5%, ..., ±50%). The results show that maximum annual spread distance, which governs long-distance dispersal, was the most sensitive of the tested parameters. At ±15% uncertainty in this parameter, mapped risk values shifted notably. No other parameter had a major effect, even at wider bounds of variation. The methods presented in this chapter are generic and can be used to assess the impact of uncertainties on the stability of pest risk maps or to identify any geographic areas for which management decisions can be made confidently, regardless of uncertainty.

Introduction to Uncertainty Analysis

This chapter describes methods for analysing uncertainty in key parameters of a spatial stochastic model used to estimate invasion risk. The model was applied to forecast the likely geographic pattern of invasion of a recently arrived forest pest, the sirex woodwasp (*Sirex noctilio* Fabricius), in eastern North America (Yemshanov *et al.*, 2009a). One primary benefit of a spatial

* Corresponding author. E-mail: fhkoch@fs.fed.us

stochastic modelling approach is that the various stages of an invasion – its arrival, spread and establishment components – can be depicted within a single modelling framework (Yemshanov et al., 2009c). Unlike other risk modelling methods that focus only on a single stage of invasion (e.g. environmental niche models, which are aimed primarily at characterizing risk of successful establishment), the approach is relatively data-intensive; in particular, it requires estimation of aspects of a pest's population or meta-population dynamics, its host distribution and its general dispersal behaviour. Nonetheless, by adopting an approach of repeated stochastic model simulations, it is possible to quantify the risk of invasion as a numeric probability (Rossi et al., 1993; Rafoss, 2003) for each geographic location (i.e. map cell) comprising the model's spatial domain (i.e. the study area). The approach also allows one to calculate the variation of the risk estimates. Thus, spatial stochastic simulation provides, for each geographic location, an estimate of both the invasion risk and the uncertainty associated with that risk estimate (Yemshanov et al., 2009a). In this case, uncertainty is synonymous with variability in model inputs (i.e. parametric uncertainty) and outputs. Uncertainties resulting from the use of imprecise terms or lack of knowledge are not addressed explicitly.

Importantly, the uncertainties associated with model outputs represent cumulative measures of uncertainties that may arise from a variety of sources; in particular, uncertainty associated with a model's parameters, its input data and possibly its structure or formulation may all contribute to the output uncertainty (Elith et al., 2002; Regan et al., 2002; Walker et al., 2003; Refsgaard et al., 2007). Additional analytical techniques are necessary to explore how uncertainty in any of these particular model elements might be influencing the outputs. The focus of this chapter is on a practical approach, first outlined in Koch et al. (2009), to examine specifically the impact of model parametric uncertainty. We use a Monte Carlo simulation approach to sensitivity analysis to do this. Monte Carlo methods involve repeated random sampling from a range of possible input values (e.g. from a probability distribution associated with a model parameter). When model simulations are completed with these randomly sampled input values, the simulation results can be compiled to obtain estimates of a phenomenon of interest.

The primary objective of sensitivity analysis in this context is to determine the relative contribution of individual model parameters to the uncertainty in the resulting outputs (Helton et al., 2006). The application of sensitivity analysis to analysing parametric uncertainty is not novel (Morgan and Henrion, 1990; Li and Wu, 2006). What makes our analysis unique is that it is applied in a spatial domain. The analysis of uncertainty is uncommon in risk maps, probably because of the perceived difficulty (Morgan and Henrion, 1990; Andrews et al., 2004; Cook et al., 2007). However, we believe that it is critical because invasions are spatial, or more precisely, spatiotemporal, processes and so it is logical to analyse the associated uncertainties from a map-based perspective.

A variety of mathematical, statistical and graphical methods have been utilized for sensitivity analysis (Frey and Patil, 2002). For a spatial stochastic model such as ours, with its capacity to generate many different realizations of the invasion process, a basic Monte Carlo approach offers a straightforward way to examine model parameter sensitivities. We begin by performing repeated simulations with the invasion model in order to get a mapped set of risk estimates based on the parameters' initial values (i.e. based on our best estimates for these values, which we determined either analytically or by consulting experts prior to running the model). The resulting set of risk estimates constitutes a baseline scenario for later comparison to the sensitivity analysis results. Because the model is stochastic, the baseline estimate for each map cell includes both the primary risk metric, P, which is a numeric probability indicating the proportion of simulation runs where successful invasion occurs, and a measure of uncertainty in that estimate, $\sigma(P)$, the standard deviation of P. Next, we identify key model parameters and vary their values within a specified set of nested bounds (±5%, ±10%, ..., ±50%), which alters the pattern of output variation. Using these nested bounds, we complete many additional simulation runs so that we can compare the parameters with one another in terms of the relative impact of uncertainty (e.g. which parameters are most sensitive at ±15% variation in their baseline values).

The analytical approach has three main objectives: (i) to identify any parameters that are highly influential and thus should be thoroughly scrutinized with respect to the consequences of uncertainty; (ii) for any influential parameters, to find the level(s) of uncertainty in inputs that dramatically change output maps (i.e. the output risk and uncertainty estimates) and so affect the maps' utility for end users (i.e. decision makers); and (iii) to determine if, even with added uncertainty, any portions of an output map remain stable enough in the presence of parametric uncertainty for an end user to utilize the map for decision support, regardless of the uncertainty.

Sensitivity analysis has a few important limitations. For instance, as applied, the approach does not address sources of uncertainty in the input data (i.e. the spatial uncertainty in particular), although a similar Monte Carlo approach can be used to vary input data for the purposes of sensitivity/uncertainty analysis (Crosetto et al., 2000; Crosetto and Tarantola, 2001). In theory, Monte Carlo techniques could also be used to compare the sensitivity of different model formulations (i.e. model formulation uncertainty) but would require significant time for computations.

Furthermore, Monte Carlo sensitivity analysis does not deal well with severe uncertainty. For models involving invasive alien species, especially recently discovered invaders, large empirical knowledge gaps may exist regarding the most important invasion drivers. This lack of knowledge makes it difficult to identify meaningful bounds in which to vary parameter values. In our case, we used a hierarchical set of percentage bounds (up to ±50%) around the values of the parameters we analysed, but in theory a parameter (e.g. the annual probability of new entries) may be off by a factor of 100, 1000 or more. Consequently, the analytical approach described here should be seen only as a local or restricted analysis of parametric uncertainty. We believe it is appropriate for the pest highlighted in this chapter (*S. noctilio* Fabricius) because the biology and behaviour of *S. noctilio* have been reasonably well documented in portions of both its native and invaded ranges. Other analytical frameworks, such as info-gap decision theory (Ben-Haim, 2006), may be better for severe parametric uncertainty.

Monte Carlo sensitivity analysis also assumes that the correct parameters to include in the model and for which to perform sensitivity/uncertainty analysis have been identified. If a model is missing key parameters or includes unnecessary or deleterious parameters, the accuracy of the output risk estimates is likely to be affected. These inaccuracies are most likely to be revealed during some sort of validation or cross-validation process with empirical data. Such data are often hard to acquire for invasive alien species. However, whether the correct set of parameters has been included in a model is really a question of model formulation uncertainty.

Because the model described in this chapter was implemented as a risk forecasting approach with essentially no opportunity for validation (i.e. *S. noctilio* was not widely distributed in North America when we performed this work), the sensitivity analysis results should be interpreted only as an assessment of the relative, and not absolute, relationships between model parameters and outputs. This is a critical point: the approach does not address the accuracy of the risk predictions, but does allow the analyst to get a basic sense of the robustness of model outputs to parametric uncertainty. Thus, while we cannot always verify or validate the accuracy of a given risk model, we can determine whether model results have any reliable decision-making value in the face of uncertainty. For instance, the technique described here allows us to identify influential parameters and, just as importantly, determine when uncertainty in a parameter starts to have a significant impact on the output risk estimates (i.e. changes a risk estimate from a relatively high value to a comparatively lower value).

Case Study: Invasion of *S. noctilio*

We modelled the potential invasion of the woodwasp species *S. noctilio* in eastern North American forests. Native to Europe, western Asia and northern Africa, *S. noctilio* has been introduced in many parts of the world and is considered an important pest of pine (*Pinus* spp.) plantations in the southern hemisphere (Carnegie et al., 2006; Corley et al., 2007). The insect has particularly impacted plantations in eastern Australia, New Zealand, South Africa and several South American countries including Brazil, Chile and Argentina. Given the insect's wide bioclimatic tolerance (Carnegie et al., 2006) and an abundance of potential hosts, *S. noctilio* was viewed as a major invasion threat to North America for several years before it was found in upstate New York in 2004 (Hoebeke et al., 2005) and southern Ontario in 2005 (de Groot et al., 2006). Essentially, it is expected to persist throughout temperate pine forests of the USA and southern Canada, although it may require 2 or 3 years to complete a generation at higher latitudes (i.e. cooler climates; Borchert et al., 2007). Pine density in the initially invaded region is comparatively low; as a result, the pest might find greater success were it to spread into the southern USA.

At the time of our analysis, *S. noctilio* had been found in more than 20 counties in New York, several counties in north-central Pennsylvania, and individual counties in Michigan and Vermont. The woodwasp had also been discovered throughout much of southern Ontario. Since that time, this insect has been discovered in additional counties in the states named above, as well as one county each in Ohio, New Jersey and Connecticut, and in western Quebec (National Agricultural Pest Information System, 2013).

With respect to hosts, *S. noctilio* has caused the greatest impacts in plantations of loblolly (*Pinus taeda*) and Monterey pine (*Pinus radiata*) in the southern hemisphere. In North America, *S. noctilio* has been confirmed as a pest on several native pines: red pine (*Pinus resinosa*), eastern white pine (*Pinus strobus*) and jack pine (*Pinus banksiana*) (Dodds et al., 2007). This insect is also found on Scots pine (*Pinus sylvestris*) in managed and unmanaged Christmas tree farms (Dodds et al., 2010). Scots pine is known as a host in the pest's native region. Evidence suggests that a number of pine species in eastern North America are suitable, if not preferred hosts. The geographic distributions of these pine species overlap and facilitate the natural spread of the insect.

S. noctilio can disperse naturally by adult flight or through human activities (e.g. movement of infested logs). It is generally known as a strong flier, but few empirical studies have been performed to quantify its dispersal capabilities (but see Corley et al., 2007). Regardless, the initial introduction of *S. noctilio* into North America was almost certainly human-mediated, likely in raw wood or solid-wood packing materials associated with international trade (Hoebeke et al., 2005).

Model and Data Sources

We used a spatially explicit, raster-based modelling framework to perform the simulations for this study. As a dynamic spatiotemporal model (Gibson and Austin, 1996; Fuentes and Kuperman, 1999), it departs from deterministic risk modelling approaches, which typically adopt the simplifying assumption that an invader's potential distribution is already in equilibrium with its environment.

The model was programmed in C++. Although the model code is not available for public use, the uncertainty analysis techniques described in this chapter can be applied to any spatial stochastic model with a similar structure. The primary requirement is that each model run yields a binary map that documents the absence (0) or occurrence (1) of some event. In our case, we simulated the annual spread and survival of *S. noctilio* over 30 years, such that each simulation run yielded a binary raster map indicating where in eastern North America the insect would have viable populations at the end of the specified time horizon. These maps are then compiled to compute a numeric probability of the event (i.e. for each map location, j, the proportion of model runs in which the event occurred, P_j) and the associated uncertainty (i.e. standard deviation of P_j, $\sigma(P_j)$) for each map location (i.e. each raster map cell). Each grid cell in the model was 5 km × 5 km, which we selected for practicality. The coarser resolution allowed us to reduce the computation time for individual simulation runs, a large number of which were necessary to analyse a set of sensitivity analysis scenarios.

In general terms, the model simulates forest growth, new arrivals, spread and survival of *S. noctilio* across eastern North America in discrete time steps (Yemshanov et al., 2009b). The model is simple with limited data requirements. Some model elements, particularly those related to dispersal, were developed from expert knowledge because empirical data were unavailable. We believe this limitation is common for many invasive alien species. For advice about eliciting expert estimates in lieu of data and then incorporating them into a model, see Morgan and Henrion (1990) and Yamada et al. (2003).

The model simulates three events for *S. noctilio*: (i) new arrivals at US and Canadian ports; (ii) spread across eastern North America; and (iii) establishment in suitable locations. Details for these simulations appear in the online supplement to Chapter 13. Additionally, Table 13.1 provides a brief summary of the main parameters and their roles within the model. Notably, we performed sensitivity analyses only for six model parameters: (i) $W_{x(t)}$, the annual probabilities of new local entries of *S. noctilio* at individual US and Canadian ports; (ii) d_{max}, the maximum annual spread distance; (iii) p_0, the local dispersal probability; (iv) s_v, the susceptibility of the host resource; (v) k, the carrying capacity for a population of *S. noctilio* at a given location; and (vi) g_v, the growth rate of the host trees. We chose this particular subset of parameters after initial model testing to limit computation time. The sensitivity analysis approach described here could certainly be applied to all of a model's parameters, but we felt total analysis was unnecessary to demonstrate the methodology. Furthermore, many models include parameters that are related to or dependent on one another in some way. For the sake of computational efficiency, it may make sense to choose one representative parameter out of related sets of parameters. However, this choice may only be practical if the analyst is familiar with a model's structure and outputs.

Although we used C++ for model development, similar models could be created in a software package like R or MATLAB that can accommodate spatial data. Both of these packages can store the map data as matrices and can implement basic statistical, mathematical and stochastic functions (e.g. testing model-derived probabilities against values from a uniform random distribution). The biggest constraint is likely to be computation time; packages like R are unlikely to be as computationally efficient as optimized code in C++, Java or similar programming languages. Because of the large number of simulation runs required for this sort of exercise, spreadsheet-based Monte Carlo software packages, such as @RISK or CRYSTAL BALL, would probably be unsuitable except at very coarse spatial scales or for small spatial areas.

Analyses

Baseline scenario

We initially used the model for a baseline scenario of *S. noctilio* spread in eastern North America over a 30-year time horizon (from 2006 to 2036). Under this baseline scenario,

Table 13.1. Parameters for components of the *Sirex noctilio* invasion model. The six parameters tested via sensitivity analysis are highlighted in bold.

Parameter	Description
All components	
t	Annual time step for the model
T	30-year time horizon for summarizing the model results (2006–2036)
New arrivals	
t_0	Used in calculating $F(t)$; corresponds to earliest year for which summary import data were available for the USA and Canada
T_{entry}	Used in calculating $F(t)$; corresponds to the presumed year that *S. noctilio* arrived in eastern North America
$F(t)$	Function describing the yearly flow of marine imports to the USA and Canada through time
$p(t)$	Total probability of successful *S. noctilio* entry into North America in year t; derived from $F(t)$
x	Individual port of entry that receives commodities associated with *S. noctilio*; total number of ports = 148
$v_{x(t)}$ ($V_{x(t)}$)	Tonnage of *S. noctilio*-associated cargo received at port x in year t, converted to a proportion, $V_{x(t)}$, of the total *S. noctilio* tonnage for the region that was received at that particular port x
$\mathbf{W_{x(t)}}$	Vector of the local probabilities of *S. noctilio* entry at each port x in year t; derived from $V_{x(t)}$
Spread	
$b(d)$	Colonization rate (i.e. rate of successful dispersal) as a function of the distance, d, from the nearest location with an established *S. noctilio* population; influenced by p_0 and d_{max}
$\mathbf{p_0}$	Local dispersal probability (i.e. probability of dispersal at a distance of 1 km)
$\mathbf{d_{max}}$	Maximum distance at which dispersing *S. noctilio* populations become established
Establishment	
$N_{j(t)}$; $N_{j(t+1)}$	*S. noctilio* population densities in grid cell j at years t and $t+1$
R	Annual *S. noctilio* population growth rate
k	Carrying capacity that constrains maximum population size
$j(t)$	Maximum volume of pine killed by *S. noctilio* in year t; depends on μ
μ	Minimum volume of pine required to support a single population unit
g_v	Function describing the age-dependent rate of host (pine) stand growth
s_v	Function describing the age- and species-dependent level of host (pine) susceptibility; influenced by a_j, a_0, a_{max} and s_{max}
a_j	Host stand age in years (i.e. the average stand age in a map cell)
a_0	Age of host stand closure (20 years)
$\mathbf{a_{max}}$	Age when host stand reaches its maximum level of susceptibility
$\mathbf{s_{max}}$	Maximum susceptibility value for ageing host stands

Please see the online supplement to Chapter 13 for more information about the model components.

model parameter values were left as originally specified during model development (i.e. as determined analytically or from expert estimates; see the online supplement to Chapter 13 for additional details). We initialized the model with a map of known *S. noctilio* infestations as of 2006, the first year when systematic field detection surveys were performed in the USA and Canada.

We generated two output metrics for this baseline scenario that we also utilized in our subsequent sensitivity analyses. The first was P_j, the probability that *S. noctilio* invades map cell j at the end of the forecast horizon. This served as our primary metric of invasion risk. The value of P_j for each map cell was calculated from the repeated model simulations:

$$P_j = \frac{\sum_{u=1}^{U} \tau_{j,u,T}}{U} \qquad (13.1)$$

where $\tau_{j,u,T}$ is a binary variable indicating presence or absence of *S. noctilio* in cell j at time horizon T for a single model replication u and U is the total number of replications for the scenario. Less than 500 replications were required for the baseline scenario to stabilize (see subsection 'Sensitivity analysis scenarios' below for further discussion about stabilization). In addition, the variation of the P_j values was characterized with a map of $\sigma(P_j)$, the standard deviation of P for each cell, which was our primary metric of the output uncertainty. The standard deviation is a commonly used metric of the uncertainty in an estimate, but it has some limitations, perhaps most notably that it can be sensitive to extreme observations. We adopted $\sigma(P_j)$ as our uncertainty metric for the sake of computational simplicity, but other uncertainty metrics such as binary entropy (MacKay, 2003) could be applied in the analyses described here.

From the maps we generated for the baseline scenario (Fig. 13.1; see colour plate section), we can make several broad forecasts regarding the expected path of the *S. noctilio* invasion in eastern North America. First, the risk of invasion (i.e. P_j; Fig. 13.1a) is expected to be high ($P_j > 0.75$) throughout the north-eastern USA, southern Ontario and Quebec, which is unsurprising because the pest is already established in this region. The area of relatively high risk also extends into the northern portion of the south-eastern USA; indeed, the southern edge of the main invasion front is expected to be near the Virginia–North Carolina border and the western edge along the eastern shore of Lake Superior in 2036. Output uncertainty (i.e. $\sigma(P_j)$; Fig. 13.1b) is generally highest near this predicted main front. Beyond the main front, the south-eastern USA contains extensive areas of medium-level ($0.25 \le P_j \le 0.75$) risk near the Atlantic and Gulf coasts (i.e. near possible ports of entry). Notably, this region contains large areas of pine forest, most of which is dominated by loblolly pine, a species understood to be highly susceptible to *S. noctilio* (see Table S13.3 in the online supplement to Chapter 13). The output uncertainty tends to be high here because the probability of a new *S. noctilio* entry at any port, and its subsequent spread and establishment, is relatively moderate compared with the probability of expansion in northern areas near existing infestations. Notably, areas of the south-eastern USA that are further inland (i.e. non-coastal) exhibit both low risk and low uncertainty, which reflects less abundant hosts and greater distance from possible sources of invaders, either ports of entry or the advancing main front.

Sensitivity analysis scenarios

We analysed the sensitivity of the invasion risk estimates using a Monte Carlo approach with four general steps: (i) defining a probability distribution for each parameter of interest; (ii) sampling from this distribution to select a value; (iii) running multiple simulations of the risk model with the parameter values sampled from the distributions; and (iv) summarizing the results from repeated realizations of this process. As acknowledged earlier, a parameter may be poorly specified due to lack of data, so its associated distribution may have to be approximated. In some cases, information about the parameter may be insufficient to characterize even the primary moments (i.e. the mean and variance) of a probability distribution. In other cases, a parameter's empirical distribution may be reasonably well fit by one of the many commonly used theoretical distribution functions (e.g. the normal, exponential or Cauchy distributions). Regardless, a simpler solution may be to assume a uniform distribution for each parameter (Morgan and Henrion, 1990). In our case, we employed a nested set of variability bounds around each tested parameter: ±5%, ±10% and so on up to ±50%. Each pair of 'plus–minus' bounds defined the end points for a symmetric uniform distribution from which we sampled values randomly. A benefit of applying the same percentage increments to all parameters (i.e. rather than changing each parameter according to an independent scale) is that comparisons of their degree of sensitivity become more straightforward.

For the sensitivity analysis, we varied one parameter at a time, while leaving all other parameters unchanged. To verify the relative impact of specific parameters, we also used the alternative approach of varying all tested parameters but one, which was kept at the baseline. Because s_v and g_v were represented as tables of values (Table 13.1; online supplement to Chapter 13), when varying either or both of these parameters, all values in the table(s) were altered identically based on the value sampled randomly from the associated uniform distribution. We must emphasize that our sampling choices (i.e. utilizing uniform distributions and sampling in ±5% increments to a maximum of ±50%) were somewhat arbitrary, although informed by our previous experience with initial testing of the model and an awareness of what would be computationally practical. In our case, these particular choices permitted us to perform a fairly comprehensive analysis of parametric uncertainty, but they may not work well for a different model applied to a different pest.

Analysts who intend to complete similar analyses should consider at least two things. First, they should decide if it is reasonable to approximate the distributions of tested parameters with something other than the uniform distribution. Certain statistical tests (e.g. Kolmogorov–Smirnov test, chi-squared test) can help determine whether a set of data are consistent with the normal distribution or some other proposed distribution, but this assumes the data for a parameter of interest are sufficient for valid testing (Morgan and Henrion, 1990). Second, as alluded to earlier in the chapter, the analysts must also identify levels of variation that will yield meaningful and interpretable results regarding the parameter sensitivities. Unfortunately, this can be a lengthy iterative process and may not be possible for species that lack even basic information for parameterizing an invasion model.

Sensitivity tests and metrics

For complex stochastic simulation models, hundreds or even thousands of replications may be necessary to stabilize the outputs (i.e. to minimize the variation in the output values that can arise simply from completing too few replications). This is especially true when a large amount of variability (i.e. uncertainty) is added to model parameters, as was the case in our sensitivity scenarios. One metric that can be used to determine the minimum number of model replications required for output map stability is S_{XY}, the sum of the squared differences in P_j map values between two trials incorporating consecutively increasing numbers of replications:

$$S_{XY} = \sqrt{\sum_{j=1}^{M}\left[\left(P_{jX} - P_{jY}\right)^2\right]} \qquad (13.2)$$

where M is the total number of map cells covering eastern North America (~156,000 cells) and P_{jX} and P_{jY} are the invasion probabilities for map cell j in trials using X and Y number of replications, $X > Y$. When S_{XY} is plotted against the number of replications Y, it depicts a declining curve; when this curve begins to flatten rather than decline, this indicates that the model has stabilized. We found that most of the sensitivity scenarios converged after 2400–2700 replications (Fig. 13.2), so we generated maps of P_j and $\sigma(P_j)$ based on 3000 model replications for each scenario. As Fig. 13.2 also suggests, other metrics, such as the square root of the total map area where P_j or $\sigma(P_j)$ is below some specified threshold, can be used to assess model stability, but they may not be as easily interpretable as S_{XY}. In our case, these latter two metrics both appear as gradually increasing curves, where the flattening that indicates model stability is not as immediately obvious as with the S_{XY} metric (Fig. 13.2).

To evaluate the effect of introduced parametric uncertainty on the output uncertainty of the risk maps (i.e. on the $\sigma(P_j)$ values), we calculated 'uncertainty ratios' for each sensitivity scenario. For any given map cell, the uncertainty ratio is the value of

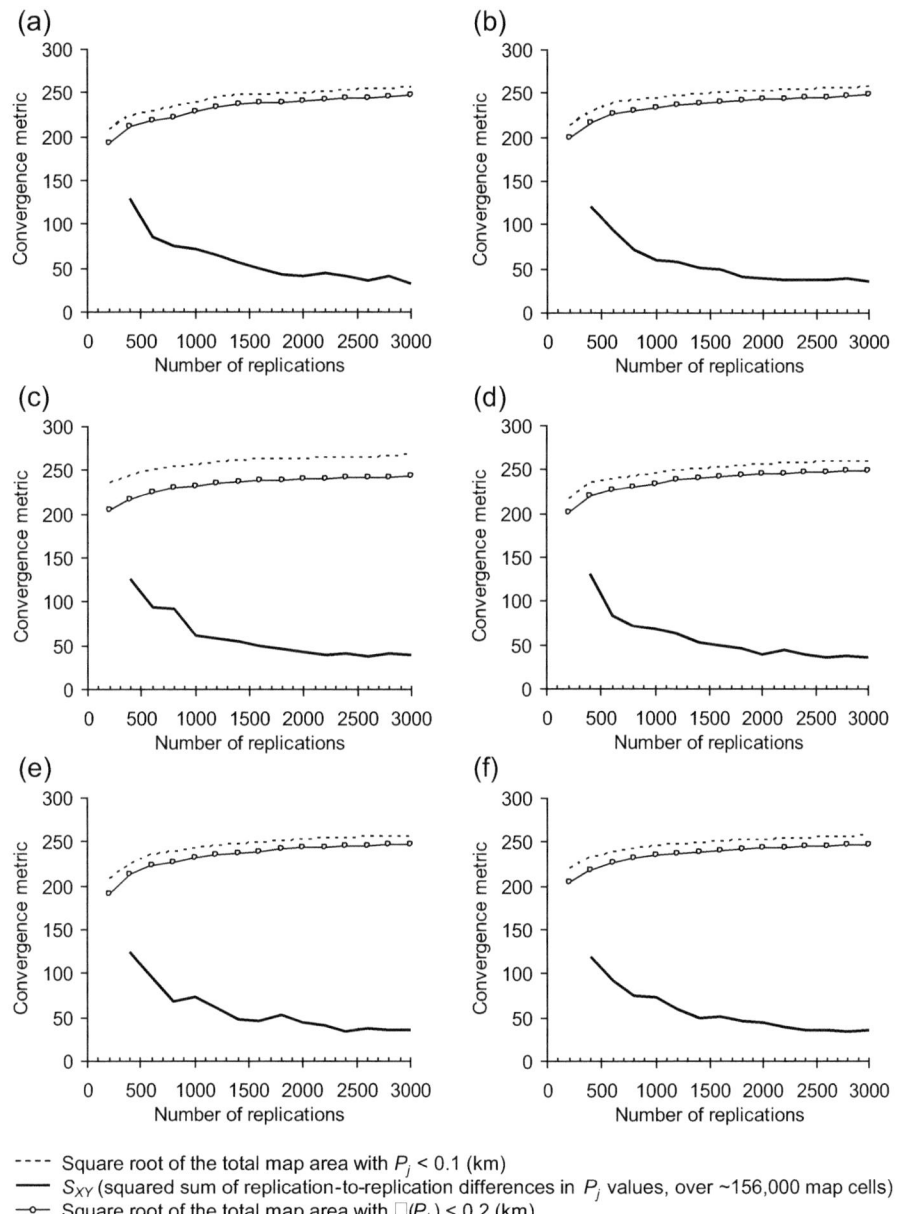

Fig. 13.2. Convergence metric values versus number of replications, at ±40% parametric uncertainty. Tested parameters: (a) local probabilities of entry at marine ports, $W_{x(l)}$; (b) population carrying capacity, k; (c) maximum annual spread distance, d_{max}; (d) local dispersal probability, p_0; (e) host susceptibility, s_v; (f) host growth rate, g_v.

---- Square root of the total map area with $P_j < 0.1$ (km)
— S_{XY} (squared sum of replication-to-replication differences in P_j values, over ~156,000 map cells)
-o- Square root of the total map area with $\square(P_j) < 0.2$ (km)

$\sigma(P_j)$ for the scenario of interest divided by $\sigma(P_j)$ for the baseline scenario. An uncertainty ratio value close to 1 indicates that varying the parameter value at the specified level does not substantially change the variability (i.e. uncertainty) of the output risk estimate. Ratio values approaching 0 indicate decreasing uncertainty in the output risk estimates, while as values progress above 1, they indicate increasing uncertainty in these estimates. Although the uncertainty ratio is a relatively simple metric, it can be mapped such that it enables straightforward visual comparison of the results of the different sensitivity scenarios, even when the compared parameters are widely different in scale or structure.

Visualizations of Uncertainty

Figure 13.3 (see colour plate section) shows uncertainty ratio maps for several of the model parameters at ±40% uncertainty. In the one-parameter-at-a-time sensitivity scenarios, changes to the maximum annual spread distance, d_{max}, increased uncertainty ratios across a large portion of the map area (Fig. 13.3a). In particular, high uncertainty ratios appeared in a broad band just beyond the estimated invasion front and in a few areas inside the front, such as coastal New England. These latter areas exhibited high risk, yet low uncertainty, under the baseline scenario. Changes to the local dispersal probability, p_0 (Fig. 13.3b), also elevated uncertainty ratio values in these locations, but to a lesser degree than observed for d_{max}. Furthermore, unlike for d_{max}, the uncertainty ratios for p_0 were < 1 in the north-western portion of our study area (i.e. the western Great Lakes region). We believe that increased uncertainty in p_0 permitted more invasion nuclei to develop in this remote area through time, which raised the invasion risk estimates, but at the same time stabilized the output uncertainties at a lower level than under the baseline scenario. This phenomenon also occurred with the other parameters besides d_{max}, as exemplified by the ratio map for the port entry probabilities parameter, $W_{x(t)}$ (Fig. 13.3c). Additionally, the map for $W_{x(t)}$ shows another phenomenon we observed for all parameters except d_{max} and p_0: high uncertainty ratios in only a small proportion of the study area, mostly near the edge of the host range (i.e. near the invasion's biological limits).

Essentially, the uncertainty ratio maps for the all-but-one sensitivity scenarios (Fig. 13.3d–f; see colour plate section) show the opposite of the one-at-a-time scenarios. When d_{max} (Fig. 13.3d) was the only parameter left fixed at its baseline value, and all other parameters were varied uniformly within a ±40% bound, the uncertainty ratios were low to moderate throughout most of the study area. In contrast, when any other single parameter was left fixed, such as the population carrying capacity, k (Fig. 13.3e), or the host growth rate, g_v (Fig. 13.3f), the uncertainty ratio values increased substantially across most of the study area. This pattern appears to confirm that d_{max} was the most influential parameter on the model outputs.

Although the uncertainty ratio maps facilitate visual comparison, they do not really quantify the differences between the sensitivity scenarios. It is possible to calculate this difference by using the S_{XY} metric (Eqn 13.2) in a second way: to compare the maps of P_j and $\sigma(P_j)$ for each sensitivity scenario, X, with the matching maps from the baseline scenario, Y. In this case, S_{XY} is calculated as the sum of the cumulative differences between the sensitivity scenario's P_j or $\sigma(P_j)$ map and the corresponding baseline scenario map of P_j or $\sigma(P_j)$. Thus, S_{XY} in this context depicts cumulative changes in the S. noctilio risk map due to the introduction of parametric uncertainty.

We cross-tabulated the S_{XY} differences for eastern North America as well as three smaller focus regions: eastern Canada, the north-eastern USA and the south-eastern USA. Figure 13.4 shows the results for all regions at ±25% and ±40% parametric uncertainty. For the entire study area, and whether calculated from the P_j or $\sigma(P_j)$ maps, the cross-tabulation results indicate that d_{max} was by far the most sensitive of the tested model parameters. The graphs of S_{XY} for P_j (Fig. 13.4a and c) suggest that p_0 was the second most-sensitive parameter when considering the entire study area. Both of these observations were also true for eastern Canada and the north-eastern USA. In

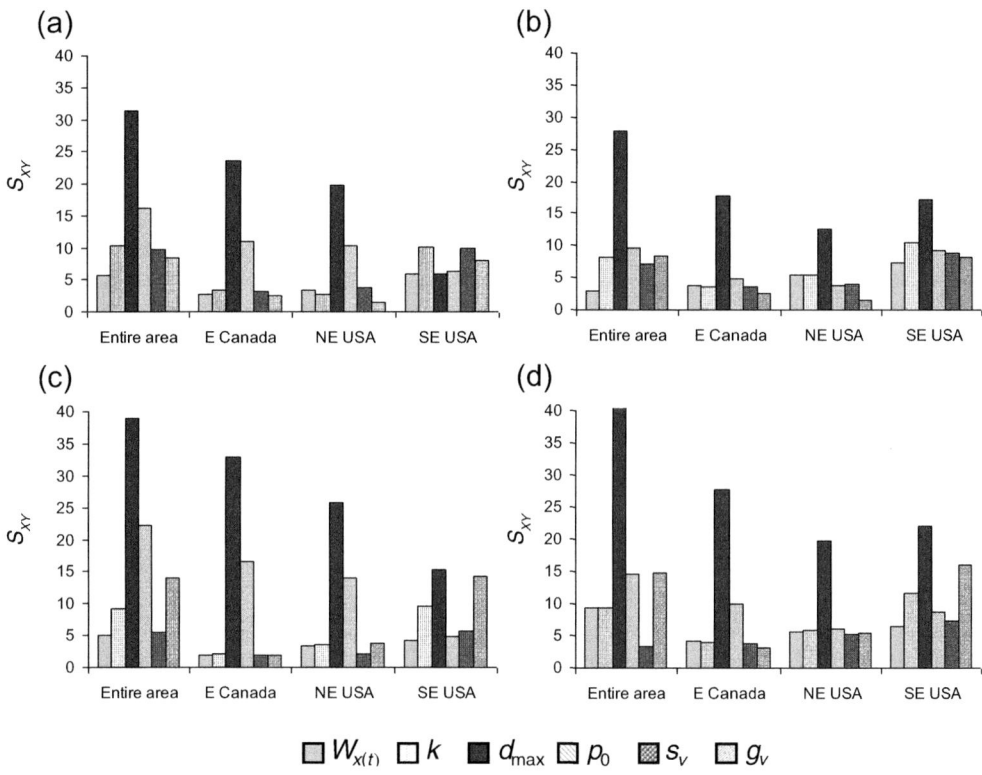

Fig. 13.4. Regional summaries of the S_{XY} metric for the sensitivity analyses. Results from one-parameter-at-a-time sensitivity analyses are presented at two variability increments: (a) for P_j at ±25% parametric uncertainty; (b) for $\sigma(P_j)$ at ±25%; (c) for P_j at ±40%; (d) for $\sigma(P_j)$ at ±40% parametric uncertainty.

contrast, the results for the south-eastern USA suggest relatively less importance for d_{max}, and greater importance for g_v, which was nearly as important as d_{max} at ±40% uncertainty (Fig. 13.4c and d). Indeed, g_v also showed moderately high sensitivity for the entire study area at ±40% uncertainty (Fig. 13.4c and d). We believe this outcome is explained by the fact that the south-eastern US region is host-rich but relatively far from currently infested locations, such that successful invasions would likely only develop from rare, and thus uncertain, new entries. In this context, a parameter governing susceptible host abundance could be nearly as important as one shaping the rate of spread.

In a final set of tests for each sensitivity scenario, we plotted the regions where introducing parametric uncertainty changed the mapped risk estimates considerably. We partitioned cells in the map of P_j for the baseline scenario into three broad classes, 'low', 'medium' and 'high' risk, corresponding to the intervals 0–0.25, 0.25–0.75 and 0.75–1, respectively. We then determined, for each sensitivity scenario, the percentage of the map area (i.e. the percentage of map cells) that moved from one risk class to another when compared with the baseline scenarios. These shifts can be portrayed in a classified map that highlights geographic locations with considerable changes in infestation risk. This classified map can also be used to determine if any geographic regions remained largely unchanged despite the introduction of parametric uncertainty at a specified level. Note that our choice of breakpoints for these risk classes was arbitrary. Certainly, other analysts might find different sets of breakpoints more meaningful or opt to use more than three risk classes. The tests described here should remain applicable, regardless of such choices.

Table 13.2 shows the percentages of the map area that shifted from one risk class to another in one-parameter-at-a-time sensitivity scenarios. Results are displayed for scenarios at small (±10%), moderate (±25%) and high (±40%) levels of introduced parametric uncertainty. At ±10% uncertainty, the sensitivity scenarios typically exhibited only modest shifts in risk class (i.e. <5% of the map area relative to the baseline scenario), except for the local probability of entry, $W_{x(t)}$, which showed a relatively large shift of 8.3% of the map area between the medium and low risk classes. We believe this shift is a consequence of the phenomenon we noted previously: added variability in $W_{x(t)}$ caused more invasion nuclei to enter geographically remote portions of our study area through time, increasing the invasion risk, P_j, but stabilizing the output uncertainty, $\sigma(P_j)$. This phenomenon likely explains similar shifts between medium and low risk for $W_{x(t)}$

(8.1% of the map area) and between high and medium risk for p_0 (8.1%) at ±40% uncertainty, as well as many of the other observed changes in risk class due to introduced parametric uncertainty. The greatest map-area shifts occurred with d_{max}, which exhibited 17.4% and 27.7% shifts from high to medium risk at ±25% and ±40% uncertainty, respectively. In addition, the results for d_{max} from the full sequence of one-at-a-time sensitivity scenarios (Table 13.3) show a 9.8% map-area shift between high and medium risk at ±15% uncertainty. This is larger than any shift observed for the other five parameters, at any variability bound increment. There was also an 8.1% map-area shift between medium and low risk at ±15% uncertainty for d_{max}.

The geographic distribution of these shifts is important. Figure 13.5 (see colour plate section) shows risk-class shifts for eastern North America for the scenarios with ±15% and ±50% uncertainty in d_{max}. At ±15% uncertainty (Fig. 13.5a), most areas within the main invasion front at the 30-year time horizon (see Fig. 13.1 and the earlier description of the baseline scenario results) did not exhibit a change in risk class, although clusters of map cells with

Table 13.2. The percentage of the map area shifting from one risk class to another when varying one parameter at a time within three different symmetric uniform ranges: ±10%, ±25% and ± 40%. Percentages are relative to the class area totals for the baseline scenario.

Shift in risk class	Model parameter					
	$W_{x(t)}$	k	d_{max}	p_0	s_v	g_v
10% added parametric uncertainty						
Low → medium[a]	0.5	4.3	3.0	1.0	1.3	1.7
Medium → low	8.3	1.2	3.5	5.4	4.4	2.6
High → medium	1.0	0.9	4.9	1.1	1.0	0.9
Medium → high	0.7	0.6	0.2	0.6	0.5	0.8
25% added parametric uncertainty						
Low → medium	2.2	2.8	3.5	2.3	2.6	2.5
Medium → low	1.7	1.3	6.6	2.0	1.6	3.7
High → medium	1.3	1.0	17.4	4.6	1.1	0.9
Medium → high	0.6	0.7	0.1	0.2	0.7	0.6
40% added parametric uncertainty						
Low → medium	0.6	3.7	4.4	1.4	2.1	5.8
Medium → low	8.1	2.2	5.7	5.5	3.5	1.1
High → medium	1.2	1.3	27.7	8.1	1.1	1.1
Medium → high	0.6	0.6	0.0	0.1	0.6	0.6

[a]Low risk, $P_j < 0.25$; medium risk, $0.25 \leq P_j \leq 0.75$, high risk: $P_j > 0.75$.

Table 13.3. The percentage of total map area shifting from one risk class to another when varying only the d_{max} parameter. Reported percentages are relative to the class area totals for the baseline scenario.

Shift in risk class	Uniform variation of d_{max}, percentage of the baseline value								
	±5%	±10%	±15%	±20%	±25%	±30%	±35%	±40%	±50%
Low → medium[a]	1.6	3.0	2.9	3.5	3.5	3.4	4.1	4.4	6.4
Medium → low	3.8	3.5	8.1	7.4	6.6	5.1	4.8	5.7	7.8
High → medium	3.4	4.9	9.8	12.1	17.4	20.6	25.2	27.7	34.9
Medium → high (×10)	2.8	2.2	1.3	1.1	0.8	0.8	0.4	0.3	0.3

[a]Low risk, $P_j < 0.25$; medium risk, $0.25 \leq P_j \leq 0.75$; high risk, $P_j > 0.75$.

high-to-medium risk shifts did appear near the front's southern and north-western edges. In the south-eastern USA, beyond the main front, shifts from medium to low risk, and vice versa, occurred primarily in coastal areas. This result likely reflects a high degree of variability in the patterns of expansion of new *S. noctilio* entries at the region's marine ports under an uncertain d_{max}, despite the fact that the region is relatively host-rich. Similar geographic patterns occurred at ±50% uncertainty (Fig. 13.5b), although more map cells were affected. Still, even at this high level of uncertainty in d_{max}, sizeable areas within and beyond the main invasion front did not display a change in risk class. In short, across much of eastern North America, the risk estimates appeared to be fairly robust to uncertainty in this highly influential parameter.

Conclusions

What did we learn about our *S. noctilio* invasion model from these analyses? Foremost, the sequence of sensitivity scenarios demonstrated that the maximum annual spread distance, d_{max}, was the most sensitive of the tested model parameters, followed to a somewhat lesser degree by p_0, the local dispersal probability. In hindsight, this result is unsurprising, since for many mechanistic models of invasion processes, model aspects governing dispersal, particularly long-distance dispersal, are the most influential and uncertain. (Parameters related to the invader's demography may also be influential and uncertain; see Neubert and Caswell, 2000 and Buckley *et al.*, 2005.) Indeed, were this or a similar modelling approach applied to another invasive pest, we could expect dispersal to figure prominently in subsequent sensitivity analyses. Of course, the results for those other models and species of interest may not be as obvious as seen here. Those results would depend substantially on the amount of interplay between the model parameters given the specific circumstances (e.g. region of interest, dispersal behaviour and population dynamics) being modelled. In our case, we modelled an invasive alien pest that is a strong flier and has hosts that are widely distributed and fairly abundant in the region of concern; hence, minimal functional connectivity between host areas may be necessary for range expansion (Minor *et al.*, 2009; Vogt *et al.*, 2009). Another species may be more constrained in a practical sense by the geographic distribution of its host(s), so uncertainty in this constrained distribution (i.e. in host-related model parameters) could have a significant impact on model projections. Nevertheless, we believe that the small suite of tests and metrics we outlined here should facilitate similar model-based analyses of invasion risks and uncertainties, even if the results end up being somewhat ambiguous. We attempted to develop a toolbox that provides analysts with a capacity to measure uncertainty quantitatively and to portray those results geographically. We believe the spatial assessment of models for alien species is critical because invasions unquestionably play out over time and space.

A particular finding of our study is that with the addition of a small degree of uncertainty for the d_{max} parameter (i.e. ±15%), a sizeable proportion of the map area displayed a drop in risk class, either from high to medium or from medium to low. Small uncertainty effects have critical implications were we to use the *S. noctilio* model to portray the forecasted invasion risks to decision makers. The most important implication is that risk estimates from the model must be interpreted cautiously, because we do not have to be very wrong about our primary dispersal assumptions before we may begin to misestimate the invasion risk, which could lead to incorrect risk management decisions. Fortunately, other model parameters that we tested only begin to affect model results when the level of uncertainty is high. We can be more confident that our risk projections are reasonably robust to uncertainty in these parameters (although see discussion below regarding the possibility of severe parametric uncertainties).

Given the known sensitivity of the model to estimates of d_{max}, how might we best proceed operationally? Two possible directions emerge from these findings. First, as Fig. 13.5 (see colour plate section) suggests, even when d_{max} is treated as highly uncertain (±50%), the risk estimates for much of the study area are fairly robust (i.e. no appreciable change occurs in the coarse, high-, medium- or low-risk rankings). In turn, these unchanged portions of the map could probably be used confidently for some decision-making tasks, such as prioritizing locations for the allocation of resources for monitoring and/or management.

Nevertheless, a decision maker may be reluctant to use a risk map if a significant portion of it has apparently been compromised by uncertainty. This predilection may just be a matter of decision makers' personal discretion or, more precisely, her or his degree of aversion to uncertainty. We accept this line of thinking. If this is the case, the sensitivity results suggest a second direction in which to proceed: determining how to resolve the lack of information about dispersal. This information gap could probably be achieved through additional research on important dispersal mechanisms of the pest of interest. Indeed, this is one of the reasons why our own research has increasingly focused on dispersal mechanisms related to human-mediated, long-distance dispersal – such as international and domestic trade (Koch *et al.*, 2011; Yemshanov *et al.*, 2012) – which are widely acknowledged as being pivotal to biological invasions yet poorly characterized. Alternatively, the dispersal modelling component for the pest of interest might also be improved via field study of its dispersal behaviour. For instance, now that *S. noctilio* is established in eastern North America, it might be possible to develop an appropriate dispersal function based on the invasive populations rather than the more indirect source of distributional observations in its native or previously invaded range. If additional research is not feasible, further review of existing literature – perhaps with guidance from experts – might uncover a species that can serve as a reasonable, if imperfect, proxy for the pest of interest when defining dispersal or other model parameters (e.g. Venette and Cohen, 2006).

In any case, our study has demonstrated that the use of sensitivity analysis techniques can reveal important sources of uncertainty (i.e. parametric uncertainty in this case) in pest risk maps and their underlying models. Still, the approach remains limited in terms of its ability to diagnose when those uncertainties might start to alter decision-making priorities. For example, while we were able to quantify when uncertainty in d_{max} began to impact the *S. noctilio* model's output risk estimates, what if we were drastically wrong about one or more of our other parameter estimates? In fact, what if one of our parameter estimates was off by a factor of 10 or more? Consequently, evaluating the parameter in question at ±50% uncertainty might generate an undeservedly optimistic impression of its robustness to uncertainty. As a possible solution, an analyst might opt to implement sensitivity scenarios with much wider uncertainty bounds (e.g. ±1000%), but for relatively fine-scale stochastic simulation models, the computation time required to complete a comprehensive analysis with very wide uncertainty bounds may make this an impractical option. This would be especially true if the analyst deemed it necessary to evaluate the sensitivity of a large number of model parameters. Sampling techniques such as Latin hypercube sampling could reduce the required number of replications (Helton and Davis, 2002; Xu et al., 2005) and thus the computation time. Yet, the best way to achieve efficiency in the face of possibly severe uncertainty may be to adopt the perspective that the impact of uncertainty is best evaluated in the context of a small set of discrete choices about how the model results will be implemented. For instance, if the results will be used to support a long-term surveillance scheme, it might be wise to lay out a few different hypothetical surveillance schemes and perform a cursory examination of how each scheme responds to various amounts of introduced parametric uncertainty. At the least, working from this perspective might help the analyst narrow down the model parameters that require particular focus, and the uncertainty bounds that should be implemented, in order to identify the most robust choice.

Ultimately, we would like to see sensitivity-based analyses of uncertainty become standard practice in pest risk modelling and mapping. Despite limitations with the approach, it can be instructive to analysts when judging the value of their outputs for decision support. For example, maps depicting shifts in risk class at various levels of added uncertainty (such as in Fig. 13.5; see colour plate section) may be paired with the 'baseline scenario' risk map outputs to communicate the potential impact of parametric uncertainty for decision making. Furthermore, while the approach does not facilitate direct incorporation of measured uncertainties into output risk products, it might possibly be used in concert with other analytical approaches that do provide this option (see Yemshanov et al., Chapter 14 in this volume).

This chapter is enhanced with supplementary resources.

To access the material please visit:

www.cabi.org/openresources/43946/

References

Andrews, C.J., Hassenzahl, D.M. and Johnson, B.B. (2004) Accommodating uncertainty in comparative risk. *Risk Analysis* 24, 1323–1335.

Ben-Haim, Y. (2006) *Info-Gap Decision Theory: Decisions Under Severe Uncertainty*, 2nd edn. Academic Press, Waltham, Massachusetts.

Borchert, D., Fowler, G. and Jackson, L. (2007) *Organism Pest Risk Analysis: Risks to the Conterminous United States Associated with the Woodwasp,* Sirex noctilio *Fabricius, and the Symbiotic Fungus,* Amylostereum areolatum *(Fries: Fries) Boidin.* US Department of Agriculture, Animal and Plant Health Inspection Service, Plant Protection and Quarantine, Center for Plant Health Science and Technology, Plant Epidemiology and Risk Analysis Laboratory, Raleigh, North Carolina.

Buckley, Y.M., Brockerhoff, E., Langer, L., Ledgard, N., North, H. and Rees, M. (2005) Slowing down a pine invasion despite uncertainty in demography and dispersal. *Journal of Applied Ecology* 42, 1020–1030.

Carnegie, A.J., Matzuki, M., Haugen, D.A., Hurley, B.P., Ahumada, R., Klasmer, P., Sun, J. and Iede, E.T. (2006) Predicting the potential distribution of *Sirex noctilio* (Hymenoptera: Siricidae), a significant exotic pest of *Pinus* plantations. *Annals of Forest Science* 63, 119–128.

Cook, A., Marion, G., Butler, A. and Gibson, G. (2007) Bayesian inference for the spatio-temporal invasion of alien species. *Bulletin of Mathematical Biology* 69, 2005–2025.

Corley, J.C., Villacide, J.M. and Bruzzone, O.A. (2007) Spatial dynamics of a *Sirex noctilio* woodwasp population within a pine plantation in Patagonia, Argentina. *Entomologia Experimentalis et Applicata* 125, 231–236.

Crosetto, M. and Tarantola, S. (2001) Uncertainty and sensitivity analysis: tools for GIS-based model implementation. *International Journal of Geographical Information Science* 15, 415–437.

Crosetto, M., Tarantola, S. and Saltelli, A. (2000) Sensitivity and uncertainty analysis in spatial modelling based on GIS. *Agriculture, Ecosystems & Environment* 81, 71–79.

de Groot, P., Nystrom, K. and Scarr, T. (2006) Discovery of *Sirex noctilio* (Hymenoptera: Siricidae) in Ontario, Canada. *The Great Lakes Entomologist* 39, 49–53.

Dodds, K.J., Cooke, R.R. and Gilmore, D.W. (2007) Silvicultural options to reduce pine susceptibility to attack by a newly detected invasive species, *Sirex noctilio*. *Northern Journal of Applied Forestry* 24, 165–167.

Dodds, K.J., de Groot, P. and Orwig, D.A. (2010) The impact of *Sirex noctilio* in *Pinus resinosa* and *Pinus sylvestris* stands in New York and Ontario. *Canadian Journal of Forest Research* 40, 212–223.

Elith, J., Burgman, M.A. and Regan, H.M. (2002) Mapping epistemic uncertainties and vague concepts in predictions of species distribution. *Ecological Modelling* 157, 313–329.

Frey, H.C. and Patil, S.R. (2002) Identification and review of sensitivity analysis methods. *Risk Analysis* 22, 553–578.

Fuentes, M.A. and Kuperman, M.N. (1999) Cellular automata and epidemiological models with spatial dependence. *Physica A* 237, 471–486.

Gibson, G.J. and Austin, E.J. (1996) Fitting and testing stochastic models with application in plant epidemiology. *Plant Pathology* 45, 172–184.

Helton, J.C. and Davis, F.J. (2002) Illustration of sampling-based methods for uncertainty and sensitivity analysis. *Risk Analysis* 22, 591–622.

Helton, J.C., Johnson, J.D., Sallaberry, C.J. and Storlie, C.B. (2006) Survey of sampling-based methods for uncertainty and sensitivity analysis. *Reliability Engineering and System Safety* 91, 1175–1209.

Hoebeke, E.R., Haugen, D.A. and Haack, R.A. (2005) *Sirex noctilio*: discovery of a palearctic siricid woodwasp in New York. *Newsletter of the Michigan Entomological Society* 50, 24–25.

Koch, F.H., Yemshanov, D., McKenney, D.W. and Smith, W.D. (2009) Evaluating critical uncertainty thresholds in a spatial model of forest pest invasion risk. *Risk Analysis* 29, 1227–1241.

Koch, F.H., Yemshanov, D., Colunga-Garcia, M., Magarey, R.D. and Smith, W.D. (2011) Potential establishment of alien-invasive forest insect species in the United States: where and how many? *Biological Invasions* 13, 969–985.

Li, H. and Wu, J. (2006) Uncertainty analysis in ecological studies: an overview. In: Wu, J., Jones, K.B., Li, H. and Loucks, O.L. (eds) *Scaling and Uncertainty Analysis in Ecology: Methods and Applications*. Springer, Dordrecht, The Netherlands, pp. 43–64.

MacKay, D.J.C. (2003) *Information Theory, Inference, and Learning Algorithms*. Cambridge University Press, Cambridge.

Minor, E.S., Tessel, S.M., Engelhardt, K.A.M. and Lookingbill, T.R. (2009) The role of landscape connectivity in assembling exotic plant communities: a network analysis. *Ecology* 90, 1802–1809.

Morgan, M.G. and Henrion, M. (1990) *Uncertainty: A Guide to Dealing with Uncertainty in Quantitative Risk and Policy Analysis*. Cambridge University Press, New York.

National Agricultural Pest Information System (2013) *Survey Status of Sirex Woodwasp – Sirex noctilio (All Years)*. US Department of Agriculture, Animal and Plant Health Inspection Service, Plant Protection and Quarantine (USDA-APHIS-PPQ) and Purdue University, West Lafayette, Indiana. Available at: http://pest.ceris.purdue.edu/map.php?code=ISBBADA&year=alltime (accessed 22 December 2013).

Neubert, M.G. and Caswell, H. (2000) Demography and dispersal: calculation and sensitivity analysis of invasion speed for structured populations. *Ecology* 81, 1613–1628.

Rafoss, T. (2003) Spatial stochastic simulation offers potential as a quantitative method for pest risk analysis. *Risk Analysis* 23, 651–661.

Refsgaard, J.C., van der Sluijs, J.P., Højberg, A.L. and Vanrolleghem, P.A. (2007) Uncertainty in the environmental modelling process – a framework and guidance. *Environmental Modelling & Software* 22, 1543–1556.

Regan, H.M., Colyvan, M. and Burgman, M.A. (2002) A taxonomy and treatment of uncertainty for ecology and conservation biology. *Ecological Applications* 12, 618–628.

Rossi, R.E., Borth, P.W. and Tollefson, J.J. (1993) Stochastic simulation for characterizing ecological spatial patterns and appraising risk. *Ecological Applications* 3, 719–735.

Venette, R.C. and Cohen, S.D. (2006) Potential climatic suitability for establishment of *Phytophthora ramorum* within the contiguous United States. *Forest Ecology and Management* 231, 18–26.

Vogt, P., Ferrari, J.R., Lookingbill, T.R., Gardner, R.H., Riitters, K.H. and Ostapowicz, K. (2009) Mapping functional connectivity. *Ecological Indicators* 9, 64–71.

Walker, W.E., Harramoës, P., Rotmans, J., van der Sluijs, J.P., van Asselt, M.B.A., Janssen, P. and Krayer von Krauss, M.P. (2003) Defining uncertainty – a conceptual basis for uncertainty management in model-based decision support. *Integrated Assessment* 4, 5–17.

Xu, C., He, H.S., Hu, Y., Chang, Y., Li, X. and Bu, R. (2005) Latin hypercube sampling and geostatistical modeling of spatial uncertainty in a spatially explicit forest landscape model simulation. *Ecological Modelling* 185, 255–269.

Yamada, K., Elith, J., McCarthy, M. and Zerger, A. (2003) Eliciting and integrating expert knowledge for wildlife habitat modelling. *Ecological Modelling* 165, 251–264.

Yemshanov, D., Koch, F.H., McKenney, D.W., Downing, M.C. and Sapio, F. (2009a) Mapping invasive species risks with stochastic models: a cross-border US–Canada application for *Sirex noctilio* Fabricius. *Risk Analysis* 29, 868–884.

Yemshanov, D., McKenney, D.W., De Groot, P., Haugen, D.A., Sidders, D. and Joss, B. (2009b) A bioeconomic approach to assess the impact of an alien invasive insect on timber supply and harvests: a case study with *Sirex noctilio* in eastern Canada. *Canadian Journal of Forest Research* 39, 154–168.

Yemshanov, D., McKenney, D.W., Pedlar, J.H., Koch, F.H. and Cook, D. (2009c) Towards an integrated approach to modelling the risks and impacts of invasive forest species. *Environmental Reviews* 17, 163–178.

Yemshanov, D., Koch, F.H., Ducey, M. and Koehler, K. (2012) Trade-associated pathways of alien forest insect entries in Canada. *Biological Invasions* 14, 797–812.

14 Making Invasion Models Useful for Decision Makers: Incorporating Uncertainty, Knowledge Gaps and Decision-making Preferences

Denys Yemshanov,[1] Frank H. Koch[2] and Mark Ducey[3]

[1]*Natural Resources Canada, Canadian Forest Service, Great Lakes Forestry Centre, Sault Ste. Marie, Ontario, Canada;* [2]*USDA Forest Service, Southern Research Station, Eastern Forest Environmental Threat Assessment Center, Research Triangle Park, North Carolina, USA;* [3]*Department of Natural Resources and the Environment, University of New Hampshire, Durham, New Hampshire, USA*

Abstract

Uncertainty is inherent in model-based forecasts of ecological invasions. In this chapter, we explore how the perceptions of that uncertainty can be incorporated into the pest risk assessment process. Uncertainty changes a decision maker's perceptions of risk; therefore, the direct incorporation of uncertainty may provide a more appropriate depiction of risk. Our methodology borrows basic concepts from portfolio valuation theory that were originally developed for the allocation of financial investments under uncertainty. In our case, we treat the model-based estimates of a pest invasion at individual geographical locations as analogous to a set of individual investment asset types that constitute a 'portfolio'. We then estimate the highest levels of pest invasion risk by finding the subset of geographical locations with the 'worst' combinations of a high likelihood of invasion and/or high uncertainty in the likelihood estimate. We illustrate the technique using a case study that applies a spatial pest transmission model to assess the likelihood that Canadian municipalities will receive invasive forest insects with commercial freight transported via trucks. The approach provides a viable strategy for dealing with the typical lack of knowledge about the behaviour of new invasive species and generally high uncertainty in model-based forecasts of ecological invasions. The technique is especially useful for undertaking comparative risk assessments such as identification of geographical hot spots of pest invasion risk in large landscapes, or assessments for multiple species and alternative pest management options.

* Corresponding author. E-mail: Denys.Yemshanov@NRCan-RNCan.gc.ca

Uncertainty in Biological Invasions

Probabilistic spatial models have been used to assess the potential for, and impacts from, ecological invasions (Rafoss, 2003; Koch *et al.*, 2009; Pitt *et al.*, 2009; Yemshanov *et al.*, 2009a,b; Prasad *et al.*, 2010; Venette *et al.* 2010; Koch and Yemshanov, Chapter 13 in this volume). Such models offer the capacity to depict fine-scale variations in key environmental and biological parameters that may influence the dynamics of invasive alien species in a landscape. However, parameters in these models require certain statistical assumptions, so when these models are extrapolated in geographical space, uncertainty that is associated with underlying model structure, parameters and data is propagated into pest risk forecasts and maps.

Description of the invasion process in probabilistic terms provides a technical means to represent uncertainties in the events that lead to invasion. Probabilistic invasion models commonly include randomization algorithms to represent the uncertain course of an invasion. For example, forecasts of where an invasive alien organism might be introduced and subsequently spread may include random elements. Alternatively, randomization algorithms can be used to draw plausible values repeatedly from statistical distributions of model inputs to measure the resultant variation in model outputs. This approach is known more generally as Monte Carlo analysis. Numerous randomized simulations of the invasion process provide a set of possible invasion outcomes (Koch *et al.*, 2009; Pitt *et al.*, 2009; Yemshanov *et al.*, 2009a). The model outputs (e.g. an invasive alien species' presence/absence or density at a site by a specified time) can be analysed statistically to determine an expected outcome or the extent of variation among outcomes. These statistics can, for example, provide important insights about managing an invasion in the face of uncertainty or targeting research to alleviate some of the uncertainty.

In this chapter, we focus on relatively simple techniques that help incorporate the uncertainty that is typically generated by probabilistic spatial models into the output risk estimates (i.e. risk maps) for invasive alien pests. Note that the 'risk' we are modelling in this example is the likelihood of the arrival of wood- and bark-boring insects without considering the level of impact. Conceptually, our approach would apply to other aspects of ecological invasions. Because our approach is from a decision maker's perspective, the final estimates should also reflect how the uncertainty in the invasion forecasts might change a decision maker's expectations of invasion outcomes.

Perceptions of Uncertainty in Model-based Pest Risk Assessments

Uncertainty is an inevitable component of invasion forecasts but can be challenging for pest risk managers (i.e. biosecurity professionals and others tasked with managing pest incursions) to factor into their decision-making processes. One of the biggest impediments has been the lack of techniques to directly incorporate uncertainty into the prioritization of risks for decision makers. Notably, human perceptions of 'more certain' versus 'less certain' outcomes are different in a decision-making context (Kahneman and Tversky, 1979; Kahneman *et al.*, 1982); a decision maker's perception of uncertainty embedded in a forecast of pest risk could change his or her priorities for action. Perceptions of uncertainty differ among people. For example, given a choice between two alternative scenarios with the same estimated probabilities of pest arrival, a cautious risk manager would assign higher priority for action to the scenario with more certainly estimated values. This type of behaviour is commonly called 'risk-averse' in economic literature: risk-averse individuals always prefer the more certain option from alternative choices with the same expected outcome (Arrow, 1971; Gigerenzer, 2002; Shefrin and Belotti, 2007). For example, the risk-averse investor would prefer a stock with a 3 ± 0.5% annual return on investment over the stock with a 3 ± 4% annual return. Conversely, 'risk-tolerant' or 'risk-seeking' investors would prefer the more variable stock because this stock has a greater potential to outperform its historical average return in the short term than the more consistent stock.

The meaning of 'risk' in investments is different from how it is typically understood in pest risk modelling (i.e. as shorthand for the likelihood of arrival, establishment and spread, and the magnitude of impacts). Indeed, risk aversion might be better characterized as aversion to uncertainty in the elements of the invasion being modelled. This terminology is admittedly ambiguous, but 'risk aversion' also implies that a pest risk manager would generally assign lower priority for action to estimates of pest invasion risk that are comparatively more uncertain. However, for information-gathering purposes, decision makers may reverse their priorities and become more risk-tolerant. For example, a pest risk model to plan a pest survey in a heterogeneous landscape may show two locations where the presence of an invasive alien species is equally probable but the estimate for one location is more uncertain than for the other. For the risk-tolerant manager, locations with more uncertain risk estimates would be assigned a higher priority for survey (Yemshanov et al., 2010) because these locations would provide more knowledge-gaining opportunities. Results from surveys in these locations also could help to reduce some of the uncertainty associated with the model parameters or inputs.

Ideally, the risk assessor would use an algorithm to consistently adjust outputs from a probabilistic invasion model to support decision making. The adjustment would depend on the amount of uncertainty in the outputs and a basic understanding of the decision maker's perceptions of uncertainty. Potential adjustments can be visualized by plotting the model-based estimates of risk in two dimensions: the mean likelihood of invasion (i.e. across all model replications) against the uncertainty of that likelihood estimate. For example, a pest's estimated mean arrival rate for each location of interest is plotted against the variance in Fig. 14.1a.

When uncertainty is ignored and the action priorities (e.g. selection of sites for surveillance) are based solely on the mean invasion likelihood values, the dividing boundaries between high- and low-priority locations can be depicted in the mean–variance space as lines with constant mean likelihood values parallel to the x-axis (Fig. 14.1b). In this case, the amount of variance in the likelihoods of pest invasion does not affect the decision-making choice. If decision makers assign higher priorities to more certain estimates of pest invasion likelihood (i.e. are 'risk-averse'), the lines that delineate action priority levels will be curved relative to the x-axis (Fig. 14.1c), so the locations with more certainly defined estimates of invasion likelihood would receive higher relative priority and vice versa. Alternatively, when a decision maker's objective is to gain more information about an invader's behaviour, or when uncertainty in the invasion likelihood values is believed to increase the overall level of decision-making priority (i.e. the decision maker is 'risk-seeking'), the locations with higher variance will be assigned a higher priority. In this case, the lines delimiting the priority levels would be curved relative to the x-axis in an opposite direction (Fig. 14.1d).

Portfolio Valuation Techniques and Pest Risk Assessment

The concepts depicted in Fig. 14.1 are strikingly similar to the financial asset valuation process which has been studied in detail and has produced a corresponding analytical framework (Arrow, 1971; Elton and Gruber, 1995). To illustrate how this framework may be translated to invasive species modelling, consider a hypothetical example of a geographical assessment of pest invasion likelihood, where a spatial invasion model has forecast the potential spread of a newly documented pest. Knowledge about the invader's behaviour in its new environment remains limited, so the

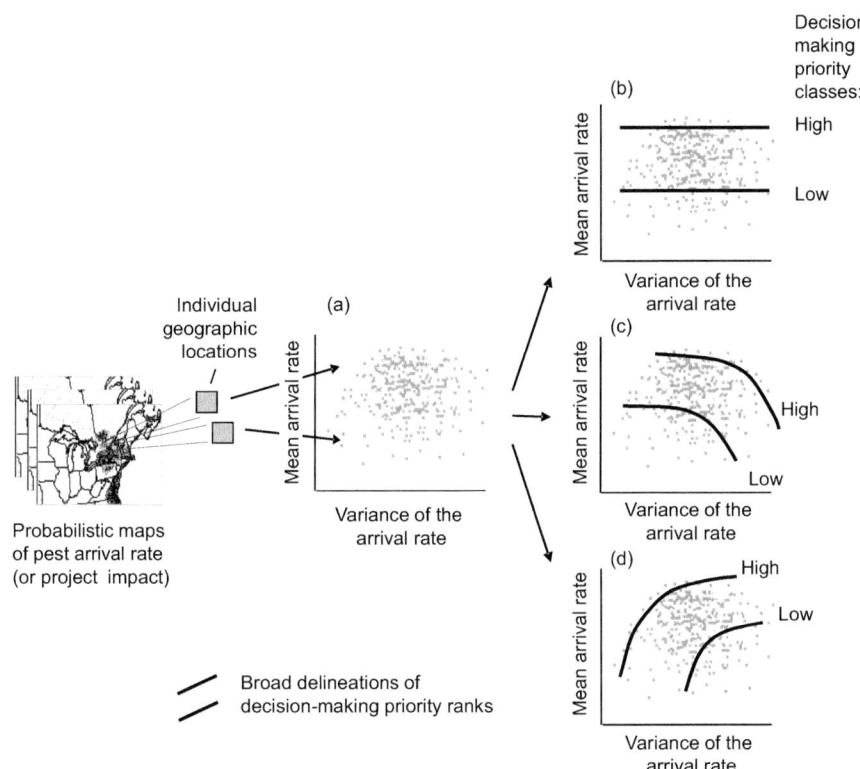

Fig. 14.1. Delineation of decision-making priorities with respect to mean pest arrival rate and uncertainty: (a) depicting individual geographical locations (i.e. map cells) as points in the dimensions of mean pest arrival rate and variance; (b) decision making is based solely on the mean values regardless of the amount of variance in the arrival rate estimates (points above the 'high' line are a high priority, those below the 'low' line a low priority and points between the lines are a medium priority); (c) decision making assigns higher priority to estimates with higher variance; (d) decision making is risk-averse, so more certain estimates of the pest arrival rate receive higher priority.

spatial model represents key parameters that characterize the species' expected behaviour in distributional form. Multiple randomized simulations generate, for each location (i.e. individual map cells, in this example), a multitude of estimated pest arrival rates within a specified time. For each location, the invader's estimated mean arrival rate and the uncertainty of that estimate (i.e. the variance of the simulated mean) can be calculated. Results for multiple locations can be plotted on a mean–variance graph (Fig. 14.1). Plots generally show the points as an amorphous cluster termed a 'cloud' (Fig. 14.1a).

Locations for resource deployment (e.g. to monitor the ongoing spread of the invader) are prioritized by finding the locations that have the 'most extreme' combinations of mean arrival rates and associated uncertainties. What constitutes 'most extreme' depends on the decision maker's perception of uncertainty. For example, uncertainty might increase (Fig. 14.1c) or decrease (Fig. 14.1d) the priority for action at a location. Regardless, the combinations of mean arrival rate and uncertainty characterize locations within the mean–variance cloud. For instance, a relatively low arrival rate that is highly uncertain would place a location on the outer (i.e. most extreme) boundary of the point cloud. Notably, this combination is analogous to finding the 'non-dominant' portfolio set in financial asset allocation when considering mean net financial returns and their volatilities (i.e. the variances of the net return values). In our case, the likelihood that the threatening pest will arrive at a previously pest-free location is analogous to the concept of 'net return', while the uncertainty of that likelihood estimate is analogous to 'volatility' (Arrow, 1971; Elton and Gruber, 1995).

In pest risk modelling, a set of individual geographical locations (e.g. map cells or polygons) within a region is analogous to a set of individual 'portfolios', where each 'portfolio' is characterized by an associated distribution of estimated net returns (i.e. analogous to model-based likelihoods of pest arrival). Commonly, we rank these locations so the highest rank would denote the most extreme combination of arrival rate and uncertainty. As we perceive the process, decision makers will continue to select the most highly ranked sites until a budget limit is reached. Lines that distinguish high-priority locations (i.e. those that need immediate management action) from lower-priority locations (i.e. where action could be deferred) can be drawn through the mean–variance cloud. These limits are represented as convex lines (i.e. 'frontiers') at user-defined boundaries within the mean–variance clouds. Figure 14.1c illustrates the general shape of these lines when the decision maker is risk-averse and Fig. 14.1d illustrates the general shape when the decision maker is risk-tolerant.

In portfolio allocation, the usual objective is to select a few portfolio combinations from a theoretically infinite set that have the desired trade-offs between net returns and their volatilities (Elton and Gruber, 1995). In our invasion risk modelling scenario, each portfolio represents a single geographical location, so their total number is finite. Under classical portfolio theory, allocation usually aims to define a single best-performing set of portfolios in financial terms (Ingersoll, 1987; Elton and Gruber, 1995). A single set is sufficient because it is assumed that any investment amount can be allocated simply in specified proportions to the set of portfolios. In geographical assessments of pest invasions, finding a single best-performing portfolio set would be analogous to identifying a small portion of the geographical region that combined the highest likelihoods of invasion and the highest variances of these estimates (or lowest, depending on decision-making goals).

In order to evaluate the rest of the map locations, we must subsequently define a hierarchy of best- to worst-performing portfolio sets for all map locations in a study area. To do this, the distribution of arrival rates for each of the n locations (i.e. square cells) in the map is evaluated to find a subset, $\aleph_1$, of locations with the most extreme combinations of arrival rate and associated uncertainty. In financial terminology, this subset is often called the 'non-dominant' or 'efficient' set. Once the non-dominant set $\aleph_1$ is found, it is assigned the highest priority rank of 1 and removed from set n temporarily. Next, a second non-dominant subset, $\aleph_2$, is determined from the rest of the map locations, $n - \aleph_1$, assigned a rank of 2, and so forth. The process is repeated until all sets of locations in the area of interest have been evaluated and assigned a priority rank. Conceptually, this technique follows an algorithm for finding nested non-dominant sets (Goldberg, 1989) and multi-attribute frontiers (Yemshanov et al., 2013).

Finding Non-dominant Frontiers

Finding nested non-dominant sets, or efficient frontiers, represents the most critical step in the analysis because the set limits must simultaneously account for a decision maker's perceptions of uncertainty and the amount of variation in the data, and yet be computationally tractable. In this chapter, we demonstrate two relatively simple approaches based on the mean–variance frontier concept and the stochastic dominance rule. We continue to draw upon the example that we established earlier. To briefly recap, an area of interest is divided into n grid cells (i.e. locations). A Monte Carlo simulation generates multiple estimates of the arrival rate of an invasive alien species into each cell. For each cell, we have an estimate of the mean arrival rate (i.e. the mean of all simulated values for that cell) and uncertainty in the arrival rate (i.e. the variance or standard deviation about the mean and the frequency distribution of arrival rates). Our goal is to help the decision maker select particular locations for action (e.g. conduct a survey for the invasive alien species).

Mean–variance frontier concept

The mean–variance frontier (MVF) concept is a visually appealing and simple technique. The mean arrival rate, $\bar{\psi}_j$, for a location j is plotted against the standard deviation of the arrival rate, $\sigma(\psi_j)$, which serves as a measure of uncertainty (Fig. 14.2a). All geographical locations are plotted on the same graph. Instead of using variance as the measure of uncertainty, we use the standard deviation because it increases monotonically with the variance, spreads points along the uncertainty axis more uniformly and facilitates frontier identification. Ultimately, the classification and ranking of locations is

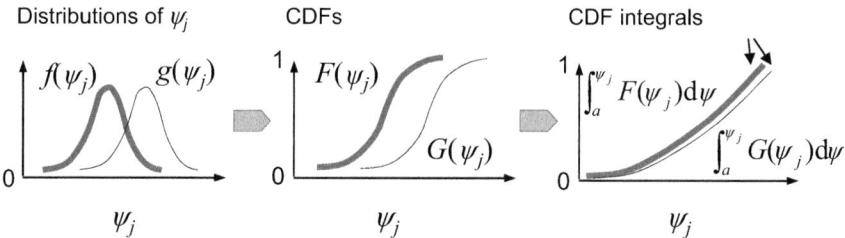

Fig. 14.2. The concepts of nested mean–variance frontiers and stochastic dominance: (a) ranking risk of invasion via nested frontiers in dimensions of mean risk and its standard deviation (i.e. risk-seeking preferences); (b) second-degree stochastic dominance rule. In (b), $f(\psi_j)$ and $g(\psi_j)$ are example distributions of pest arrival rates at two corresponding map locations, f and g; $F(\psi_j)$ and $G(\psi_j)$ are the cumulative distribution functions (CDFs) of $f(\psi_j)$ and $g(\psi_j)$; $\int_a^{\psi_j} F(\psi_j) d\psi$ and $\int_a^{\psi_j} G(\psi_j) d\psi$ are the integrals of the CDFs. The 'SSD' label highlights the second-order stochastic dominance condition (i.e. $\int_a^{\psi_j} F(\psi_j) d\psi$ and $\int_a^{\psi_j} G(\psi_j) d\psi$ do not cross each other).

similar whether the variance or standard deviation is used. The point, and corresponding map location, in the outer boundary of the cloud (i.e. the furthest from the origin) are assigned a priority rank of 1, the highest decision-making priority. Points that are an equivalent distance from the origin are also given a rank of 1. This set of priority 1 locations is removed from the mean–variance cloud, and the next set of points that are most distant from the origin in the new cloud are found and assigned rank 2, and so on. Each frontier represents a layer that has one-point width. The estimation of frontiers proceeds inwards one by one, like peeling an onion, as depicted in Fig. 14.2a, until all points are evaluated and assigned a corresponding priority rank. For the risk-seeking decision maker, the highest-priority locations are in the upper-outermost convex frontier of the mean–variance cloud (Fig. 14.2a). For risk-averse decision makers, the highest-priority locations are in the upper-innermost convex boundary where points exhibit extreme combinations of low variance and/or high mean values (Arrow, 1971; Elton and Gruber, 1995).

Stochastic dominance

Another popular technique to identify best-performing sets of portfolios is based on the second-degree stochastic dominance (SSD) rule. The SSD is a pair-wise ordering rule for distributions of observations. The SSD rule compares two distributions based on their cumulative distribution functions or CDFs (Fishburn and Vickson, 1978; Whitemore and Findlay, 1978; Levy, 1998). In our case, we compare two geographical locations, f and g. For each location, the assemblage of model-based invasion outcomes (i.e. across all of the model simulations) is described by the distribution, $f(\psi_j)$ or $g(\psi_j)$, of the pest arrival rate ψ_j, which can vary within an interval of values $[a; b]$. For simplicity, we consider a range of ψ_j values between 0 and 1, the absolute minimum and maximum values possible for this variable (Fig. 14.2b), but acknowledge that the range a–b can be broader if other metrics for pest risk are used. The SSD rule compares the distributions of f and g as represented by the integrals of their respective cumulative distribution functions:

$$\int_a^{\psi_j} F(\psi_j) d\psi \quad \text{and} \quad \int_a^{\psi_j} G(\psi_j) d\psi$$

where

$$F(\psi_j) = \int_a^{\psi_j} f(\psi_j) d\psi \quad \text{and}$$

$$G(\psi_j) = \int_a^{\psi_j} g(\psi_j) d\psi$$

Location g dominates the alternative f by second-degree stochastic dominance if

$$\int_a^{\psi_j} [F(\psi_j) - G(\psi_j)] d\psi \geq 0 \text{ for all } \psi_j \text{ and}$$

$$\int_a^{\psi_j} [F(\psi_j) - G(\psi_j)] d\psi > 0 \text{ for at least one } \psi_j$$

(14.1)

Note that the integrals of the CDFs are calculated for the entire range of ψ_j values. For the SSD condition to be met, the integrals of the CDFs for $F(\psi_j)$ and $G(\psi_j)$ (i.e. lines depicting CDF integrals in Fig. 14.2b) must not cross. The SSD rule does not require $f(\psi_j)$ or $g(\psi_j)$ to be normally distributed. The rule also meets the assumptions of the risk-averse decision maker: given two choices with the same expected value, the more certain choice is always preferred (Levy, 1992; Levy and Levy, 2001). Because $G(\psi_j)$ and $F(\psi_j)$ represent the full distributions of model-based pest arrival rates at locations f and g, uncertainty in the ψ_j values may cause the dominance conditions to fail. Because g may not dominate f by the SSD rule it will end up with a comparatively lower rank due to the uncertainty (Levy, 1998).

For our example, potential differences in pest arrival rates ψ_j among locations were analysed by multiple, pair-wise SSD tests. For the test, one location was designated f and the other, g. By comparing all possible pairs of map locations (i.e. $n(n-1)$ pairs), we identified a subset ψ_1 that could not be dominated by any element in the rest of the set, $n - \aleph_1$, according to the SSD rule (i.e. that would make the dominance conditions in Eqn 14.1 fail). After the first non-dominant subset ψ_1 was found, assigned a risk rank of 1 and removed from set n, the next non-dominant subset, $\aleph_2$, was found, assigned a risk rank of 2 and so on until all elements of n that represented the map locations were evaluated.

An Example with a Pathway-based Pest Invasion Model

We illustrate the portfolio-based methodology with a case study involving potential invasive alien species in Canadian forests. We use a relatively simple probabilistic model to estimate the likelihood that invasive forest insects may be carried in commercial freight transported to major Canadian municipalities via the North American road network. For simplicity, our model does not consider local pest spread by biological means or the pests' population dynamics. The choice of a particular model is not critical in demonstrating the portfolio-based approach. However, we have chosen a pathway-based model over more common spatial spread models because of its capacity for predicting human-assisted movement of invasive pests over long distances, a phenomenon that most spread models cannot predict well (Andow et al., 1990; Buchan and Padilla, 1999; Melbourne and Hastings, 2009). Pathway-based models describe the spread of a species through a network. The network is composed of nodes (e.g. a set of parks, campgrounds or cities) and connections between nodes. The network analysis prioritizes the degree of connectivity between the nodes (e.g. number of truck trips) over distance. So, for example, if 500 truck trips occur per day between cities A and C and only ten truck trips occur per day between A and B, a species is more likely to be moved from city A to city C than to city B even if city B is only 50 km from A and city C is 500 km from A. In general, pathway-based models can predict low-probability long-distance dispersal events better than common spatial spread models.

In our case, we associated the long-distance dispersal of invasive alien insect pests of forests with the movement of traded commodities via trucks on the North American road network. Traded commodities have been recognized as a reasonable predictor of the human-mediated movement of invasive species (e.g. Tatem et al., 2006; Hulme et al., 2008; Floerl et al., 2009; Hulme, 2009; Kaluza et al., 2010; Koch et al., 2011). We used a Commercial Vehicle Survey (CVS) maintained by Transport Canada as our primary data source to forecast movement of wood-boring forest pests with commodities and freight (Yemshanov et al., 2012a,b). We included commodity categories that involve raw wood products or are associated with significant quantities of wood packing materials (Table 14.1). These materials are acknowledged as a potential source of invasive alien forest pests, despite the implementation of International Standard for Phytosanitary Measures No. 15 (ISPM 15), which stipulates treatment of these materials to reduce the risk of pest introduction via international trade (USDA APHIS, 2000). The effectiveness of ISPM 15 has been questioned (Reaser and Waugh, 2007; Reaser et al., 2008), implying that such measures cannot completely eliminate risk (Haack and Petrice, 2009; Liebhold, 2012).

CVS data were collected during a 2005–2007 survey at truck weigh stations across Canada. Each CVS record summarized a single shipment route reported by a driver. The summary included the route origin, destination, (if applicable) the location(s) where the route crossed the US–Canadian border and a description of the cargo (i.e. type and tonnage). We selected records from the CVS database with commodity types that are commonly associated with invasive alien forest pests (Table 14.1). We reformatted the CVS data into a list of 'origin–destination' network segments so the location at the beginning of each segment (i.e. a part of a route from the CVS) was treated as an 'origin' and the location at the end of a segment as a 'destination' (Fig. 14.3). The network included about 11,000 individual locations in total.

We used the 'origin–destination' network to estimate the rate of transmission of

Table 14.1. Commodity categories from the US–Canadian Standard Classification of Transported Goods (SCTG) commonly associated with transport of bark- and wood-boring forest insects.

Commodity category (SCTG code)
Monumental or building stone (10)
Logs and other wood in the rough (25)
Wood products (26)
Non-metallic mineral products (31)
Base metal in primary or semi-finished forms and in finished basic shapes (32)
Articles of base metal (33)
Machinery (34)
Electronic and other electrical equipment and components and office equipment (35)
Motorized and other vehicles, including parts (36)
Transportation equipment, not elsewhere classified (37)
Precision instruments and apparatus (38)
Furniture, mattresses and mattress supports, lamps, lighting fittings (39)
Miscellaneous manufactured products (40)

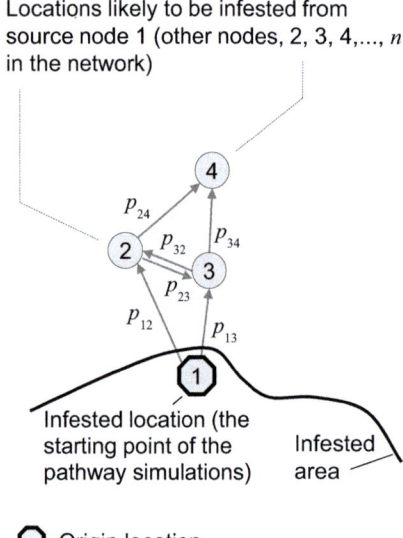

Fig. 14.3. Schematic representation of pest invasion spread in a network-based pathway model. The parameters p_{ij} represent the probability of movement from location i to location j.

invasive forest pests through an individual pathway segment between two given locations, i and j, in the network. We first summed the tonnages of forest pest-associated commodities recorded in the CVS data with one sum per directional pathway between locations i and j, which were designated m_{ij} and m_{ji} (note $m_{ij} \neq m_{ji}$). For each route segment ij, the rate, p_{ij}, of a forest pest being moved from i to j over the survey period (2005–2007) was estimated from the total tonnage of freight shipments of forest-pest-associated commodities from i to j:

$$p_{ij} = 1 - \exp(-m_{ij}\lambda_t) \qquad (14.2)$$

where λ_t is the likelihood of a pest being moved with one tonne of relevant commodities over the survey period t. Essentially, λ_t is a multiplier that converts the tonnage value into a rate estimate. Given the scope of our case study (i.e. modelling the potential human-assisted movement of an entire class of forest pests) it is impossible to calculate an exact λ_t. However, because our primary focus was to prioritize locations in terms of their relative potential to receive invasive forest pests from elsewhere (i.e. to establish a partial order among geographical locations in the transportation network), precise knowledge of λ_t was not critical (Yemshanov et al., 2012b). We used recent records of the spread of the emerald ash borer, a significant forest pest, along the most prominent vector of its expansion in Ontario, Canada (i.e. along the Highway 401 corridor between Windsor and Toronto), as well as the corresponding tonnages of relevant commodities moved through this corridor, to estimate λ_t (Yemshanov et al., 2012b). Briefly, we estimated λ_t via a series of iterative pathway model simulations that were intended to match the model-estimated and known rate of emerald ash borer spread along the Highway 401 corridor in southern Ontario given relevant trucking statistics.

Essentially, the model was comprised of a pathway matrix where each element p_{ij} estimates the rate of an invasive forest pest being moved with commercial truck transport from one geographical location, i, to another, j. The rows of the matrix represent the starting points of individual pathway segments, while the columns represent the segments' end points. The matrix also has an extra column that denotes the probability, p_{i0}, that the invasive organism fails to arrive at any location j from location i:

$$\mathbf{P}_t = \begin{bmatrix} 0 & p_{12} & \cdots & p_{1n} & p_{10} \\ p_{21} & 0 & \cdots & p_{2n} & p_{20} \\ \vdots & \vdots & \vdots & \vdots & \vdots \\ p_{n1} & p_{n2} & \cdots & 0 & p_{n0} \end{bmatrix} \quad (14.3)$$

where

$$p_{i0} = 1 - \sum_{j=1}^{n} p_{ij}$$

The introduction of the last column was required because the transmission probability values p_{ij} in a given row do not always sum to 1. Since the CVS data did not document the duration of stay at intermediate locations during transit, the diagonal elements of the matrix, p_{ii}, were not estimated. We left the p_{ii} values at zero and instead added a column with the p_{i0} values which bring the sum of each row to 1.

We used the pathway matrix to generate transmissions of a hypothetical invasive forest pest through the transportation network. The model generated discrete transmission pathways via multiple iterations. For each iteration, the model generated a single pathway route that started from an origin location i and passed through a number of destination locations until a location with no outgoing paths or the termination state based on the p_{i0} value was chosen. As depicted in Fig. 14.3, the model simulated the movement of the pest from each point of 'origin' i to other locations j by extracting the associated vector of probabilities p_{ij} (i.e. the row of matrix values associated with i) from the pathway matrix and using it to select the next pathway point. Finally, a rate of pest arrival was estimated from the number of times the pest arrived at j from i over multiple pathway simulations:

$$\varphi_{ij} = J_i / K \quad (14.4)$$

where J_i is the number of individual pathway simulations that started at location i and passed through location j. K is the total number of individual simulations of pathway spread from i. The value of K (i.e. 2×10^6 simulations for each origin point in this study) was limited by the available computing capacity.

The arrival rates were then rearranged so each j 'destination' in the transportation network was characterized by an empirical distribution, ψ_j, of pest arrival rates φ_{ij} from all other nodes i, $i \neq j$ (i.e. $n - 1$ locations). This distribution described the location's potential to receive a forest pest with commercial freight transported from elsewhere. Because each destination had a distribution of arrival rates simulated from one origin at a time, the level of uncertainty associated with each location j depends on the connectivity of the pathway network and the variation in commodity flows along the pathway segments. In order to generate a geographically continuous coverage, we aggregated point-based arrival rates into a 15 km × 15 km grid map by combining the ψ_j

values from each location into one common distribution. Each map cell was characterized by a distribution of ψ_j pest arrival rates from other map cells in Canada. We then applied our portfolio-based techniques, the mean–variance frontier and the stochastic dominance rule, to these distributions of arrival rates to rank geographical locations.

While we used portfolio-based techniques in our analyses for all of Canada, we only show the region in southern Ontario and Quebec with the highest traffic flows and road densities so that patterns can be illustrated more clearly (Fig. 14.4; see colour plate section). We express the risk ranks, $l_{j\ SSD}$ and $l_{j\ MVF}$, generated with the mean–variance and stochastic dominance rules, relative to the largest rank possible for each respective method. We have calculated the relative rankings, $r_{j\ MVF}$ and $r_{j\ SSD}$, as:

$$r_{j\ SSD} = 1 - \{(l_{j\ SSD} - 1) / \max[l_{j\ SSD}]\};$$
$$r_{j\ MVF} = 1 - \{(l_{j\ MVF} - 1) / \max[l_{j\ MVF}]\}$$
(14.5)

where $\max[l_{j\ SSD}]$ and $\max[l_{j\ MVF}]$ represent the maximum rank values in the SSD and MVF classifications. The relative rankings have values from 0 to 1, so the relatively highest-priority ranks are close to 1 and lowest are close to 0. These relative rankings are ordinal values and are meant to facilitate comparisons of MVF and SSD results.

To explore the geographic patterns of risk identified by the ranking methods we have further divided the ranks into broad classes with the arbitrary ranking thresholds of 0.4, 0.6, 0.75, 0.9 and 0.95 (Fig. 14.4; see colour plate section). Many of the highest ranked cells in eastern Canada ($r_{j\ MVF}$, $r_{j\ SSD}$ > 0.75) are associated with major transportation arteries in southern Ontario and Quebec, which suggests that these corridors could serve as key pathways for new pest arrivals. The two ranking methods identified similar highest-priority areas (Fig. 14.4a and b); however, the map based on the SSD rule had more high-ranking sites (i.e. $r_{j\ SSD}$ > 0.9) than the map based on MVF. Table 14.2 illustrates the agreement between the priority rankings among the highest-ranked locations. (Note that each map cell was assigned a location name corresponding to the nearest large municipality.)

While the highest-ranked lists generated with the MVF and SSD rules are relatively close, they differed substantially from the ranks based only on the mean arrival rates.

We also compared maps of risk ranks derived with the MVF and SSD rules with the geographical distribution of the standard deviation of the arrival rates, $\sigma(\psi_j)$ (Fig. 14.4c; see colour plate section). Uncertainty, represented by standard deviation, figured prominently in location characterizations based on the MVF and SSD rules (Fig. 14.4a and b; see colour plate section); the highest-ranked locations typically included areas with high variability in pest arrival rates.

The impact of uncertainty on relative priority ranks, r_j, is even more evident when individual map locations are plotted as points on a graph of the mean pest arrival rate, $\bar{\psi}_j$, against the standard deviation, $\sigma(\psi_j)$ (Fig. 14.5). In Fig. 14.5a, the idealized boundaries between these general risk classes, as delineated with the MVF rule, are tilted at an angle, $\beta > 90°$, because the frontiers were selected starting with the upper-outermost boundary of the mean–variance cloud. This pattern shows that for any two locations with equal mean arrival rates, the location with higher variability would be assigned a higher priority rank, just as we had intended.

For delineations based on the SSD rule, the tilt angle β of the boundaries between the risk classes was below 90° (Fig. 14.5b). This result implies that for any two locations with equal mean arrival rates, the location with the more certain estimate (i.e. with lower variability) would be assigned a higher rank.

In summary, the two approaches perform similarly in terms of the highest- and lowest-ranked locations, but for moderate risk ranks, the methods place differing levels of emphasis on certainty in the arrival rate estimate (i.e. the MVF approach emphasizes areas of high

Table 14.2. Location rankings for Canadian cities based on mean arrival rate, mean–variance frontiers (MVF) and second-degree stochastic dominance (SSD) methods. A ranking based only on the locations' mean arrival rate values is included for comparison.

Location name[a]	Mean arrival rate ψ_i (× 10^{-5})	f_i[b]	Ranking method MVF Rank	MVF f_i	SSD Rank	SSD f_i	Location name (continued)	Mean arrival rate ψ_i (× 10^{-5})	f_i	Ranking method MVF Rank	MVF f_i	SSD Rank	SSD f_i
Cornwall, ON	15.9	145*	1	1	16*	1	Gananoque, ON	35.0	2	20	0.91	32	0.94
Toronto, ON	65.3	1	2	0.995	1	1	Ottawa, ON	7.7	38	32	0.88	18	0.96
Windsor, ON	29.8	3	4	0.98	1	1	Calgary, AB	21.4	8	14	0.94	56	0.9
Kitchener, ON	21.2	9	3	0.99	6	0.99	Oshawa, ON	23.9	5	26	0.89	31	0.94
Drummondville, QC	7.6	39	5	0.98	11	0.98	Nobleton, ON	2.3	80	45	0.86	15	0.96
Trois-Rivieres, QC	8.9	30	7	0.97	7	0.99	Sorel, QC	2.0	89	41	0.86	20	0.96
Iroquois, ON	6.0	45	8	0.96	5	0.99	White Rock, BC	22.6	7	21	0.91	60	0.89
Moncton, NB	9.1	28	6	0.97	10	0.98	Niagara Falls, ON	5.5	13	33	0.87	34	0.93
Ste Madeleine, QC	11.1	22	10	0.95	8	0.99	Abbotsford, BC	19.9	11	23	0.91	68	0.88
Quebec, QC	8.3	34	15	0.93	1	1	Kingston, ON	13.8	18	31	0.88	47	0.91
Montreal, QC	15.8	15	9	0.95	9	0.98	Sault Ste Marie, ON	12.5	20	40	0.86	42	0.92
Sarnia, ON	23.4	6	13	0.94	14	0.97	Orono, ON	14.0	17	50	0.85	44	0.92
Lacolle, QC	4.8	52	11	0.94	12	0.97	Sparwood, BC	10.4	24	25	0.9	76	0.86
London, ON	26.8	4	12	0.94	19	0.96	Ingersoll, ON	19.8	12	36	0.87	63	0.89
Hamilton, ON	14.8	16	18	0.92	13	0.97	Thunder Bay, ON	9.1	27	44	0.86	140	0.79
St Georges, QC	2.6	73	16	0.93	16	0.96	Halifax, NS	1.6	102	42	0.86	36	0.93
Napanee, ON	21.0	10	17	0.92	23	0.95	Winnipeg, MB	8.4	32	30	0.88	49	0.91
Windsor, QC	7.3	42	22	0.91	17	0.96	Edmonton, AB	8.2	35	54	0.84	87	0.85
North Bay, ON	13.5	19	19	0.92	24	0.95	Vancouver, BC	6.4	43	57	0.83	96	0.84

The list is sorted by the sum of $f_{i\text{MVF}}$ and $f_{i\text{SSD}}$ values.
[a] Location name based on nearest large municipality at 15 km spatial resolution: AB, Alberta; BC, British Columbia; MB, Manitoba; NB, New Brunswick; NS, Nova Scotia; ON, Ontario; QC, Quebec.
[b] f_i, relative rank denotes the location's rank order relative to the total number of ranks for that metric. Values nearest 1.0 are the highest priority.

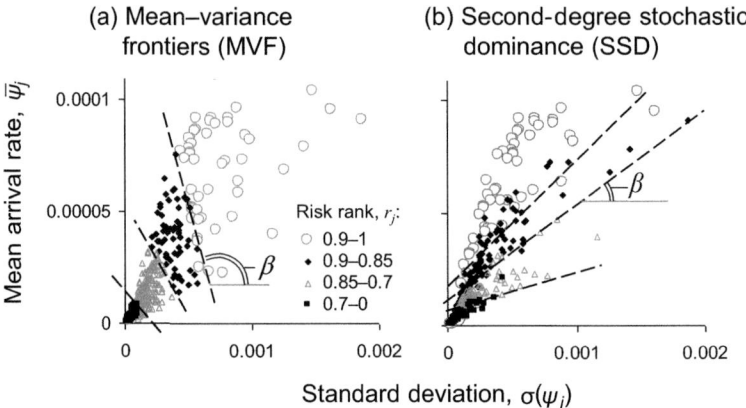

Fig. 14.5. Relative pest risk ranks, r_j, for each location, j, plotted with respect to mean pest arrival rate, $\bar{\psi}_j$, and its standard deviation, $\sigma(\psi_j)$. β denotes the tilt angle between the idealized boundaries of coarsely defined risk classes in the mean–variance cloud and the line indicating constant mean arrival rate ($\bar{\psi}_j$ = const). Different symbols delineate broad classes of the risk ranks and dashed lines depict idealized boundaries between broad classes of risk ranks.

uncertainty more). This distinctive behaviour occurs because the delineation of the non-dominant sets in the MVF and SSD classifications effectively starts from the opposite sides of the mean–variance cloud and thus the two methods offer different treatments of uncertainty.

Assessing the Utility of Portfolio-based Techniques for Pest Risk Assessment

Historically, the adoption of portfolio-based techniques in economic studies was driven by the need to find the best possible allocation of financial assets in the context of uncertain returns (Levy and Markowitz, 1979; Götze et al., 2008). The idea of incorporating decision-making priorities as a component of financial predictive models applies well to other decisions that have to be made under uncertainty. In our model-based assessment of the likelihood of pest arrival, portfolio-based approaches provided a tractable way to incorporate uncertainty into quantitative assessments of the relative risk of pest arrival among several sites that is consistent with decision-making priorities and to communicate those differences in a single decision support product (i.e. a map of relative priority ranks in our study).

In economics, if the objective is to find the smallest set of best-performing investment portfolios, the SSD and MVF approaches have been criticized as too coarse to be practical (Hardaker et al., 2004; Hardaker and Lien, 2010). However, we found the discriminatory power of SSD and MVF rules to be sufficient for our pest risk modelling case. Since our study required ranking of all geographic locations in the map and the total number of locations was large, SSD and MVF rules were sufficient to identify geographical hot spots as small as a few adjacent map cells. Also, the magnitude of the variation in the pest arrival rate values was considerably larger than the typical volatility associated with investment analyses; hence differences between the CDF integrals in the case of the SSD rule and convex mean–variance frontiers in the case of the MVF approach were more discernible.

In our example, we demonstrated how the incorporation of uncertainty via the MVF and SSD methods can change the interpretation of geographical estimates of the pest arrival risk. The output pest arrival risk was a dimensionless priority rank aimed to assist decision makers with allocations of resources for pest surveillance and control. Maps of risk ranks can be used as inputs for further economic assessments of survey costs which may include optimization of resources for survey and pest control efforts. The methodology might be used with a cost-based metric that estimates potential benefits of successful detections or management actions in response to an outbreak, culminating in an optimal cost allocation study.

The capacity of the portfolio-based techniques to account for uncertainty also improves the utility of stochastic invasion models (Demeritt *et al.*, 2007) as risk assessment and decision support tools. We illustrated the methods with a relatively simple invasion model, but the approach can be linked to more complex, probabilistic models where uncertainty in model predictions is expected to be high. Similarly, the approach could be used with spatial models to generate plausible invasion outcomes at multiple geographical locations. For example, stochastic cellular automata and gravity models (e.g. Haynes and Fotheringham, 1984; Muirhead *et al.*, 2006; Pitt *et al.*, 2009; Yemshanov *et al.*, 2009a) could generate multiple maps, each representing a possible outcome of the invasion process. Multiple maps can be rearranged so each geographical location is characterized by a distribution of invasion likelihoods or impact metrics which could be used with one of the portfolio-based techniques presented here.

Incorporating decision makers' perceptions of uncertainty

The MVF and the SSD techniques can address decision-making strategies where uncertainty is treated differently. The MVF approach, as implemented in this study, may be suitable when the uncertainty about a pest is a factor that might increase the priority for decision makers. In particular, the MVF concept is useful when both the central tendency (i.e. the mean) and variability (e.g. the standard deviation) of the risk metric represent critical decision-making variables. For instance, the MVF approach could be applied in model-based assessments for early detection of invasive organisms, when the need to gain more information and reduce uncertainty about the invader's presence or absence is paramount. In such cases, the model-based rate of pest arrival may not sufficiently characterize information gains from unexpected events such as detections of a pest in low-probability locations. Consequently, the uncertainty of the arrival rate estimate becomes a distinctly important variable, so the prioritization should include arrival rate and its variance. Alternatively, delineation based on the SSD rule may be more suitable for assessments that support costly and irreversible decisions such as restricting trade or imposing a regulation, or when decision makers are risk-averse.

Ultimately, the applicability of a particular portfolio-based method can be affected by the type of pest invasion model and the nature of the model output used in the assessment. One analytical strategy that may be worth considering is to estimate the priority ranks with more than one algorithm and then undertake an extra analysis step of aggregating the multiple sets of output ranks into a single-dimensional decision priority metric using multi-criteria aggregation techniques that do not require prior setting of the criteria weights (e.g. the multi-attribute frontier aggregation described in Yemshanov *et al.*, 2013).

Computational remarks

Both the MVF and SSD techniques perform a delineation of nested efficient frontiers for a particular map (i.e. a particular set of spatial elements), which means they only rank those elements relative to one another. In order to compare the maps of multiple scenarios or different geographical regions, the priority ranks would need to be remapped to a new common scale. The simplest approach would be to combine all geographical sets or scenarios into a single superset that includes all alternative maps or scenario data sets, and then assign the priority ranks with respect to all possible invasion outcomes and scenarios that could be found in this superset.

The MVF and SSD techniques use somewhat different approaches to generate the rank values. The MVF approach essentially 'peels' the mean–variance cloud of points, starting from the outermost layer. Alternatively, the stochastic dominance approach ranks the geographical locations via multiple pair-wise SSD tests. As a consequence, the two techniques typically yield different numbers of locations at the highest ranks. For example, Table 14.2 lists eight locations with priority ranks above 0.95 for the MVF-based classification versus 20 locations for the SSD-based approach. The allocation of the frontiers in the MVF approach can also be influenced by local variations of the point density in the mean–variance space: more frontiers can be delineated in regions of the mean–variance space with higher point density. Also note that the MVF rule assumes that the mean values and the variance provide an adequate representation of the distribution as a whole (Gandhi and Saunders, 1981). Alternatively, the stochastic dominance approach evaluates the entire cumulative distribution of expected outcomes and does not require prior evaluation of the distribution shape (Fishburn and Vickson, 1978).

The application of either the SSD or MVF approach to large geographical data sets, such as high-resolution outputs of invasion and dispersal models with a large number of spatial elements (i.e. map cells or polygons), is computationally demanding[1]. For example, the SSD test has a computational complexity on the order of $n(n-1)/2$ and the most basic algorithm to find two-dimensional convex MVF frontiers has a complexity on the order of n^2. For very large data sets, further reduction of the computing time is possible by implementing more efficient convex frontier delineation algorithms (Porter et al., 1973; Kung et al., 1975; Rhee et al., 1995; Papadias et al., 2003).

Acknowledgements

The authors extend their gratitude and thanks to Kirsty Wilson and Marty Siltanen (Natural Resources Canada, Canadian Forest Service) for technical support and diligence with preparing the CVS data set for this case study. The participation of Denys Yemshanov was supported by an interdepartmental NRCan–CFIA Forest Invasive Alien Species initiative. The participation of Mark Ducey was supported by the Agriculture and Food Research Initiative Competitive Grants Program Grant No. 2010-85605-20584 from the National Institute of Food and Agriculture.

Notes

[1] Software that performs basic MVF and SSD rankings can be obtained from the lead author at Denys.Yemshanov@NRCan-RNCan.gc.ca.

References

Andow, D.A., Kareiva, P.M., Levin, S.A. and Okubo, A. (1990) Spread of invading organisms. *Landscape Ecology* 4, 177–188.

Arrow, K.J. (1971) *Essays in the Theory of Risk Bearing*. Markham, Chicago, Illinois.

Buchan, L.A.J. and Padilla, D.K. (1999) Estimating the probability of long-distance overland dispersal of invading aquatic species. *Ecological Applications* 9, 254–265.

Demeritt, D., Cloke, H., Pappenberger, F., Theilen, J., Bartholmes, J. and Ramos, M.-H. (2007) Ensemble predictions and perceptions of risks, uncertainty, and error in flood forecasting. *Environmental Hazards* 7, 115–127.

Elton, E.J. and Gruber, M.J. (1995) *Modern Portfolio Theory and Investment Analysis*, 5th edn. Wiley, New York.

Fishburn, P.C. and Vickson, R.C. (1978) Theoretical foundations of stochastic dominance. In: Whitemore, G.A. and Findlay, M.C. (eds) *Stochastic Dominance: An Approach to Decision-Making under Risk*. Lexington Books, D.C. Heath and Co., Lexington, Massachusetts, pp. 39–144.

Floerl, O., Inglis, G.J., Deym K. and Smith, A. (2009) The importance of transport hubs in stepping-stone invasions. *Journal of Applied Ecology* 46, 37–45.

Gandhi, D.K. and Saunders, A. (1981) The superiority of stochastic dominance over mean variance efficiency criteria: some clarifications. *Journal of Business Finance & Accounting* 8, 51–59.

Gigerenzer, G. (2002) *Calculated Risks: How to Know When Numbers Deceive You*. Simon & Schuster, New York.

Goldberg, D.E. (1989) *Genetic Algorithms in Search, Optimization, and Machine Learning*. Addison-Wesley, Reading, Massachusetts.

Götze, U., Northcott, D. and Schuster, P. (2008) *Investment Appraisal – Methods and Models*. Springer, Berlin.

Haack, R.A. and Petrice, T.R. (2009) Bark- and wood-borer colonization of logs and lumber after heat treatment to ISPM 15 specifications: the role of residual bark. *Journal of Economic Entomology* 102, 1075–1084.

Hardaker, J.B. and Lien, G. (2010) Stochastic efficiency analysis with risk aversion bounds: a comment. *Australian Journal of Agricultural and Resource Economics* 54, 379–383.

Hardaker, J.B., Richardson, J.W., Lien, G. and Schumann, K.D. (2004) Stochastic efficiency analysis with risk aversion bounds: a simplified approach. *Australian Journal of Agricultural and Resource Economics* 48, 253–270.

Haynes, K.E. and Fotheringham, A.S. (1984) *Gravity and Spatial Interaction Models*. Sage, Beverly Hills, California.

Hulme, P.E. (2009) Trade, transport and trouble: managing invasive species pathways in an era of globalization. *Journal of Applied Ecology* 46, 10–18.

Hulme, P.E., Bacher, S., Kenis, M., Klotz, S., Kuhn, I., Minchin, D., Nentwig, W., Olenin, S., Panov, V., Pergl, J., Pysek, P., Roques, A., Sol, D., Solarz, W. and Vila, M. (2008) Grasping at the routes of biological invasions: a framework for integrating pathways into policy. *Journal of Applied Ecology* 45, 403–414.

Ingersoll, J.E. Jr (1987) *Theory of Financial Decision Making*. Rowman & Littlefield, Lanham, Maryland.

Kahneman, D. and Tversky, A. (1979) Prospect theory of decisions under risk. *Econometrica* 47, 263–291.

Kahneman, D., Slovic, P. and Tversky, A. (1982) *Judgment Under Uncertainty: Heuristics and Biases*. Cambridge University Press, New York.

Kaluza, P., Kolzsch, A., Gastner, M.T. and Blasius, B. (2010) The complex network of global cargo ship movements. *Journal of the Royal Society Interface* 7, 1093–1103.

Koch, F.H., Yemshanov, D., McKenney, D.W. and Smith, W.D. (2009) Evaluating critical uncertainty thresholds in a spatial model of forest pest invasion risk. *Risk Analysis* 29, 1227–1241.

Koch, F.H., Yemshanov, D., Colunga-Garcia, M., Magarey, R.D. and Smith, W.D. (2011) Establishment of alien-invasive forest insect species in the United States: where and how many? *Biological Invasions* 13, 969–985.

Kung, H.T., Luccio, F. and Preparata, F.P. (1975) On finding the maxima of a set of vectors. *Journal of the Association for Computing Machinery* 22, 469–476.

Levy, H. (1992) Stochastic dominance and expected utility: survey and analysis. *Management Science* 38, 555–593.

Levy, H. (1998) *Stochastic Dominance: Investment Decision Making under Uncertainty*. Kluwer Academic Publishers, Dordrecht, The Netherlands.

Levy, H. and Markowitz, H. (1979) Approximating expected utility by a function of mean and variance. *The American Economic Review* 69, 308–317.

Levy, M. and Levy, H. (2001) Testing for risk aversion: a stochastic dominance approach. *Economics Letters* 71, 233–240.

Liebhold, A.M. (2012) Forest pest management in a changing world. *International Journal of Pest Management* 58, 289–295.

Melbourne, B.A. and Hastings, A. (2009) Highly variable spread rates in replicated biological invasions: fundamental limits to predictability. *Science* 325, 1536–1539.

Muirhead, J.R., Leung, B., van Overdijk, C., Kelly, D.W., Nandakumar, K., Marchant, K.R. and MacIsaac, H.J. (2006) Modelling local and long-distance dispersal of invasive emerald ash borer *Agrilus planipennis* (Coleoptera) in North America. *Diversity and Distributions* 12, 71–79.

Papadias, D., Tao, Y., Fu, G. and Seeger, B. (2003) An optimal and progressive algorithm for skyline queries. *Proceedings of the 2003 Association for Computing Machinery Special Interest Group on Management*. International Conference on Management of Data, San Diego, California, pp. 467–478.

Pitt, J.P.W., Worner, S.P. and Suarez, A.V. (2009) Predicting Argentine ant spread over the heterogeneous landscape using a spatially explicit stochastic model. *Ecological Applications* 19, 1176–1186.

Porter, R.B., Wart, J.R. and Ferguson, D.L. (1973) Efficient algorithms for conducting stochastic dominance tests on large numbers of portfolios. *Journal of Financial and Quantitative Analysis* 8, 71–81.

Prasad, A.M., Iverson, L.R., Peters, M.P., Bossenbroek, J.M., Matthews, S.N., Syndor, T.D. and Schwartz, M.W. (2010) Modeling the invasive emerald ash borer risk of spread using a spatially explicit cellular model. *Landscape Ecology* 25, 353–369.

Rafoss, T. (2003) Spatial stochastic simulation offers potential as a quantitative method for pest risk analysis. *Risk Analysis* 23, 651–661.

Reaser, J.K. and Waugh, J. (2007) *Denying Entry: Opportunities to Build Capacity to Prevent the Introduction of Invasive Species and Improve Biosecurity at US Ports*. International Union for Conservation of Nature (IUCN), Gland, Switzerland.

Reaser, J.K., Meyerson, L.A. and Von Holle, B. (2008) Saving camels from straws: how propagule pressure-based prevention policies can reduce the risk of biological invasion. *Biological Invasions* 11, 193–203.

Rhee, C., Dhall, S.K. and Lakshmivarahan, S. (1995) The minimum weight dominating set problem for permutation graphs is in NC. *Journal of Parallel and Distributed Computing* 28, 109–112.

Shefrin, H. and Belotti, M.L. (2007) Behavioral finance: biases, mean–variance returns, and risk premiums. *CFA Institute Publications* 2007 (June), 4–12.

Tatem, A.J., Rogers, D.J. and Hay, S.I. (2006) Global transport networks and infectious disease spread. *Advances in Parasitology* 62, 293–343.

USDA APHIS (2000) *Pest Risk Assessment for Importation of Solid Wood Packing Materials into the United States*. US Department of Agriculture, Animal Plant Health Inspection Service, Plant Protection Quarantine, Riverdale, Maryland.

Venette, R.C., Kriticos, D.J., Magarey, R., Koch, F.H., Baker, R.H.A., Worner, S.P., Gómez, N.N., McKenney, D.W., Dobesberger, E.J., Yemshanov, D., De Barro, P.J., Hutchison, W.D., Fowler, G., Kalaris, T.M. and Pedlar, J. (2010) Pest risk maps for invasive alien species: a roadmap for improvement. *BioScience* 60, 349–362.

Whitemore, G.A. and Findlay, M.C. (1978) Introduction. In: Whitemore, G.A. and Findlay, M.C. (eds) *Stochastic Dominance: An Approach to Decision-Making under Risk*. Lexington Books, D.C. Heath and Co., Lexington, Massachusetts, pp. 1–36.

Yemshanov, D., Koch, F.H., McKenney, D.W., Downing, M.C. and Sapio, F. (2009a) Mapping invasive species risks with stochastic models: a cross-border United States–Canada application for *Sirex noctilio* Fabricius. *Risk Analysis* 29, 868–884.

Yemshanov, D., McKenney, D.W., De Groot, P., Haugen, D.A., Sidders, D. and Joss, B. (2009b) A bioeconomic approach to assess the impact of an alien invasive insect on timber supply and harvests: a case study with *Sirex noctilio* in eastern Canada. *Canadian Journal of Forest Research* 39, 154–168.

Yemshanov, D., Koch, F.H., Ben-Haim, Y. and Smith, W.D. (2010) Detection capacity, information gaps and the design of surveillance programs for invasive forest pests. *Journal of Environmental Management* 91, 2535–2546.

Yemshanov, D., Koch, F.H., Lyons, B., Ducey, M. and Koehler, K. (2012a) A dominance-based approach to map risks of ecological invasions in the presence of severe uncertainty. *Diversity and Distributions* 18, 33–46.

Yemshanov, D., Koch, F.H., Ducey, M. and Koehler, K. (2012b) Trade-associated pathways of alien forest insect entries in Canada. *Biological Invasions* 14, 797–812.

Yemshanov, D., Koch, F.H., Ben-Haim, Y., Downing, M. and Sapio, F. (2013) A new multicriteria risk mapping approach based on a multiattribute frontier concept. *Risk Analysis* 33, 1694–1709.

15 Assessing the Quality of Pest Risk Models

Steven J. Venette

Department of Communication Studies, The University of Southern Mississippi, Hattiesburg, Mississippi, USA

Abstract

The quality of pest risk models depends on their reliability and validity. Reliability refers to the model's trustworthiness and dependability, but the concept is extended to consistency and stability. Approaches to help establish and test consistency are described. Validity assures that models actually portray the expectations or assertions of the model design. Elements of validity can be characterized as face, criterion, content, concurrent and construct validity; and the merits of these elements are presented. In the social sciences, factor analysis has become the favoured method for evaluating reliability and validity. Factor analysis is especially useful when models group variables into indices. This chapter discusses the rationale and interpretation of components of factor analysis and suggests how this analysis might be incorporated into pest risk analysis in the future. Studies in communication reveal that non-technical end users can better understand and trust risk assessments when the assessments contain the best information presented in real language rather than mathematical expressions.

Introduction and Scope

The scientific quality of any risk model depends on its validity and reliability. While requirements and best practices for validity and reliability are discussed in the research literature, even in the context of risk analysis generally (Aven and Heide, 2009), less attention has been given to risk modelling, especially models of pest invasion. Inherently, risk modelling and verification are subject to 'standards of proof that are to some extent arbitrary, disputable, and subjective' (Graham, 1995). Regardless, evidence-based risk models, when valid and reliable, significantly improve risk analysis. Many researchers carefully present arguments supporting the validity and reliability of their models; others leave their work far more open to criticism. Therefore, this chapter offers concepts and tools from the social sciences that should be valuable when constructing arguments about the efficacy of a model. Additionally, this chapter offers advice on assessing the validity and reliability of pest risk analysis (PRA) models.

Clearly, not all pest risk models are likely to be equally effective at characterizing pest risk. Some weaknesses in risk assessment, unfortunately, tend to appear with some consistency. For example, some risk assessments have been 'used to predict success and failure in treatment programs' even though the tool employed was designed for needs assessment (Baird, 2009, p. 4). Other models may include factors that are only marginally correlated with pest risk; perhaps because as items are added, coefficients of reliability tend to rise. Still other models may be based on the assumption that the perception of a risk is

* Email: Steven.Venette@usm.edu

linear when, in reality, it may be better described logarithmically. Some researchers may believe that highly detailed indices are necessarily more effective than brief, simple models. Finally, a multivariate index might be constructed assuming that each subscale has an equal influence on the underlying latent variable. Each of these examples highlights a potential threat to the structure and application of a model to generate pest risk maps because of weaknesses in reliability or validity.

The goal of reliability and validity analysis depends in large measure on the type of risk assessment. When relevant empirical data are available, two major approaches to estimate the probability of an event are applicable: relative frequency and Bayesian (Aven and Heide, 2009). Frequency-based approaches rely on estimating the proportion of times that a certain event would occur given particular conditions. In this approach, the actual probability is unknown or uncertain, but model parameters come from observed data. Frequency-based risk assessment relies on traditional statistical methods to evaluate reliability and validity, and uncertainties are often communicated using point estimations and confidence intervals. Bayesian methods use probability as a measure of uncertainty based on previous data (Fabre et al., 2006). The analysis focuses on estimating or predicting parameters and assessing their uncertainties. Regardless of approach, researchers are obligated to present cogent, evidenced arguments that they have met the standards of both reliability and validity.

Reliability

In the social sciences, reliability is understood commonly as trustworthiness and dependability. For pest risk modelling, reliability is consistent with this definition but, more precisely, describes consistency and stability of research methods and findings. Reliability is established when two overarching criteria, consistency and stability, have been met.

Consistency of measurement

Analysts must show that variables have been measured in a consistent manner, often referred to as measurement reliability. Classical test theory explains that observed measurements are the combination of the actual true score and an error score (Carmines and Zeller, 1979; Spector, 1992). Risk modellers should estimate the proportion of variation that is attributable to random error, as opposed to measurement error. This hypothetical example helps distinguish between random and measurement error: a derived model consistently predicts that 132 of 500 insect larvae will survive 48 h of exposure to −30°C, but well-controlled experiments repeatedly show that 100 actually survive; then the model's predicted outcome contains the 'true' value of 100 larvae, plus an error of 32 larvae. In this example, the measurement error is consistent; thus, the proportion of error that is due to random variation of measurement would be zero. The model's scale would not be considered accurate, but it is reliable. It is measuring in a consistent manner – each prediction is off by 32 insects. However, if the freezer was defective and could not hold a consistent temperature, the larvae may spend varying amounts of time above and below the critical temperature during each iteration of the experiment. Because the time spent below −30°C would not be consistent, at least a portion of the error would be random.

The ratio of random error to measurement error may be difficult to calculate. Several statistical tools exist to help researchers estimate the proportion of random error and ultimately to evaluate overall reliability. The relevance of particular tools depends on different assumptions or types of data. Kuder and Richardson (1937) present several ways to estimate reliability. For example, their formula KR20 allows for the use of dichotomous data when calculating a reliability coefficient. One of the most common ways to evaluate internal consistency is Cronbach's (1951) coefficient α, derived from Kuder and Richardson (1937). Cronbach's α is used when multiple items are combined to create a single variable. For example, in PRA, risk models frequently use several factors to create an index that represents environmental suitability, say. Cronbach's α is a function of total variance of all the items measured, the sum of non-communal variance and the number of variables used to create the composite. While α is a measure of internal consistency, it is more specifically the ratio of common source variation to total variation. Common source variation is non-random error. Cronbach's α coefficient ranges from 0, when all of the variance is attributable to random error and reliability is theoretically perfect, to 1, when none of the variance appears to be random and the model is completely unreliable.

The use of Cronbach's α as a measure of reliability is beneficial for multiple reasons. First, α is flexible. It can be used with different scales, indices or levels (i.e. ordinal, nominal, integral and ratio) of data. For example, α is commonly used for summated rating scales. These latent variables may be a combination of continuous data, Likert-type scales or even dichotomous items such as semantic differentiation scales.

Second, α can be rearranged to provide researchers an estimate of how many items must be added to achieve a desired reliability. Remembering that coefficient α is a function of the number of variables used in an index, if variance remains constant, adding items increases the coefficient value. In essence, consistency over a greater number of items is evidence of higher reliability. The Spearman–Brown prophesy formula allows a researcher to calculate a multiplier, i, of the current number of factors in the model to reach a desired reliability; the inverse of the observed α coefficient, $1 - \alpha$, is multiplied by the desired reliability, $\hat{\alpha}$, and that number is divided by the product of the observed coefficient and the inverse of the desired coefficient; i.e. $i = [\hat{\alpha}(1-\alpha)] / [\alpha(1-\hat{\alpha})]$ (Spector, 1992). For example, if a researcher finds that a composite of eight factors yields α of 0.65, but a coefficient of 0.7 is desired, the existing number of items (8) would need to be increased by a factor of 1.3. That multiplier is equal to 0.35 (i.e. the inverse of the observed α) multiplied by 0.7 (the desired coefficient), which is then divided by the product of 0.65 (the observed α) and 0.3 (the inverse of the desired coefficient). In this case, 0.245 divided by 0.195 equals approximately 1.3. Since the eight original items multiplied by 1.3 yields an estimate of 10.4, the next whole number, 11, is the predicted number of items needed to achieve an α of 0.7. Of course, the notion that using more variables yields increased reliability assumes that added items will be as 'equally reliable' as the others.

Third, common statistical software packages can calculate Cronbach's α and report how α would change if certain factors were eliminated. Hypothetically, if 11 factors produce a coefficient of 0.7, the software could show that eliminating the third factor, for example, would increase the coefficient to 0.75. Item analysis in this manner aids data reduction and promotes simplification of models. Additionally, this analysis allows for a fairly easy way to identify 'clunkers' (i.e. items that are not working as intended and do not benefit the reliable measurement of the composite variable).

Fourth, Cronbach's α is becoming ubiquitous, particularly in the social sciences. Sijtsma (2009) supposes that 'probably no other statistic has been reported more often as a quality indicator of test scores' (p. 107). Other scholars, editors and reviewers generally understand and accept α as an estimate of reliability. In the least, α allows researchers a means of presenting quantitative data to support arguments that a model is reliable.

Although the use of Cronbach's α can provide a consistent and fairly clear way to identify the proportion of non-random variance in a measure, several disadvantages to its application also exist. Cronbach and Shavelson (2004) acknowledged that 'coefficients are a crude device that does not bring to the surface many subtleties implied by variance components' (p. 394). At times, α underestimates true reliability; and if the coefficient is used to correct correlations for attenuation, the true correlation would be overestimated. Likely the largest problem with the use of α is that it is often reported devoid of context (Schmitt, 1996). The coefficient only provides an estimate of internal consistency. Researchers should argue contextually whether that level of consistency is sufficient.

Stability of measurement

The test–retest method is common in the social sciences as an assessment of change over time (i.e. stability). Variation is usually due to some intervention such as training, therapy or communication, so instability may be desirable. For risk modelling, measuring the same variable more than once may also be appropriate. In this case, researchers are not looking for changes in responses, but rather stability in findings. Lack of differences points to stability in that the measurement is consistently yielding the same (or at least similar) results.

Risk models are often tested against a second, independent set of data and are considered 'validated' if they can accurately forecast those observed responses. This practice is sound and laudable. 'Measure twice and cut once' is not just good advice for carpenters; it should remind modellers that relying on unvalidated models can cause problems. For example, King (2013) discovered that a model created using one data set explained a large amount of the variance; however, that model failed to produce good results with any other data. In the sciences, the benefits of test replication to establish research findings are appreciated; replication to evaluate consistency of measurement can also be helpful.

Another approach to establish stability is the alternate-form method. This technique requires researchers to use at least two different ways to arrive at what should be highly similar results. For example, an instructor might use several versions of the same examination. The distribution of test scores should be consistent across versions. If this is true, the scores should be highly correlated. However, such a test cannot account for variation due to aptitude of the individual student, so correlation co-efficients will likely be lower than with test–retest methods.

Reliability can also be evaluated by comparing one portion of a data set with another. Split-half reliability, as the name implies, compares one half of the data with the other. The division can be made in any number of ways. First-half last-half reliability simply looks for consistency between the first half of the data and the last. Odd–even splits, obviously, compare results from the odd-numbered data points in an ordered list with the even-numbered points. Random-halves reliability divides the data set into two groups through random assignment; because the items are selected randomly, tests can be run on the same data set many times. An average of coefficients can be reported along with a confidence interval.

When dividing data into groups to assess reliability, the degrees of freedom are halved. The correlations can be corrected for attenuation due to the decrease in degrees of freedom. Just what type of correction to use, if any at all, is controversial (Borsboom and Mellenbergh, 2002). Correction increases the coefficient of agreement but is not a particularly conservative approach. Researchers will have to evaluate the context of the study to determine how conservative the approach should be. For example, exploratory studies allow for flexibility, while confirmatory experiments usually are approached conservatively.

Additional tools

Additional tools may prove useful when establishing reliability. These approaches use Pearson's product moment correlation as the reliability coefficient, making calculation fairly parsimonious. In fact, the interpretation of the coefficient is the same as a correlation. While no universal standards exist, the following ranges should be appropriate in most contexts: 0.0 to 0.2 indicates no or very low reliability; 0.2 to 0.4 is low reliability; 0.4 to 0.7 is substantial reliability; and 0.7 to 1.0 would be high to perfect reliability.

Validity

Validity is a more nebulous concept than reliability. In the context of pest risk modelling, the central question for validity is, 'Does the model actually model what it is attempting to model?' Primarily, validity depends on the way constructs (i.e. ideas with multiple components) are defined. The constructs being modelled must be very clear and if multiple constructs are being modelled, then the connections between them must be solid. The definitions and connections depend on conceptualization based on the review of literature. Validity is supported when the concepts are defined in a manner consistent with the extant literature. Because risk models are often complex structures, constructs can be divided into simpler constructs (i.e. sub-constructs) that can be more clearly established as valid. However, care should be taken to avoid unnecessary complexity by subdividing too finely. Additionally, Venette *et al.* (2010) warn 'as the degree of the [pest risk] model's structural freedom increases, it becomes impossible to make generalizations about the applicability or performance of a model for a new [invasive alien] species based on previous experiences without a deeper understanding of the specific problem formulation' (p. 355). Ultimately, the validity of a risk model depends largely on its structure.

Researchers must construct arguments supporting the validity of their selected or derived model. Claims of validity are supported by evidence of internal and external accuracy. Internal validity is a reflection of the methodological approach that leads to accurate findings. Modellers must be able to demonstrate that they have measured what they have intended to measure. Four basic types of validity are evoked to accomplish this end: (i) face validity; (ii) criterion validity; (iii) content validity; and (iv) construct validity.

Face validity

Face validity is perhaps the weakest form as it is a subjective evaluation of the extent to which the model accurately does what it is designed to do. To strengthen the argument for face validity, researchers can use a panel of experts to examine the model and to provide feedback about whether the model appears valid. In the least, this approach suggests that the modellers have followed best practices or met field expectations. Expert-driven, rule-set modelling functionally relies on face validity.

Criterion validity

Criterion validity is a reflection of how effectively variables, or the entire model, classify a case or predict future results. For example, in psychology, a new scale for depression can demonstrate criterion validity experimentally using a sample of people who previously have been diagnosed with depression and a same-sized group of people confirmed as not having depression. If the test correctly discriminates between people who are and who are not depressed (i.e. individual cases are correctly classified), then criterion validity is supported (Cronbach, 1960). A valid risk model should similarly be able to classify threats appropriately. If a model predicts that a pest is likely to establish in a location because climatic conditions in a species' native range and an area of concern are similar, and the pest is found in that location, then some evidence would exist for the criterion validity of that model.

A type of criterion validity, predictive validity, is not based on the model's ability to correctly classify known cases, but rather is defined by the accuracy of predicting occurrence. Predictive validity comes from a model being able to predict future cases, but waiting for a predicted invasion to actually occur in order to evaluate validity is nonsensical. Rather, predictive validity could be evaluated using previously studied outbreaks. As an example, if the parameters of a fruit fly invasion model were set to parallel April of 1990, would the model have predicted the outbreak of Mediterranean fruit flies in southern California? Similarly, if a model forecasts the spread of emerald ash borer in the USA, would the model correctly identify the states that are already classified as having emerald ash borer? The extent that actual events or outcomes are predicted is the level of predictive validity. A correlation can be calculated using the data from the event and the predicted data from the model. The correlation can be reported as a coefficient of validity.

Another type of criterion validity is called concurrent validity, when a new model's results compare favourably with those of a previously validated model.

Content validity

Content validity is often established through the review of literature. The question being answered is, 'How can the end user be confident that the material included in the model accurately reflects current field information?' A model built on the premise that a particular insect cannot thrive in arid environments is invalidated by the discovery of that insect thriving in a desert. The model would no longer accurately reflect the state of the extant content knowledge. More often, the problem with a model's content validity is due to the failure to account for sufficient facets of what is being modelled. A model of pest invasion would lack content validity if only weather data are considered.

Statistically, Cronbach's α is an indirect indicator of content validity. While α is not a test of unidimensionality, a high coefficient may be the result of each variable measuring the same concept. Just what that concept is, is unknown. Of course, researchers should examine the inter-item correlation matrix to ensure that each item is working as intended.

Construct validity

Construct validity is a reflection of the way concepts are connected in the model itself. In a flow chart, content validity deals with what is in each node, individually; construct validity is an evaluation of the arrows connecting those concepts along the path of analysis (Cronbach and Meehl, 1955). For example, one step of a model might be a prediction of the number of an invasive plant species that could be introduced into an area and a second step could be an estimate of the proportion of the introduced plants that could reproduce in that environment. That order makes sense insofar as the number of plants that could reproduce is only relevant in the context of how many plants could exist there. When concepts are independent, entry into a model does not logically need to follow any particular sequence. One concept does not precede the other. In sum, construct validity seeks to establish that concepts are logically connected and that the sequencing is appropriate. Venette et al. (2010) support the notion that the construction or choice of a reliable model is important in the context of PRA by explaining that 'indiscriminate model selection may cause incorrect estimates of invasive alien species' potential ranges and inaccurate assessments of pest risk. Consequently, decision makers may select the wrong mitigation measure and over- or underinvest in that strategy' (p. 350).

Uncertainty

Ultimately, if a researcher is unable to articulate a model's validity and degree of reliability, then subsequent risk estimates should be viewed as speculative. People and organizations that rely on that tool's estimates for policy formation or risk communication must be made aware of the uncertainty so that appropriate rhetorical strategies can be used to interpret and communicate the information. Policy makers and communicators deal with uncertainty constantly. For example, communicators strategically use qualifiers and limitations as rhetorical devices to articulate ambiguity. The problem is not that the people who use the models' results do not understand uncertainty; rather, risk analysts often are unable to communicate uncertainty effectively simply because ambiguity and imprecision often give scientists epistemological angst. Risk scientists need increased training specifically in how to articulate uncertainty in a meaningful way.

Factor Analysis

Factor analysis has become the gold standard for evaluating the validity and reliability of structures (i.e. indices) used to represent complex events or trends. In this chapter, the term 'factor analysis' defines a group of statistical tools used to identify underlying dimensions of groups of variables that are combined to represent a latent variable. Factor analysis can be particularly useful when evaluating the structure of models, such as FloraMap, that use principal component analysis, which is closely related to factor analysis, or any model that groups variables into indices, similar to how CLIMEX uses a composite index based on species' responses to climate data.

Two general types of factor analyses exist: exploratory and confirmatory. Exploratory analysis is used to uncover factors in data when the structure has not been previously established. For more on exploratory factor analysis (EFA), see Costello and Osborne (2005). Conversely, confirmatory analysis, as the name implies, is designed to determine whether a pre-established structure can be replicated with new data. For a detailed discussion of confirmatory factor analysis (CFA), see Gorsuch (1983) and Kline (1994). If a model assumes that an index based on the same weather variables is equally informative for different insects, for example, CFA might be used to test that assumption.

Exploratory factor analysis

EFA begins with the calculation of a correlation matrix for the applicable variables. Bartlett's test of sphericity is used to ensure that the variables are sufficiently correlated to continue with the factor analysis. More specifically, the null hypothesis states that no relationship exists between the variables. The test produces a χ^2 statistic with a corresponding P value. With Bartlett's test, a low P value (i.e. < 0.05) is an indication that the variables are correlated and that factor analysis can continue to the next step.

Next, the Kaiser–Meyer–Olkin (KMO) measure is used to determine sampling adequacy. KMO is an index that compares the magnitudes of the observed item correlations to the magnitudes of the partial correlation coefficients. Because partial correlations report the strength of the association between two variables when the other variables are held constant, a large coefficient indicates that the relationship between two variables cannot be adequately explained by the other variables; in other words, potential common factors do not exist. Conversely, a large value for KMO (i.e. >0.5) suggests that one or more potential factors can be identified among the variables. Thus, if KMO is high, moving to the next step is warranted.

After KMO gives an indication that a potential underlying structure exists, the actual factors need to be extracted. Many methods exist for extraction; the specific details of each are beyond the scope of this chapter. However, a common approach is to use principal component analysis. The initial solution reports communalities, or the proportion of variance that can be attributed to the common factors. Eigenvalues are also calculated. An eigenvalue is the total amount of variance accounted for by a factor. Generally, potential factors that have eigenvalues < 1 are considered unimportant as the amount of explained variance is small. Another way to determine how many meaningful factors exist is to look at a scree plot of the eigenvalues. This plot is a line graph connecting the variances explained by each factor. The researcher tries to identify the number of meaningful factors by looking for an 'elbow' in the graph. So in a hypothetical case where four potential factors exist, the factor that explains the most variance might have an eigenvalue of 7.5, the second might be 6.2, the third could be 1.2 and the fourth 1.0. If just eigenvalues are considered, all four would be included in continued analysis. In this case, using the scree plot, a clear elbow exists at the third factor (Fig. 15.1). This finding suggests that a two-factor model might explain nearly as much of the variance as a model with four factors, but would be simpler and easier to interpret.

To more easily interpret the structure of the data, the factor solution is rotated. Rotation seeks to maximize the loading of each variable on one factor. 'Loading' is a term used to describe the relevance of the variable to the factor. Again multiple methods exist to rotate the factors. Varimax rotation, for example, yields orthogonal factors, meaning that they are not correlated with each other; alternatively, promax produces factors that can be inter-correlated. Regardless of the method used, the rotated matrix is used to identify which variables load on each of the factors. A variable can fairly conservatively be said to load at a value of 0.4; 0.3 may also be acceptable if a less conservative approach is justified. Variables that load on more than one factor are said to be 'complex' and can make interpretation of the structure more difficult.

Luckily, the entire process of determining the number of factors and interpreting loadings is made far simpler when a single factor adequately describes the structure of the data. When only a single eigenvalue is greater than 1, or if a clear elbow exists after the first factor, then a single factor can be extracted (Fig. 15.2). When only one factor is extracted, rotation

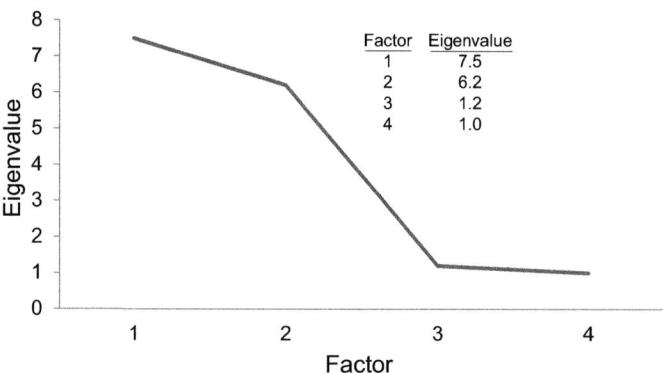

Fig. 15.1. A scree plot with a clear elbow after the second factor.

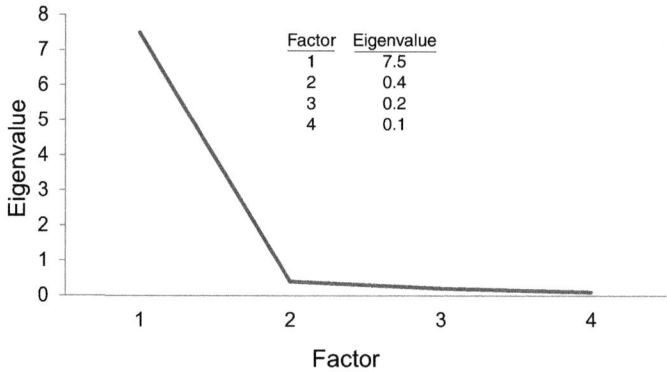

Fig. 15.2. A scree plot with an elbow after the first factor, indicating a single-factor solution.

is unnecessary because all items load on that single factor. Interpretation could not be easier.

A hypothetical example of the use of EFA in pest risk modelling might be helpful at this point. A researcher might be constructing a model of the threat posed by a particular invasive alien species if a new agricultural product is allowed to be imported into a country. Part of that model might consider whether suitable climate exists for the pest to establish, as many pest risk models do. Many variables can be combined to create a single index of climate suitability. For the sake of the example, the modeller believes that climate can be represented by average temperature, temperature variation, average precipitation and average relative humidity. Appropriate geographic subsections of a country (i.e. grid cells) will be assigned a score on a scale of 0 to 3 for each of the variables where 0 means 'not at all conducive to pest invasion' and 3 represents 'very conducive to pest invasion'. So the first grid cell might be rated as 3 for average temperature, 2 for temperature variation, 3 for average precipitation and 2 for relative humidity. Using a summated rating scale (Spector, 1992), each score would be added together to create the data point for the climate suitability index. Grid cell one would have a score of 10 out of a maximum score of 12. Grid cell two could be rated as a 1, 2, 1 and 1, for a total score of 5. The process would continue until a score for each grid cell had been tabulated.

Now the researcher is curious to see whether the scale is working as intended. Because no previous research has indicated that a particular factor structure exists for those variables, EFA would be appropriate. The correct commands are entered into the statistical software (e.g. in SPSS software, the command FACTOR) and the results appear in seconds. Bartlett's test of sphericity is significant ($P < 0.05$), so items do appear to be inter-correlated. Moving to step two is reasonable. The KMO measure is 0.8 and because it is greater than 0.5, moving forward with the analysis is justified again. When the factors are extracted, only one has an eigenvalue greater than 1.0. A clear elbow exists after the first factor (Fig. 15.2). The loadings table shows that average temperature loads at 0.72, temperature variation at 0.91, average precipitation at 0.68 and relative humidity at 0.45 (Table 15.1). Each variable is contributing to the amount of variance explained by the factor model. These results are quite positive. The researcher can use these findings in support of an argument that combining the four observed variables into one latent variable such as 'climate' is a valid way to reduce data and simplify the risk analysis model.

An example of a less positive result might be helpful as well. Using the same context as the previous example, again with a significant test of sphericity and a high KMO measure, but this time the eigenvalues are different. One dimension has a value of 6.5, the second at 5.6 and a third at 0.3. Examination of the scree plot confirms that the elbow exists after the second factor (Fig. 15.1). Two factors are extracted and the model is rotated. The loadings table (Table 15.2) shows that average temperature loads on the first dimension at 0.87, but not the second (loading of 0.12). Temperature variation also loads on the first at 0.65, but

Table 15.1. Example of a factor component matrix with all items loading on the first factor.

Variable	Factor	
	1	2
Average temperature	0.72	0.17
Temperature variation	0.91	0.01
Average precipitation	0.68	0.16
Relative humidity	0.45	0.25

Table 15.2. Example of a factor component matrix with items loading on two factors.

Variable	Factor	
	1	2
Average temperature	0.87	0.12
Temperature variation	0.65	0.23
Average precipitation	0.29	0.59
Relative humidity	0.41	0.58

not the second (0.23). Average precipitation loads on the second (0.59) but not the first (0.29). Relative humidity loads on the first factor (0.41), but loads more on the second (0.58). Looking at the structure, it appears that the first dimension is identifying a latent variable associated with temperature and the second seems to be associated with moisture. This interpretation would also account for relative humidity loading on both dimensions, as relative humidity is a function of temperature and moisture. The results in this example do not support the claim that the four climate variables can be combined. Creating separate temperature and moisture variables appears to be a valid approach, but additional exploratory analysis is needed to determine the extent that this two-factor model explains the variance in the data.

Confirmatory factor analysis

CFA aims to determine how well the data fit a predetermined structure. Using structural equation modelling, CFA assesses the congruence between the predicted structure and the factor structure in the experimental data (e.g. Gorsuch, 1983). A high degree of fit supports the conclusion that the pre-existing structure is generalizable.

CFA is a fairly complex type of analysis. Similar to the previous discussion of EFA, a simplified explanation is presented. Again, the purpose of this chapter is not to replace a statistics text, but rather to explain how the tests aid the assessment of validity and reliability. CFA reflects that theories and data are interconnected. Theory provides an explanation for observed phenomena. This explanation is grounded through the analysis of data. Models are simplified representations of the relationships or predicted relationships between concepts, which are operationalized using variables. These connections between variables form constructs and constructs come together to support theory. CFA provides a statistical means to test the risk analysis model structure (Vehkalahti, 2007).

CFA assumes that the constructs composing the model have linear relationships. The relationship could be direct (i.e. a basic correlation), but constructs may also have complex relationships (Baron and Kenny, 1986). For instance, one or more concepts might affect the direction or strength of the relationship between two other variables. As an example of this moderator relationship, a strong positive relationship might exist between average rainfall and damage caused by a plant disease, but that relationship may only be seen when it is moderated by temperature. In other words, the disease thrives in wet conditions, but only when temperatures are warm.

As previously mentioned, CFA begins with a model that has been previously established through the analysis of one or more data sets. CFA tests the extent to which the new data fit that model. If the fit is not close, the model is rejected. In logic, the claim that all swans are white, for example, is invalidated upon the discovery of one black swan. The argument that a risk model accurately represents 'reality' is disconfirmed when that model fails upon the introduction of new data. The model as currently constructed cannot be considered valid. If the data do fit the model well, the researcher cannot technically say that the model is validated in the same way that hypotheses are never proven. The researcher simply reports the findings of the CFA and concludes that the model continues to work well in light of the new data.

Regardless of the approach, risk analysts are responsible for articulating how their models are constructed. If the relationships between variables emerge from theory deductively, those relationships should be tested using the data for the particular context being analysed. In this case, CFA would be most helpful. If the relationships arise inductively from the particular case data, EFA helps justify the researcher's structuring of the model. Regardless, users cannot accurately evaluate the validity of any model if decisions about structure are not reported transparently and tested systematically.

Conclusion

Because pest risk estimates play a major role in policy formation and enactment, end users must have the best information available to make important decisions. Scientists often fear that their information must be 'dumbed down' for end users to understand. Over-simplification is anathema to best information. However, a statement such as 'we are 90% confident that our model is reliable at a 0.81 level' is unlikely to appeal to many end users. Careful translation to real language that does not depend on mathematical expression is difficult but vital to the message's reception.

This chapter presents recommendations for evaluating the validity and reliability of pest risk models with the goal of helping risk analysts and end users assess the worth of those models. The tools offered may aid the development of 'best practices' that ultimately could improve the quality of the science. Such best practices should provide clear standards for reporting how validity and reliability were assessed. Risk assessments should not be based on models that are arbitrary, disputable or subjective.

References

Aven, T. and Heide, B. (2009) Reliability and validity of risk analysis. *Reliability Engineering and System Safety* 94, 1862–1868.

Baird, C. (2009) *A Question of Evidence: A Critique of Risk Assessment Models Used in the Justice System.* National Council on Crime and Delinquency, Madison, Wisconsin.

Baron, R.M. and Kenny, D.A. (1986) The moderator–mediator variable distinction in social psychological research: conceptual, strategic, and statistical considerations. *Journal of Personality and Social Psychology* 51, 1173–1182.

Borsboom, D. and Mellenbergh, G.J. (2002) True scores, latent variables, and constructs: a comment on Schmidt and Hunter. *Intelligence* 30, 505–514.

Carmines, E.G. and Zeller, R.A. (1979) *Reliability and Validity Assessment.* Sage, Newbury Park, California.

Costello, A.B. and Osborne, J.W. (2005) Best practices in exploratory factor analysis: four recommendations for getting the most from your analysis. *Practical Assessment, Research and Evaluation* 10, 1–9.

Cronbach, L.J. (1951) Coefficient alpha and the internal structure of tests. *Psychometrika* 16, 297–334.

Cronbach, L.J. (1960) *Essentials of Psychological Testing.* Harper Press, New York.

Cronbach, L.J. and Meehl, P.E. (1955) Construct validity in psychological tests. *Psychological Bulletin* 52, 281–302.

Cronbach, L.J. and Shavelson, R.J. (2004) My current thoughts on coefficient alpha and successor procedures. *Educational and Psychological Measurement* 64, 391–418.

Fabre, F., Pierre, J.S., Dedryver, C.A. and Plantegenest, M. (2006) Barley yellow dwarf disease risk assessment based on Bayesian modelling of aphid population dynamics. *Ecological Modelling* 193, 457–466.

Gorsuch, R.L. (1983) *Factor Analysis.* Lawrence Erlbaum Associates, Hillsdale, New Jersey.

Graham, J.D. (1995) Verifiability isn't everything. *Risk Analysis* 15, 109–109.

King, M. (2013) Power of speech style: a relational framing perspective. MS thesis, University of Southern Mississippi, Hattiesburg, Mississippi.

Kline, P. (1994) *An Easy Guide To Factor Analysis.* Routledge, London.

Kuder, G.F. and Richardson, M.W. (1937) The theory of the estimation of test reliability. *Psychometrika* 2, 151–160.

Schmitt, N. (1996) Uses and abuses of coefficient alpha. *Psychological Assessment* 8, 350–353.

Sijtsma, K. (2009) On the use, the misuse, and the very limited usefulness of Cronbach's alpha. *Psychometrika* 74, 107–120.

Spector, P.E. (1992) *Summated Rating Scale Construction.* Sage, Newbury Park, California.

Vehkalahti, K. (2007) Factor analysis and the reliability of measurement scales. *Proceedings of the International Meeting of the Psychometric Society.* Tokyo, Japan, 9–13 July 2007.

Venette, R.C., Kriticos, D.J., Magarey, R.D., Koch, F.H., Baker, R.H.A., Worner, S.P., Gomez Raboteaux, N.N., McKenny, D.W., Dobesberger, E.J., Yemshanov, D., De Barro, P.J., Hutchison, W.D., Fowler, G., Kalaris, T.M. and Pedlar, J. (2010) Pest risk maps for invasive alien species: a roadmap for improvement. *BioScience* 60, 349–362.

Index

accuracy and precision 224
active dispersal 55, 57
addition rule matrix 25
aerial dispersal 9, 49–61
alien species, definition 2
alternate-form method 226
aphids, windborne dispersal 52, 54, 56, 57, 58–59
ArcGIS software 71
area at highest risk 19, 20–21, 23–24, 26, 28
    see also endangered area maps
area of potential establishment 19, 22–23, 26, 28, 30
arrival of invasive alien species 8–9
    in economic impact models 149–151
    human transportation 9, 35–47, 111, 213–218
    windborne 9, 49–61
Asian gypsy moth (*Lymantria dispar asiatica*) 89–90
Asian longhorned beetle (*Anoplophora glabripennis*) 37
asparagus pests 9
assessors (risk assessors) 5, 6–8, 14
    communication of uncertainty 228–229, 233
AUC (area under ROC curve) 14, 72, 76
Australia
    Chilean needle grass invasion 173–185
    mimosa eradication 146–158
    nematode invasion 107–108
    windborne insect pests 52–53, 57, 58–60
average history maps 87

Bactrocera dorsalis (Oriental fruit fly) 92
bad neighbours 157
BAMM tool (NAPPFAST) 88
Bartlett's test of sphericity 229
Bayesian networks 174, 176–178
best case scenario 25, 118, 123
bioclimatic envelopes *see* species distribution models
biosecurity 1–2, 3, 8
    economic impact assessment of eradication 145–158
    international agreements 15, 18–19
    US port inspections 36
bird cherry-oat aphid (*Rhopalosiphum padi*) 52, 54, 56, 57, 58–59
biting midge (*Culicoides imicola*) 52–53, 56, 57, 59–60
bluetongue virus, vectors 52–53, 59–60
boundary displacement method for estimating spread rates 138–139, 141
Burmese python (*Python molurus bivittatus*) 66–67, 68–77

C++ programs 56, 193
CABI databases 100
Canada, invasion of forest pests 213–218
causal (process-based) models 54, 171–186
CFSR (Climate Forecast System Reanalysis) database 84
Chilean needle grass (*Nassella neesiana*) 173–185
citizen scientists 133

citrus black spot (*Guignardia citricarpa*) 87
classification of variables 5, 24, 25
climate change 127
climatic data
    in the MAXENT model 68, 72, 79
    in the NAPPFAST model 83–84, 94
    in spread models 117, 121, 127
CLIMEX model 10, 117, 121–122, 127
cluster analysis 111
    hierarchical clustering 109–110, 111
    k-means 108–109, 111
    self-organizing maps 10, 98–108, 110–111
cold exclusion model in NAPPFAST 91
common ground-dove (*Columbina passerina*) 166–168
communication of uncertainty 228–229, 233
computational power
    cluster analysis 108
    process-based analysis and GIS software 180–181
    uncertainty analysis 193, 203, 220
concurrent validity 228
conditional probability tables (CPT) 176–178
confirmatory factor analysis 229, 232
confusion matrices 13
consistency (reliability) of models 224–226
construct validity 228
content validity 228
cooperation (or its lack) in pest management 157–158
correlative model (MAXENT) 65–79
covariance 77, 79
Cox's risk matrix 26
CPT (conditional probability tables) 176–178
criterion validity 227–228
Cronbach's α 224–226, 228
Culicoides imicola (biting midge) 52–53, 56, 57, 59–60

data sources
    in estimation of spread rate 133–134, 139, 140
    freight movements (North America) 36–37, 40, 45–46, 213
    pest species databases 88, 100, 110–111, 133–134
    in process-based models 174–175
    weather 83–84
decision makers (risk managers) 5, 6
    NAPPFAST model 89

perception of uncertainty 202, 207–208, 219, 228
decision support schemes/systems (DSS) 20–31
deductive models 10
    NAPPFAST 20, 82–94
degree-day template in NAPPFAST 85, 89–90
deposition in aerial transport process 52
Diabrotica virgifera (western corn rootworm) 24, 122
    DSS maps of risk 27–29
    spread models 122–126
DISMO R software 67, 68
dispersal
    dispersal kernels 120, 126, 179
    international trade 35–47, 213–218
    in process-based models 176, 178
    stratified 132, 151–152
    uncertainty in model predictions 201–202, 213–218
    windborne 9, 49–61
    see also spread
distance regression method (spread rate estimation) 137–138, 139, 141
distribution models *see* species distribution models
DSS (decision support schemes/systems) 20–31

Ecoclimatic Index (EI) 5, 27, 117, 121, 127
ecological impact *see* environmental impact
ecological niche models *see* species distribution models
economic impact 2, 12–13
    area of highest risk 26
    cost/benefit analysis of eradication 145–158
    economic injury level 19–20, 26–27
    spread models 117, 118, 122, 123
EI (Ecoclimatic Index) 5, 27, 117, 121, 127
eigenvalues 229–230
EIL (economic injury level) 19–20, 26–27
emerald ash borer (*Agrilus planipennis*) 37, 140
endangered area maps 19, 21, 26–27
ENMTOOLS software 67, 73
entry *see* arrival of invasive alien species
environmental data
    in the MAXENT model 68, 72, 79
    in the NAPPFAST model 83–84, 94
    in spread models 117, 121, 127

environmental impact 2, 24
    process-based (causal) models 171–186
    spatial modelling 162–169
EPT tool (NAPPFAST) 89
eradication, economic impact assessment 145–158
error
    random or measurement 224–226
    type I/type II 13
    see also uncertainty
establishment of invasive alien species 9–11
    area of potential establishment 19, 23, 26
    hierarchical clustering 109–110, 111
    k-means clustering 108–109, 111
    MAXENT model 65–79
    NAPPFAST model 20, 82–94
    prioritization 97–98
    in process-based modelling 176
    self-organizing maps 98–108, 110–111
Europe
    cluster analysis 109–110
    PRATIQUE project 19, 20–31
    spread models 121, 122–126
European corn worm (*Ostrinia nubilalis*) 107
European gypsy moth (*Lymantria dispar dispar*) in the USA 132–133, 134–140
European and Mediterranean Plant Protection Organization (EPPO) 23, 100
evolution 15
expert elicitation 148, 149
    face validity 227
    participatory modelling 175, 178, 181–182, 183–184
exploratory factor analysis 229–232

face validity 227
factor analysis 229–232
false positives/false negatives 13–14
financial impact *see* economic impact
Finland, SOM analysis 106–107
fire ant (*Solenopsis invicta*) 164–169
flight duration/termination in insects 57–58
forest pests
    in Canada 213–218
    urban forests in the USA 37–47
free riders 157
Freight Analysis Framework (FAF) database 36–37
freight transportation 9, 35–47, 111
    portfolio-based analysis of uncertainty 213–218

GARP model 10
GBIF upload tool (NAPPFAST) 88
Generic Pest Forecast System (GPFS) (NAPPFAST) 83, 84, 86, 92
GENSIM program 56
geographic information systems (GIS) 180
Geospatial Modelling Environment (ARCGIS) 71
GI (Growth Index) 117, 119, 121, 127
giant sensitive plant (*Mimosa pigra*) 146–158
Global Biodiversity Information Facility 68, 88
Global Pest and Disease Database (GPDD) 100
GPFS (Generic Pest Forecast System) (NAPPFAST) 83, 84, 86, 92
Growth Index (GI) 117, 119, 121, 127
gypsy moth (*Lymantria dispar*)
    L. d. asiatica in Japan 89–90
    L. d. dispar in the USA 132–133, 134–140

habitat distribution in spread models 117, 121, 122
habitat suitability in process-based models 171–185
    see also species distribution models
habitat susceptibility in process based models 171–185
    see also dispersal
hazard index 164, 167
hierarchical clustering 109–110, 111
high risk matrix 25–26
history maps 87
human-mediated introduction 9, 35–47, 111
    portfolio-based analysis of uncertainty 213–218
HYSPLIT model 9, 50, 52, 55–56, 57, 59

impact of invasive alien species 2, 12–13
    economic 13, 19–20, 26–27
        cost/benefit analysis of eradication 145–158
    environmental 24
        process-based (causal) models 171–186
        spatial modelling 162–169
import *see* trade and transportation
individual-based models 12, 58
inductive models 10–11
    MAXENT 65–79
    NAPPFAST global analyst tool 88
    self-organizing maps 98–108, 110–111
infection template in NAPPFAST 86, 90–92

info-gap decision theory 191
insect pests
    cluster analysis 107, 109
    distribution map of San Jose scale 3–4
    impact assessment (red imported fire ant) 164–169
    NAPPFAST model 85, 89–90, 92
    spread models 122–126, 132–133, 134–140
    uncertainty analysis (woodwasp) 192–203
    windborne dispersal models 9, 49–61
integrative models 12–13
International Pest Risk Mapping Workgroup (IPRMW) recommendations 14–15
International Standards for Phytosanitary Methods (ISPMs) 18–19
international trade
    introduction of alien species 9, 35–47, 111, 213–218
    SPS Agreement 15
introduction of invasive alien species *see* arrival of invasive alien species
invasive alien species, definition 2

Japan, Asian gypsy moth 89–90
Japanese beetle (*Popillia japonica*) 85

k-means clustering 108–109, 111
Kaiser–Meyer–Olkin (KMO) measure 229

limiting factor matrix 25
logistic regression 11, 165, 167

maize pests
    European corn worm 107
    western corn rootworm 24, 27–29, 122–126
maps and mapping 2–3
    area definitions 19–20
    NAPPFAST output 87
    pest risk maps 181
    PRATIQUE project DSSs 5–8, 20–31
    SOM output 102–103
matrix rules for combining pest risk maps 25–26
MAXENT model (maximum entropy) 10, 65–79
maximum rule matrix 25

MCAS-S (Multi-Criteria Analysis Shell for Spatial Decision Support) 23
mean–variance frontier (MVF) concept 211–212, 216–220
mechanistic models of dispersal 54, 61, 201
MESS maps (multivariate environmental similarity surface) (MAXENT) 72
Mimosa pigra (giant sensitive plant) 146–158
minimum convex polygon (MCP) (MAXENT) 71–72, 75
minimum rule matrix 25
minimum training presence (MTP) (MAXENT) 77
models and modelling 2, 4–5, 6, 13–15
    types 8–13
    see also specific models
modified average matrix 25
Monte Carlo analysis 9, 190, 191, 207
MVF (mean–variance frontier) concept 211–212, 216–220

NAPPFAST model 10, 20, 82–94
nascent foci of infestation 152
negative externality 157
nematode pests, in Australia 107–108
NETICA™ software 177
neural networks 98–108, 110–111
New Zealand 98, 103, 104
non-dominant sets 210–213

Old World bollworm (*Helicoverpa armigera*) 55
Oriental fruit fly (*Bactrocera dorsalis*) 92
Ostrinia nubilalis (European corn worm) 107
overfitting in models 14, 72–74, 75

Parthenium hysterophorus (feverfew) 185
partial budget models 145–146, 153
participatory modelling methods 175, 178, 181–182, 183–184
pathway models 9, 213–218
Pearson's product moment correlation 226
pest risk analysis (PRA)
    mapping of endangered/high risk areas 18–31
    production of models/maps 4–8
    purposes 3, 21
    quality *see* quality of models/maps
    types of model 8–13
pheromone traps 140
phytosanitary measures *see* biosecurity

plants
   as hosts
      area at highest risk 23–24, 26
      EIL 19–20, 26–27
      NAPPFAST model 82–94
      SOM analysis (Finland) 106–107
   as pests 8–9
      Chilean needle grass invasion 175–185
      mimosa eradication 146–158
PMTRAJ model 9, 50, 52, 56–57, 58–59
point data 133, 136–140
polygonal data 133, 135–140
population density model of spread 116, 119–120, 124–126
population dynamics models 20, 54
portfolio analysis 208–220
positive externality 157
PRATIQUE project 19, 20–31
precision and accuracy 224
predictive validity 227–228
presence-and-absence modelling 66, 116
presence-only modelling 66, 71–72, 77
price of commodities 152–153
principal component analysis 229
probability maps 87, 91
process-based (causal) models 54, 171–186
Python molurus bivittatus (Burmese python) 66–67, 68–77

quality of models/maps 13–15, 223–224
   acknowledgement of uncertainty 228–229, 233
   economic impact assessments 155–156
   factor analysis 229–232
   MAXENT 75–77
   NAPPFAST 84, 88
   process-based pest risk models 174, 181–185
   reliability 224–226
   self-organizing maps 102–103, 104–105, 110
   sensitivity analysis 104–105, 155–156, 191
   validity 4–5, 8, 227–228
quarantine pests 2, 133–134

R software 120–121, 193
radial range expansion 116, 118–119, 123–124
rain splash 86

receiver-operating characteristic (ROC) curve 14, 72, 76
red imported fire ant (*Solenopsis invicta*) 164–169
regularization value (MAXENT) 73–74
reliability of models 224–226
resolution of maps 5
Rhopalosiphum padi (bird cherry-oat aphid) 52, 54, 56, 57, 58–59
risk analysis *see* pest risk analysis
risk assessors 5, 6–8, 14
   communication of uncertainty 228–229, 233
risk-aversion/risk-tolerance 207–208, 212
RMTA (Real-Time Mesoscale Analysis) database 84
ROC (receiver-operating characteristic) curve 14, 72, 76
rule-based models 12

SAHM software 77
sampling bias, in MAXENT 66, 71–72, 75
San Jose scale (*Quadraspidiotus perniciosus*) 3–4
satellite infestation sites 152
self-organizing maps (SOM) 10, 98–108, 110–111
sensitivity 13–14
sensitivity analysis
   economic impact assessments 155–156
   self-organizing maps 104–105
   uncertainty in a spatial stochastic model 189–203
sine curve method 85
sink habitats 77
Sirex noctilio (woodwasp) 192–203
SOM (self-organizing maps) 10, 98–108, 110–111
spatial modelling
   impact of alien species on native species 162–169
   spatial stochastic model in uncertainty analysis 189–203, 207
Spearman–Brown prophecy formula 225
species distribution models 10–11
   hierarchical clustering 109–110, 111
   k-means clustering 108–109, 111
   MAXENT 65–79
   NAPPFAST 10, 20, 82–94
   self-organizing maps 10, 98–108, 110–111
specificity 13, 14

split-half reliability method 226
spread of invasive alien species 11–12, 128, 131–132
    in economic impact models 151–152
    generic models 115–127
    in process-based models 172, 178, 179
    spread rate estimation 132–141
SPS Agreement (Application of Sanitary and Phytosanitary Measures) (WTO) 15
square-root area regression method 135–137, 139, 141
SSD (stochastic dominance) technique 212–213, 216–220
stability
    of measurement 226
    of model outputs 196
stakeholders 6
    cooperation 157–158, 175
standard deviation 195
stochastic dominance (SSD) technique 212–213, 216–220
stratified dispersal 132, 151–152
sudden oak death (*Phytophthora ramorum*) 86, 90–92
swallow-tailed kite (*Elanoides forficatus*) 166–168

t-distribution 120
temperature
    NAPPFAST cold exclusion model 91
    NAPPFAST degree-day template 85, 89–90
10-percentile training presence 77
test–retest method 226
total investment benefit 152–153
trade and transportation 9, 35–47, 111
    portfolio-based analysis of uncertainty 213–218
trait-based screening assessments 8–9
trajectory modelling of windborne dispersal 9, 49–61
transferability of models 66
transportation of freight 9, 35–47, 111
    portfolio-based analysis of uncertainty 213–218
type I/type II errors 13

uncertainty 8, 207
    best- and worst-case scenarios 25, 118, 123
    communication of 228–229, 233
    as perceived by decision makers 202, 207–208, 219, 228
    portfolio valuation techniques 208–220
    sensitivity analysis 189–203
        SOMs 104–105
USA
    distribution of San Jose scale 3–4
    establishment of the Burmese python in Florida 66–67, 68–77
    impact of the red imported fire ant 164–169
    introduction of species via freight transport 9, 35–47
    NAPPFAST model 20, 82–94
    pest species databases 100, 111
    quarantine 133–134
    spread of the gypsy moth 132–133, 134–140
    uncertainty analysis of *S. noctilio* invasion 192–203

validity/validation of models 4–5, 8
    factor analysis 229–232
    process-based pest risk models 174, 181–185
    types of validity 227–228
    see also quality of models/maps
Venn tool (NAPPFAST) 88
visual appearance of maps 5

water hyacinth 29–30
weather
    input into MAXENT model 68, 72, 79
    input into NAPPFAST model 83–84, 94
weeds 8–9
    economic impact of mimosa eradication 145–158
    process-based risk mapping of Chilean needle grass 175–185
western corn rootworm (*Diabrotica virgifera virgifera*) 24, 122
    DSS maps of risk 27–29
    spread models 122–126
windborne dispersal 9, 49–61
'Wombling' 141
wood-infesting pests 35–36, 39, 133
    uncertainty analysis of invasion from USA to Canada 213–218
woodwasp (*Sirex noctilio*) 192–203
World Trade Organization (WTO) 15, 19
worst case scenario 25, 118, 123

Printed and bound by CPI Group (UK) Ltd, Croydon, CR0 4YY
23/03/2026

14847628-0001